Applied Tribology: Lubrication

Applied Tribology: Lubrication

Edited by Bernice Wyong

Clanrye International
www.clanryeinternational.com

Clanrye International,
750 Third Avenue, 9th Floor,
New York, NY 10017, USA

ISBN: 978-1-63240-929-4

Cataloging-in-Publication Data

Applied tribology : lubrication / edited by Bernice Wyong.
p. cm.
Includes bibliographical references and index.
ISBN 978-1-63240-929-4
1. Tribology. 2. Lubrication and lubricants. 3. Bearings (Machinery). 4. Lubrication systems. I. Wyong, Bernice.
TJ1075 .A67 2020
621.89--dc23

For information on all Clanrye International publications visit our website at www.clanryeinternational.com

CLANRYE INTERNATIONAL

Printed in China.

Contents

Preface

The main aim of this book is to educate learners and enhance their research focus by presenting diverse topics covering this vast field. This is an advanced book which compiles significant studies by distinguished experts in the area of analysis. This book addresses successive solutions to the challenges arising in the area of application, along with it; the book provides scope for future developments.

The field that deals with interacting surfaces in relative motion is known as tribology. It is a multidisciplinary subject that applies principles of various academic fields such as physics, chemistry, biology, mathematics, engineering and materials science. The study and application of the fundamentals of friction, wear and lubrication are fundamental to this field. Lubrication is the technique of using a lubricant to lessen the friction or wear present between any surfaces. Such lubricants are the fluid materials that are characterized by viscosity. Fluidostatic lubrication and Fluid-fluid lubrication are the two types of lubrication used in tribology. It reduces the rate of wear and stress at bearings. This book contains some path-breaking studies in the field of applied tribology. While understanding the long-term perspectives of the subject, the book makes efforts in highlighting their impact as a modern tool for the growth of the discipline. The readers would gain knowledge that would broaden their perspective about the discipline.

It was a great honour to edit this book, though there were challenges, as it involved a lot of communication and networking between me and the editorial team. However, the end result was this all-inclusive book covering diverse themes in the field.

Finally, it is important to acknowledge the efforts of the contributors for their excellent chapters, through which a wide variety of issues have been addressed. I would also like to thank my colleagues for their valuable feedback during the making of this book.

Editor

1

Correlation Study of Physicochemical, Rheological, and Tribological Parameters of Engine Oils

Prashant Thapliyal[1,2] and G. D. Thakre[3]

[1] *Department of Physics, Army Cadet College, Indian Military Academy, Dehradun 248007, India*
[2] *Uttarakhand Technical University, Sudhowala, Dehradun 248007, India*
[3] *Tribology and Combustion Division, CSIR-Indian Institute of Petroleum, Mohkampur, Dehradun, India*

Correspondence should be addressed to Prashant Thapliyal; prashant.197104@gmail.com

Academic Editor: Meng Hua

The physicochemical and tribological studies of mineral and synthetic commercial engine oils have been carried out to investigate their performance variability and to propose generalized relationship among different physicochemical and performance parameters. Physicochemical parameters have been determined using standard test procedures proposed in ASTM and Indian Standards (BIS). The rheological parameters of these lubricants have been investigated to identify the flow behavior. The tribological performance in terms of their antifriction and antiwear properties has been studied using four-ball tribotester. Correlation and regression analysis has been performed to ascertain relationship among physicochemical and tribological parameters and the causes of performance variability are highlighted. An empirical relation to calculate coefficient of friction as a function of physicochemical properties has been established using regression analysis. The developed relation has fair degree of reliability, as percentage of deviation is less than 20%.

1. Introduction

Lubricants play a vital role in present day automotives. The engine oils in particular lubricate all the critical parts of IC engines. They not only reduce friction and wear between the moving parts but also dissipate frictional heat generated between the contacting parts of the engines [1]. The engine oils are basically formulated using base oil and additive package. The chemistry of formulated engine oil by and large determines its physicochemical properties and also the in situ tribo performance behavior. The physicochemical properties like viscosity, density, TAN (total acid number), TBN (total base number), and sulphated ash are considered to be important characteristic properties of the engine oils. These properties provide information on general applicability of engine oils. Along with physicochemical properties, the flow behavior of engine oils is also an important aspect. The flow behavior of engine oil is dependent on the rheology of the oil and, therefore, it is very much essential to have a thorough knowledge of the rheological behavior of the lubricants [2].

The lubricants on the basis of their rheological behavior are characterized as Newtonian and non-Newtonian fluids. The fluids with molecular mass less than 1000 kg/mol show Newtonian behavior at low pressure and shear stress [3]. Recently it has been reported that the non-Newtonian behavior of lubricants results into improved load carrying capacity and reduced contact friction in hydrodynamic porous journal bearings [4]. The engine oils exhibit viscoelasticity under non-Newtonian flow condition and show time dependent strain [5]. Viscoelasticity results in shear thinning of lubricants. Therefore, the viscosity of engine oil is considered to be one of the major rheological parameters that have profound influence on the lubricant performance behavior. Thus, the physicochemical properties and rheological and tribological behavior of engine oils are interdependent.

Tribology is the study of friction and wear of the machine parts. Lubricating oil forms a thin film between the surfaces

Table 1: Lubricants selected for the study.

Sl. number	Lubricant code	SAE grade	Base oil	Application
1	L_1	SAE-40	Mineral	Diesel engine
2	L_2	SAE20W-50	Mineral	Diesel/gasoline engine
3	L_3	SAE20W-50	Mineral	Gasoline engine
4	L_4	SAE5W-40	Synthetic	Diesel/gasoline engine
5	L_5	SAE5W-40	Synthetic	Diesel/gasoline engine

that separates adjacent moving parts and minimizes the direct contact between them. As a result of this, the heat generated due to frictional heating decreases. Efficient lubrication aids in wear reduction, thus protecting the engine components from frequent failures. On the basis of the ratio of lubricant film thickness to the composite surface roughness of the contacting surfaces, different lubrication regimes ranging from boundary to hydrodynamic lubrication may occur. These lubrication regimes have dependency on the contact pressure and on the surface velocity of the surfaces in contact [6]. In this context, experimental-statistical methods have been extensively used to characterize friction in dry contact and second-order polynomial equation established for coefficient of friction [7]. In another attempt differentiation between API GL performance levels of automotive gear oils was conducted using tribological tests on four-ball and crossed-cylinder wear test apparatus. The statistical analysis performed has revealed the differentiation in performance levels of automotive gear oils [8].

Interrelationship among various physicochemical and tribological parameters can be an effective tool to understand the behavior and performance variability of lubricants. Various attempts have been made to establish empirical relations among physicochemical parameters using mathematical/statistical techniques. In this context, variation in tribo performance of commercial engine oils was studied and correlations between tribological parameters like friction and wear with physicochemical properties were established [9]. Similarly empirical relations were established between temperature and absolute viscosity for lubricants derived from seed oils [10]. In order to predict the tribological properties an algorithm called phenomenological and predictive model was developed for organic sulphide based lubricants. The model was validated using experimental data on weld load using four-ball machine [11]. Over the years, it has been observed that the theoretical models have been used to rationalize experimental data on the physicochemical properties of binary mixtures of vegetable oils with various classes of mineral base oils [12]. Moreover, multivariable statistical analyses techniques have been used for predicting the pressure viscosity coefficient of lubricants using NMR experiments [13].

A number of studies have been carried out in the past on determining and establishing the dependency among different lubricant parameters since Barus established a relation between viscosity and pressure by introducing pressure viscosity coefficient "α" [14]. In the recent past comprehensive characterization of lubricating fluids of same viscosity but different additive and base stock compositions was performed to investigate the frictional behavior, thermophysical and rheological properties, and mechanical efficiency in hydraulic motors [15]. The lubricant viscosity is temperature dependent. Studies have been carried out to establish the dependency of temperature and other parameters on viscosity of engine oil. Relationship between shear dependent viscosity, temperature, and pressure for polymer thickened lubricants was also established [16]. It has been reported that better rheological behavior with temperature leads to better tribological performance [17].

On the basis of literature review carried out it is observed that attempts have been made to develop dependencies among various characteristic properties of the lubricants. However, no comprehensive dependency in the form of empirical relations between the physicochemical properties and tribological performance of the engine oil exists. Hence, in the present work attempts have been made to investigate the relations between the physicochemical properties and the tribological performance of engine oils. The study has been performed on commercial engine oils and the characteristic properties pertaining to physicochemical, rheological, and tribological performance determined. The performance parameters were then correlated using correlation and regression analysis to establish dependency relations among them. The study will aid the lubrication and maintenance engineers in selecting appropriate parameters for the successful operations of the engines.

2. Experimental

2.1. Lubricant Selection. In this study, five different commercial engine oils coded as $L_1, L_2, \ldots, L_5$ have been considered. The details of the selected lubricants are given in Table 1. The motive behind selecting the said lubricants is to understand the performance behavior of lubricants presently available in the market and establish relations among their characteristic properties and performance behavior.

2.2. Lubricant Characterization. The selected lubricants have been characterized for their physicochemical properties, rheological behavior, and tribological performance. The physicochemical properties provide the basic qualitative information on the products selected while the rheological and tribological behaviors provide the information on performance of the lubricants. The TAN measures the presence of organic and strong inorganic acids in the oil and is an indicator of oil oxidation that may lead to corrosion of the components. TBN being the measure of basic components represents the ability of oil to neutralize acids produced in it during normal use. Similarly, sulphated ash represents the amount of metallic elements derived from the detergent and antiwear additives of the oil. The additive packages contain elements like calcium, magnesium, zinc, molybdenum, phosphorus, and so on that help in enhancing the performance of the engine oil.

2.2.1. Physicochemical Properties. The physicochemical properties such as density, viscosity, viscosity index, sulphated ash, total acid number (TAN), and total base number (TBN) have been determined using standard test procedures proposed in ASTM and Indian Standards (BIS). The metallic elements present in the additive package have been determined using the Inductively Coupled Plasma Atomic Emission Spectrometer (ICPAES), model: PS 3000 UV (DRE), Leeman Labs Inc. (USA).

2.2.2. Rheology. The variation of rheological parameters (viscosity, shear stress, and torque) with temperature has been investigated using RHEOPLUS/32 MCR 302 from Anton Paar Austria. The rheometer capable of performing rheological studies in rotational or oscillatory mode consists of an EC motor with a torque range of 10–200 mNm. The experiments have been performed using concentric cylinder geometry as shown in Figure 1. The clearance space between the concentric cylinders was filled with the lubricant to be tested and the inner cylinder was rotated with the help of a spindle at the desired speed. Two different sets of experiments were performed to determine the variation of coefficient of viscosity with temperature and the shear rate. The first set of rheology experiments was performed at a constant shear rate of 10/s and the temperature was varied from 20 to 50°C with variation rate of 4°C per minute. The variation in coefficient of viscosity with temperature was monitored and recorded. In yet another experiment, the shear rate was varied from 1 to 100/s at room temperature and the variation in coefficient of viscosity with shear rate was monitored and recorded.

2.2.3. Tribology. Tribological performance tests have been conducted on four-ball tribotester (FBT) using the standard wear test procedure as mentioned in ASTM D: 4172B. The FBT used in the present study is shown in Figure 2.

(1) Friction Analysis. The FBT machine assesses antiwear and antifriction properties of lubricants. For this the FBT utilizes a four-ball sliding contact geometry formed in between four balls each of diameter 12.7 mm. The four balls are assembled in a tetrahedron with bottom three balls fixed in the ball pot while the fourth ball mounted on the vertical shaft is free to rotate at predefined spindle speed. The lubricant to be tested is introduced in the stationary ball pot forming thin lubricating film between the bottom three and the top ball. The contact friction in terms of friction torque is continuously recorded during the entire test duration.

(2) Wear Analysis. The contact wear in terms of wear scar diameter is measured at the end of the test using an industrial apochromatic microscope. The friction torque is later converted into coefficient of friction using empirical relations.

TABLE 2: Experimental test condition.

Parameter	Value
Load	40 kgf
Temperature	75°C
Speed	1200 rpm
Test duration	1 hr

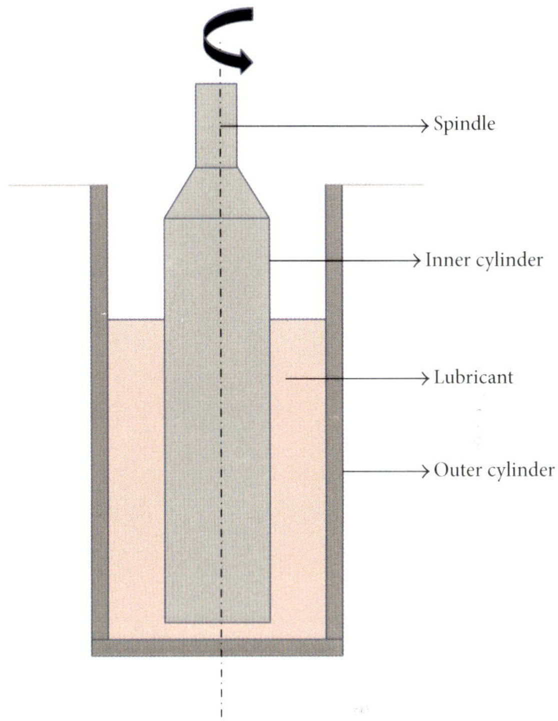

FIGURE 1: Concentric cylindrical geometry of rheometer.

Each lubricant is tested twice and the wear scar diameter (WSD) along the vertical and horizontal axis is measured for all the bottom three balls, thus providing 12 readings for a given lubricant. The average of the 12 readings is reported as the wear scar diameter.

The experiments are performed on balls made up of AISI chrome alloy standard steel number E- 52100, grade 25 EP (extra polish). The test conditions used are given in Table 2.

Postexperimental investigations on the used test specimens were performed to investigate the mode and mechanism of wear. Further, the capability of additives to form boundary layers on the test surface was investigated using scanning electron microscopy (SEM) using FESEM from FEI Netherlands model Quanta 200F fitted with EDX system.

3. Results and Discussion

3.1. Physicochemical Analysis of Lubricants. The results for the measurements carried out on the physicochemical properties of the lubricants are given in Table 3.

It is evident from Table 3 that commercial engine oils are nearly similar in their physicochemical characteristics. The density of these lubricants is of the order of 0.8 g cm^{-3},

Table 3: Physicochemical properties of the lubricants.

Sl. number	Characteristics	Lubricant name				
		L_1	L_2	L_3	L_4	L_5
1	Density at 15°C (g cm^{-3})	0.8711	0.8910	0.8695	0.8655	0.8526
2	Kinematic viscosity (mm^2/s) @ 40°C	123.06	166.71	154.93	83.68	79.82
3	Kinematic viscosity (mm^2/s) @ 100°C	14.17	17.75	17.93	13.28	13.05
4	Viscosity index (VI)	115	117	118	162	166
5	TAN (mg KOH/g)	0.44	1.93	0.93	2.13	2.00
6	TBN (mg KOH/g)	11.16	11.09	9.65	14.41	14.25
7	Sulphated ash %wt	1.06	0.77	0.93	0.80	1.10

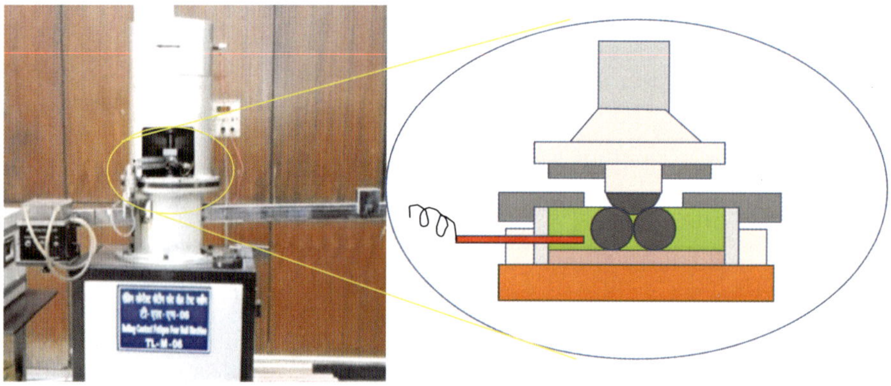

Figure 2: Four-ball tribotester (FBT).

irrespective of the brand of lubricant and the nature of base oil (mineral/synthetic). The tested lubricants have VI > 110. However, the synthetic lubricants have very high VI of above 160. A high VI is very much desired so that there is lesser variation in viscosity with change in temperature. The TAN, TBN, and sulphated ash are higher for synthetic oils. This may be due to the presence of higher concentrations of additives in them. The TAN values are fairly in the range of 0.5–2.25 for the selected lubricants. The synthetic lubricants with low viscosity at 40 and 100°C possess a very high VI. This may be due to the presence of VI modifiers in the oil. The TBN values for the oils are in the range of 9–15 mg KOH/g, with synthetic oils having high TBN values. The sulphated ash content is almost similar around 1% wt for all the selected lubricants.

The results for the trace metal analysis are given in Table 4. The results reveal the presence of very high concentrations of extreme pressure additives containing elements like zinc, phosphorous, and molybdenum. The synthetic oils show high concentrations of Zn and P and almost negligible Mo. Among the selected lubricants, L_2 has highest additive concentration with Zn = 977, Mo = 93, and P = 894 mg/l. The presence of zinc, molybdenum, and phosphorus has a direct influence on the friction and wear behavior of the lubricants.

Table 4: Trace metal analysis.

Sl. number	Lubricant code	Element (mg/l)		
		Zn	Mo	P
1	L_1	549.10	36.60	512.30
2	L_2	977.10	93.30	893.50
3	L_3	724.60	50.00	677.60
4	L_4	907.10	1.00	857.90
5	L_5	924.60	<1.00	877.90

3.2. Rheological Investigations

3.2.1. Variation of Viscosity with Temperature. The variation in dynamic viscosity with temperature is shown in Figure 3. It is observed that the coefficient of viscosity decreases monotonically with increase in temperature. As shown in Figure 3 the decrease is not linear; however, it is in tandem with the general trends of variation of lubricant viscosity with temperature. Lubricant L_2 has the highest value for coefficient of viscosity, that is, 0.5 Pa·s at 293 K. It has the largest negative gradient with temperature indicating that it is more susceptible to temperature variations. L_4 and L_5 have smaller values of dynamic viscosity being synthetic lubricants. But these lubricants show better stability as compared to the mineral based lubricants as they have smaller negative temperature gradient of viscosity.

The variation of viscosity with temperature for the selected lubricants with the help of curve fitting technique is found to obey Reynolds' equation [18] given by

$$\mu_0 = be^{-aT_A}, \tag{1}$$

where μ_0 is the dynamic viscosity at atmospheric pressure and T_A is absolute temperature.

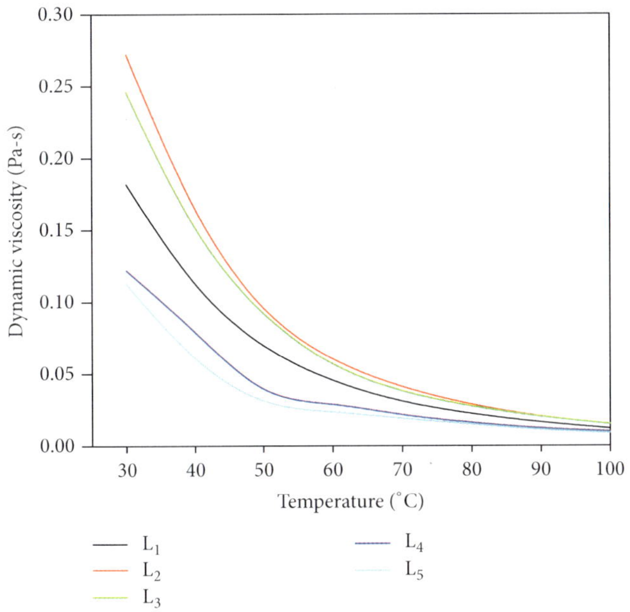

FIGURE 3: Variation of viscosity with temperature.

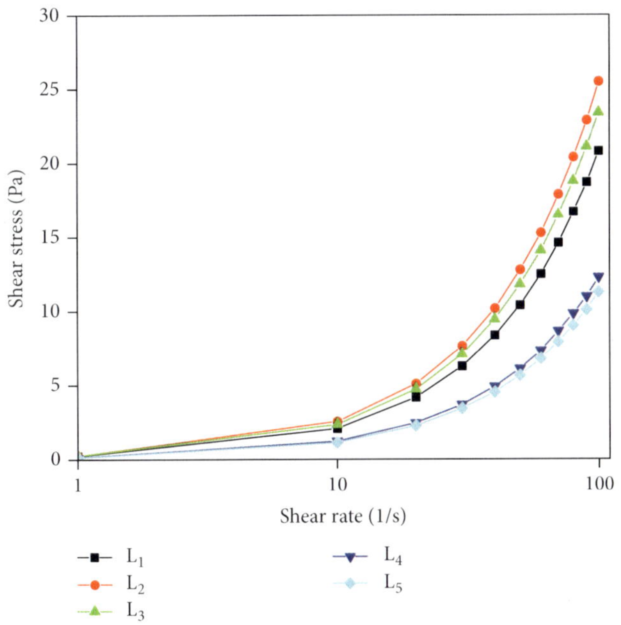

FIGURE 4: Variation of shear stress with shear rate for the lubricants.

3.2.2. *Variation of Shear Stress with Shear Rate.*

The variation of shear stress/shear rate is shown in Figure 4. As observed from Figure 4, all the selected lubricants describe nonlinear behavior representing non-Newtonian behavior indicating presence of viscoelasticity. All of them have yield stress showing viscoplastic nature with L_2 having highest value.

Utilizing the experimental data represented in Figure 4, curve fitting was performed with the help of curve fitting tool box in MATLAB software. The best cure fit equation thus obtained is given by

$$\tau = m\left(\frac{\partial u}{\partial y}\right)^n. \quad (2)$$

TABLE 5: Power law index of the lubricants.

Sl. number	Lubricant code	Power law index
1	L_1	0.9967
2	L_2	0.9969
3	L_3	0.9916
4	L_4	0.9940
5	L_5	0.9998

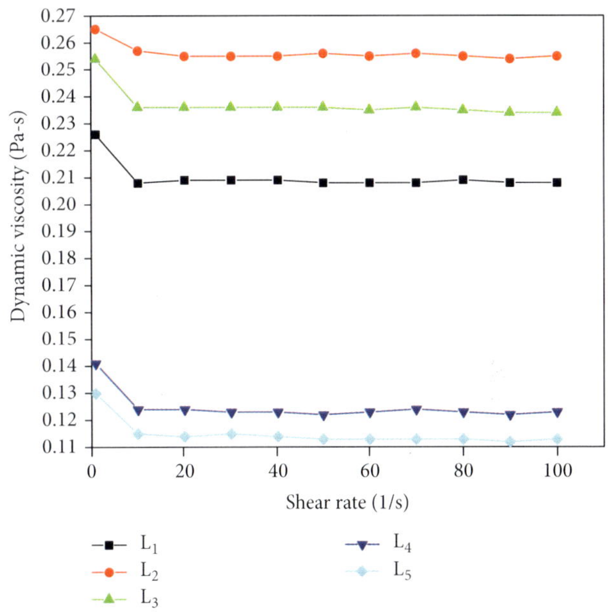

FIGURE 5: Variation of viscosity with shear rate for the lubricants.

Equation (2) represents the flow behavior of power law fluid. The value of power law index "n" signifies the Newtonian and non-Newtonian behavior of lubricants. "n" < 1 represents shear thinning behavior, "n" > 1 represents shear thickening, and $n = 1$ represents the Newtonian fluid. The power law index values as obtained from the curve fitting procedure are given in Table 5. The values of "n" for the lubricants are close to less than 1 representing shear thinning behavior of the lubricants. This further confirms that the lubricants represent non-Newtonian behavior.

3.2.3. *Variation of Viscosity with Shear Rate.*

The variation of viscosity with shear rate is shown in Figure 5. As observed from Figure 5 the viscosity initially decreases with increase of shear rate. This behavior is observed at lower shear rates, that is, shear rate < 10/s. At higher shear rates there is no significant variation and the coefficient of viscosity is almost constant over the entire range of shear rate. The lubricant L_1 shows the largest variation of viscosity with shear rate and L_2 the least. Beyond shear rate of 10/s the viscosity is almost independent of shear rate. Decrease of viscosity with shear rate is more pronounced for L_1 showing more shear thinning and hence more viscoelastic behavior. Small values of dynamic viscosity for L4 and L5 are attributed to their synthetic origin and having SAE grade 5W-40.

TABLE 6: Tribological performance of lubricants.

Sl. number	Lubricant code	Coefficient of friction	Average wear scar diameter (mm)
1	L_1	0.1429	0.710
2	L_2	0.1155	0.746
3	L_3	0.1416	0.676
4	L_4	0.0890	0.391
5	L_5	0.0881	0.446

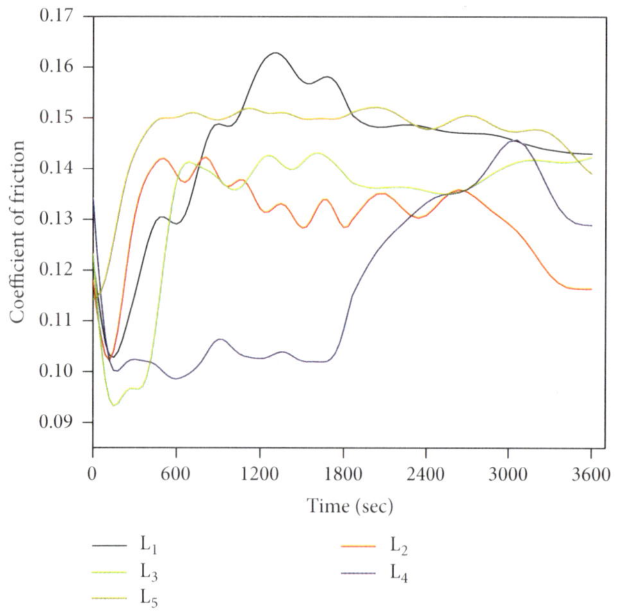

FIGURE 6: Friction coefficient versus time in seconds for L_1, L_2, L_3, L_4, and L_5.

3.3. Tribological Investigations. The tribological performance of lubricants is defined in terms of their friction and wear behavior.

3.3.1. Friction Behavior. Figure 6 shows the variation in coefficient of friction for the lubricants over the entire experimental duration. It is observed from Figure 6 that the coefficient of friction increased during the early stage of experiment and later on remained almost constant. Due to the initiation of the wear scar the coefficient of friction increased in the early stage; later on due to the rubbing wear the coefficient of friction became almost constant. The kinetic friction, that is, the coefficient of friction at the end of the test, is the highest for lubricant L_1 (μ = 0.1429) and lowest for lubricant L_2 (μ = 0.1155). This behavior of lubricants can be attributed to the presence of extreme pressure and antifriction additives. Lubricant L_1 has relatively lower concentrations of Zn and P as evident from Table 4, while L_2 has got the highest concentration of these elements. In case of synthetic base lubricants, the friction coefficient of $\mu = 0.0890$ and $\mu = 0.0881$ is observed for lubricants L_4 and L_5, respectively. Though synthetic oils possess a very low viscosity, yet higher concentrations of Zn and P present in them enhance the film forming capability of these oils at the given test load, hence decreasing the friction coefficient when compared with the mineral based lubricants. The lubricants are often blended with zinc dialkyl dithiophosphate (ZDDP) as multifunctional additive. The Zn and P resent in this additive form the polar moieties which are able to adhere on the steel surface and protect the surface from damage. This adsorbed layer of additive is known as boundary film which under pressure (applied load) gets strengthened, thereby reducing the friction and wear.

3.3.2. Wear Behavior. The wear scars as observed on the ball test specimens are shown in Figure 7. The morphology of wear scar reveals normal rubbing wear within the contact. The rubbing marks are distinctly visible along the sliding direction.

For a better comparison of test results, the coefficient of friction and WSD is tabulated in Table 6. The lubricant L_4 reported the best antiwear performance with WSD of 0.391 mm, while the lubricant L_2 reported the poorest performance with WSD of 0.746 mm.

3.3.3. Postexperimental Analysis. Figure 8 shows the SEM micrographs of used ball test specimens. The SEM micrographs reveal that the wear surfaces have undergone normal rubbing wear under the influence of load in the sliding direction. The wear tracks observed are parallel to the direction of sliding. The lubricant L_4 and L_5 show a smoother surface with some edge serrations along the wear track. The surface smoothening must have occurred due to the rubbing of the asperities under the applied load. Similarly the micrograph for L_2 lubricated specimen shows severe wear with small micropits. It also shows scuffing on the surface. The L_1 lubricated specimen has smooth wear tracks representing smoothening of surface asperities. Also the wear associated with this lubricant is very low due to this smoothening action. The L_3 lubricated specimen shows rigorous scuffing on the steel surface. The scuffing marks are deeper and hence a larger wear scar diameter is observed with this lubricant. The SEM micrograph for the L_3 lubricated specimen reveals some surface distress with scuffing in the direction of sliding. The surface damage is observed in the form of nonuniform removal of material from the surface.

The EDX analysis for the specimens reveals the presence of elements like zinc, sulphur, phosphorus, and so on, which signifies that a thin boundary layer of lubricant is formed on the steel surfaces. The boundary films formed with the help of extreme pressure additives help in protecting the surfaces from further damage.

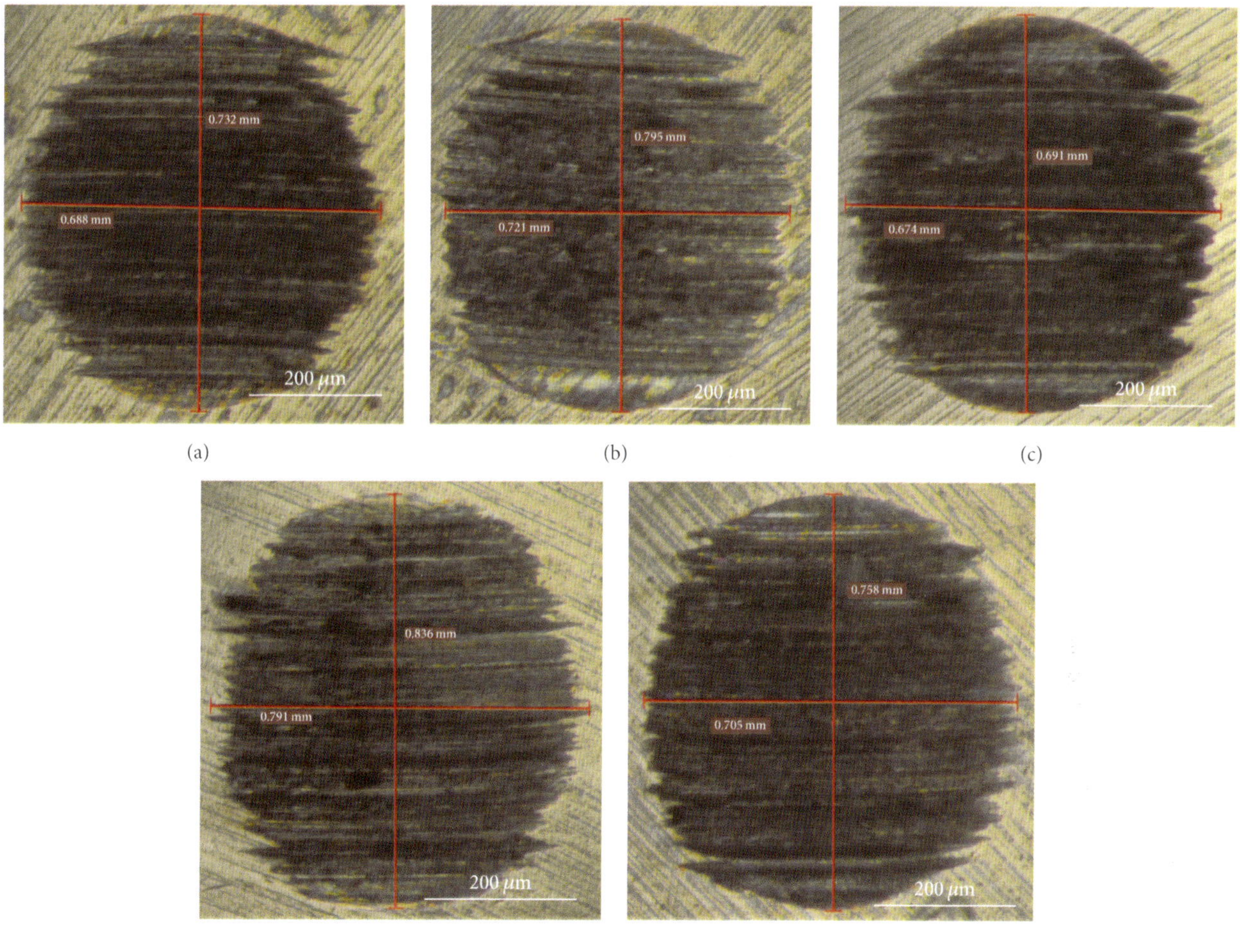

FIGURE 7: Wear scar diameters for the lubricants (a) L_1, (b) L_2, (c) L_3, (d) L_4, and (e) L_5.

3.4. Correlation Analysis. Correlation analysis predicts the association between two or more variables and infers the strength of the relationship among them. The value of correlation coefficient "r" reflects the extent to which the two individual variables are related [19]. The value of r ranges between −1 and +1. +1 value indicates perfectly positive correlation while −1 indicates perfectly negative correlation. The "r" is determined with the help of (i) covariance $\mathrm{Cov}(x, y)$ between any two variables that measures the variability of the (x, y) pairs around the mean of x and mean of y and (ii) sample variances of x and y, that is, S_x^2 and S_y^2 that represent the variability of x-scores and y-scores around their respective sample means $\overline{X}$ and $\overline{Y}$, respectively. Thus, the "r" is calculated using the formula

$$r = \frac{\mathrm{Cov}(x, y)}{\sqrt{S_x^2 S_y^2}}. \quad (3)$$

The lubricant properties given in Table 3 and the performance characteristics given in Table 6 have been therefore used to determine the correlation coefficients. Table 7 gives the correlation coefficients obtained using correlation analysis among different physicochemical and tribological parameters.

On examining the correlation coefficients of physicochemical and tribological properties, it is observed that the kinematic viscosity at 40°C has a positive correlation coefficient of 0.83 indicating that density affects viscosity directly. Positive correlation of 0.92 between metal additive Mo and density and of 0.95 between Mo and kinematic viscosity at 40°C shows that Mo affects the density and kinematic viscosity of the lubricant positively. Very high positive correlation coefficient of 0.94 between VI and TBN is a clear indicator that more neutralization of acid produced improves the VI of oil, thus prolonging the operational life. Trace metals Zn and P have very high value of correlation coefficients 0.96 and 0.98, respectively, with TAN indicating that though they improve the performance of the oil, yet they cause increase in lubricant acidity. This subsequently leads to increase in friction as interaction between the surfaces enhances oxidation and oxides in general get adsorbed on the surface [20]. Negative correlation of significance between WSD and TAN with value −0.55 and between COF and TAN with value −0.84 indicates that the increase in the value of TAN does not affect COF and WSD as TBN also has strong negative correlation of −0.92 with COF and −0.91 with WSD. It means the formation of acids in the process is combated by

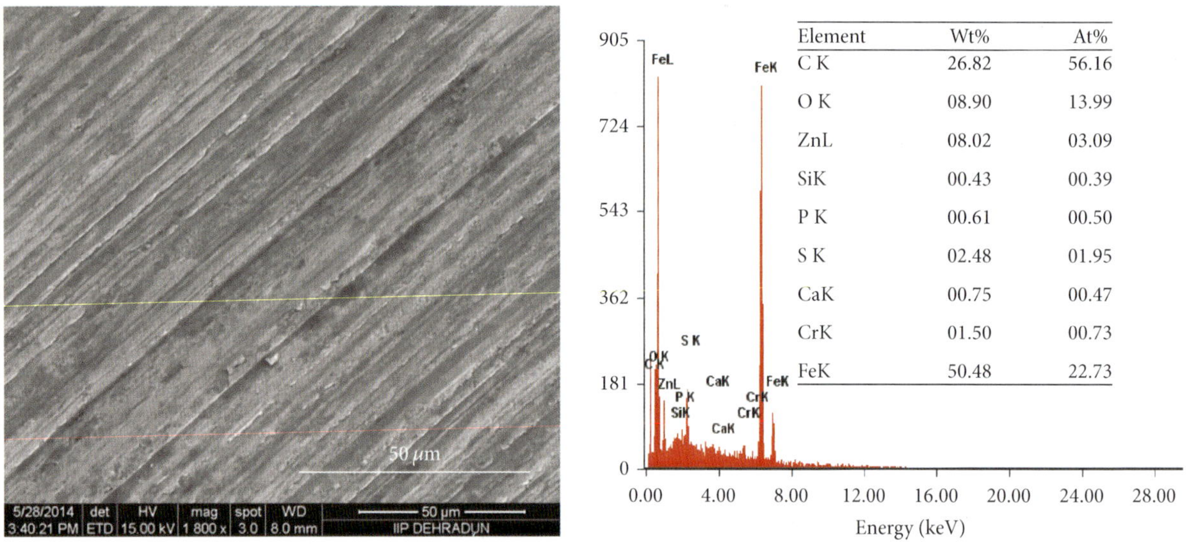

Element	Wt%	At%
C K	26.82	56.16
O K	08.90	13.99
ZnL	08.02	03.09
SiK	00.43	00.39
P K	00.61	00.50
S K	02.48	01.95
CaK	00.75	00.47
CrK	01.50	00.73
FeK	50.48	22.73

(a)

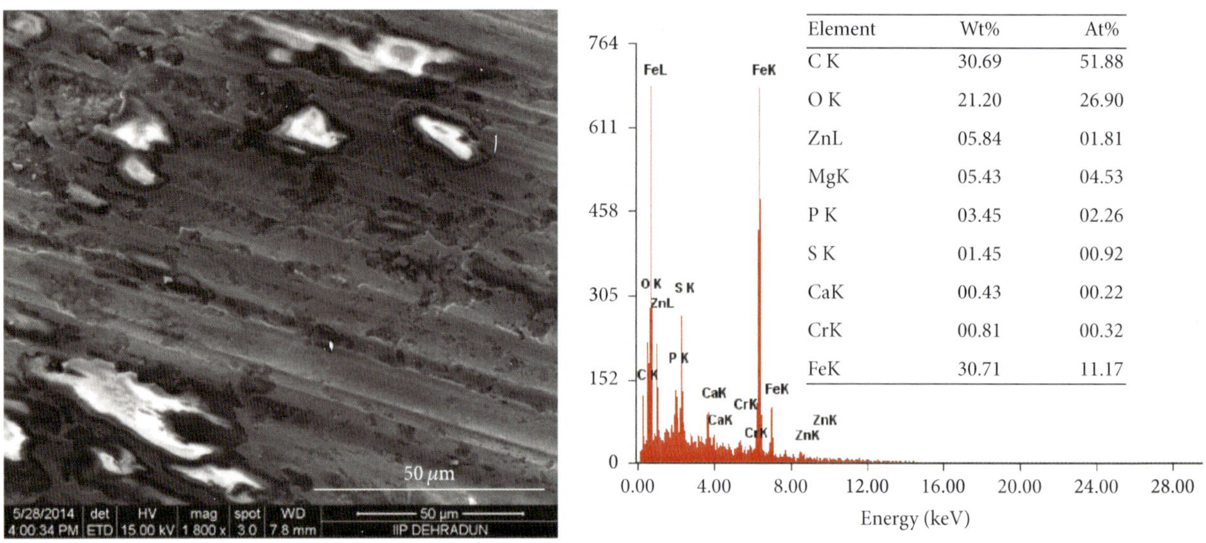

Element	Wt%	At%
C K	30.69	51.88
O K	21.20	26.90
ZnL	05.84	01.81
MgK	05.43	04.53
P K	03.45	02.26
S K	01.45	00.92
CaK	00.43	00.22
CrK	00.81	00.32
FeK	30.71	11.17

(b)

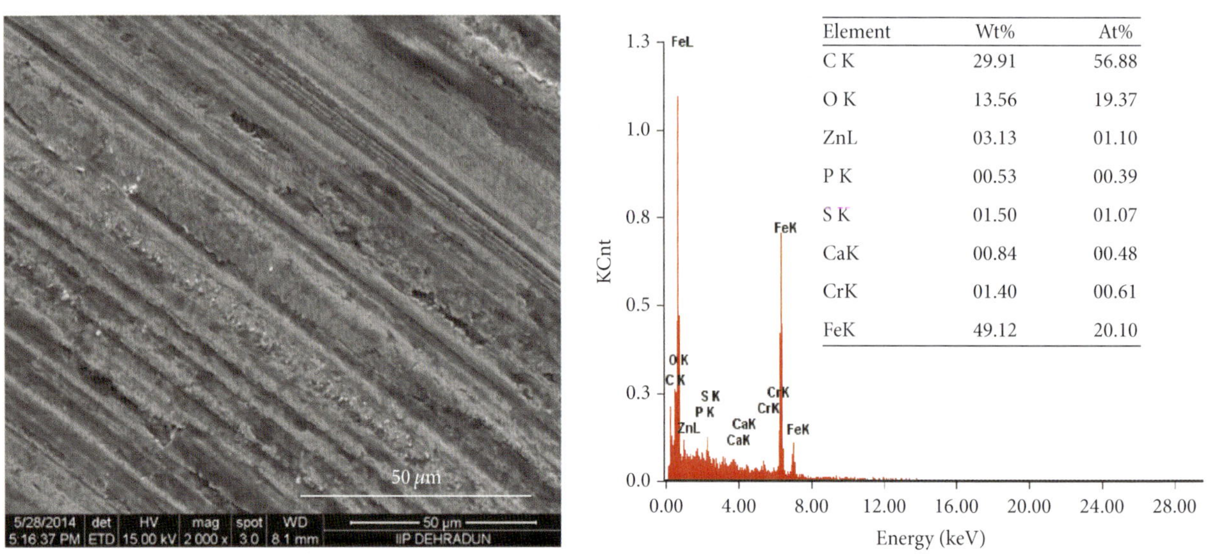

Element	Wt%	At%
C K	29.91	56.88
O K	13.56	19.37
ZnL	03.13	01.10
P K	00.53	00.39
S K	01.50	01.07
CaK	00.84	00.48
CrK	01.40	00.61
FeK	49.12	20.10

(c)

FIGURE 8: Continued.

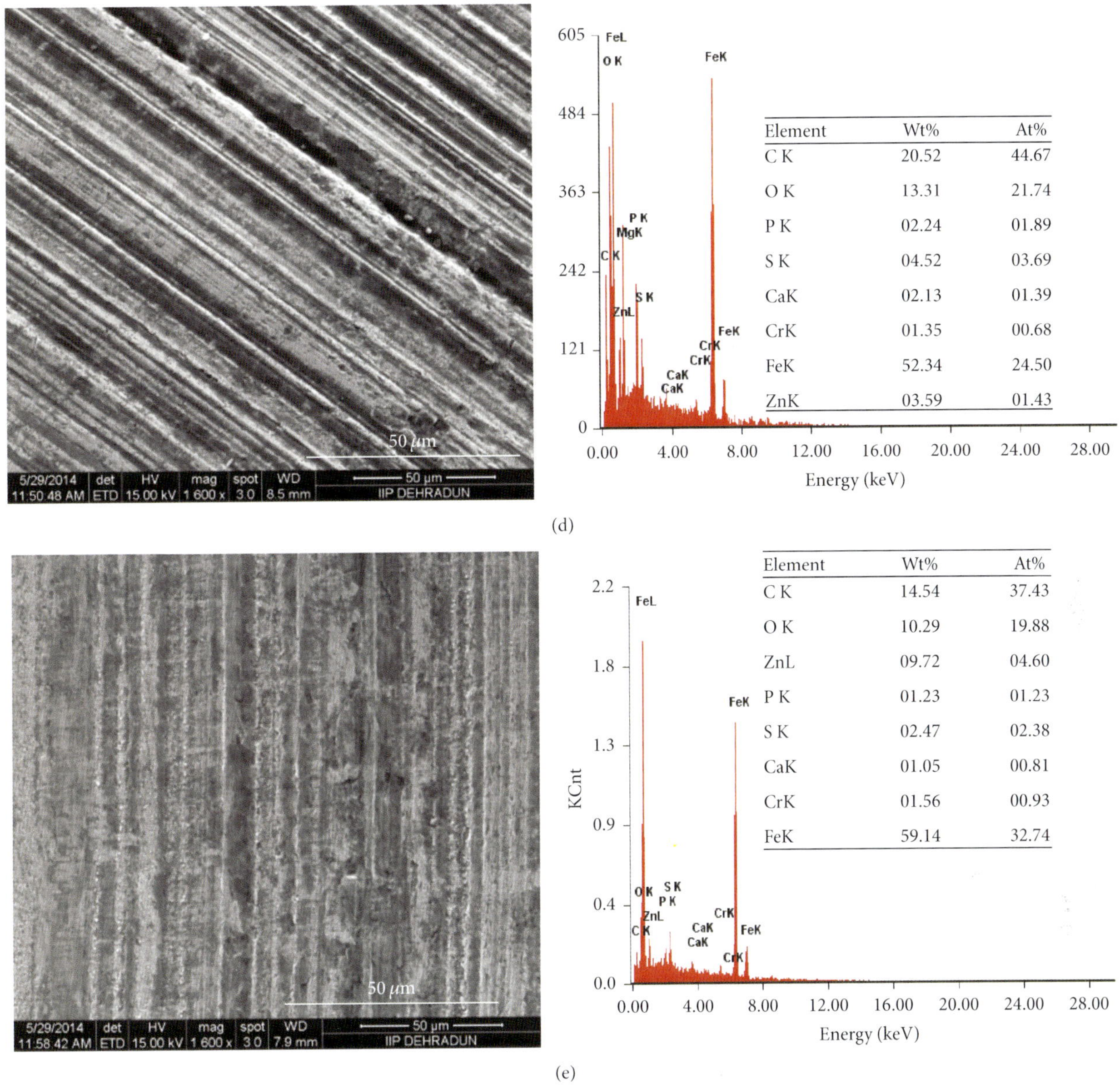

FIGURE 8: SEM/EDX micrographs for the lubricants (a) L_1, (b) L_2, (c) L_3, (d) L_4, and (e) L_5.

the presence of bases in the additive package. Strong positive correlation of 0.82 is incidental as it is a proven fact that they are almost unrelated as some lubricants give antifriction properties while others only give antiwear properties.

3.5. Regression Analysis for Lubricant Properties. Regression analysis has been performed for estimating the causal relationships for coefficient of friction and WSD with the physicochemical characteristic properties. Linear regression is the technique used for establishing causal relationship between a dependent variable and two or more independent variables. This helps to establish a relationship between the parameters of interest. The dependent variable, coefficient of friction (μ), and independent variables, density @ 15°C (ρ), kinematic viscosity @ 40°C (ν_{40}), and TAN, for the selected lubricants are given in Table 8.

The first-order multiple regression model was implemented on the data given in Table 8 and the regression statistics was established. The regression statistics involved determination of the values for correlation coefficient (R^2) and the standard error (E). The values obtained in the present analysis are $R^2 = 0.82$ and $E = 0.01$ representing strong relationship among the variables. Subsequently ANOVA (analysis of variance) was performed to determine the level of variability within the regression model. The significance and parameters, namely, degrees of freedom (df), sum of squares (SS), and the mean square (MS), obtained from ANOVA are given in Table 9.

Inference for multiple regression was later drawn by fitting a linear equation to the observed data. The least square fit was assumed and the line residuals were determined. The test statistics, that is, the ratio of slope and standard deviation

Table 7: Correlation coefficient matrix for lubricant properties (physicochemical and tribological).

	ρ @ 15°C g-cm^{-3}	ν @ 40°C	ν @ 100°C	VI	TAN	TBN	Sulphated ash	Zn	Mo	P	COF	WSD
ρ @ 15°C g-cm^{-3}	1*	**0.83**	0.73	0.73	0.08	0.59	−0.68	0.08	**0.92**	0.01	0.41	**0.75**
ν @ 40°C	**0.83**	1*	.91*	0.90	**−0.53**	0.91	**−0.50**	−0.15	**0.95**	0.22	**0.73**	**0.78**
ν @ 100°C	0.73	.91*	1*	0.76	−0.34*	0.85	−0.46*	0.02	**0.88**	−0.04	**0.68**	**0.81**
VI	−0.73	−0.90	−0.76	1*	0.71	**0.94**	0.13	0.53	−0.83	0.60	**−0.92**	**−0.97**
TAN	−0.08	**−0.53**	−0.34*	0.71	1*	0.70	0.33	**0.96**	−0.20	**0.98**	**−0.84**	**−0.55**
TBN	−0.59	−0.91	−0.85	0.94	0.70	1*	0.08	0.50	−0.76	0.56	**−0.92**	**−0.91**
Sulphated Ash	−0.68	**−0.50**	−0.46*	0.13	0.33	0.08	1*	0.51	−0.46	−0.47	−0.31	−0.45
Zn	0.08	−0.15	0.02	0.53	**0.96**	0.50	−0.51	1*	0.03	0.99	**−0.79**	−0.42
Mo	**0.92**	**0.95**	**0.88**	**0.83**	0.20	0.76	−0.46	0.03	1*	−0.04	0.55	**0.88**
P	0.01	−0.22	−0.04	0.60	**0.98**	0.56	−0.47	0.99	−0.04	1*	**−0.83**	−0.49
COF	0.41	**0.73**	**0.68**	**−0.84**	**−0.84**	**−0.92**	−0.31	**−0.79**	0.55	**−0.83**	1*	0.82*
WSD	**0.75**	**0.91**	**0.76**	**−0.97**	**−0.55**	**−0.91**	−0.45	**0.88**	−0.35	0.65	0.82*	1*

*Not relevant.

Table 8: Data for regression analysis.

Sl. number	COF (μ)	Density, (ρ) @ 15°C	Kinematic viscosity, ν_{40} @ 40°C	TAN
1	0.1429	0.8711	123.06	0.44
2	0.1155	0.8910	166.71	1.93
3	0.1416	0.8695	154.93	0.93
4	0.0890	0.8655	83.68	2.13
5	0.0881	0.8526	79.82	2

Table 9: Analysis of variance (ANOVA).

Source	df	SS	MS	F	Significance F
Regression	3	0.003263	0.001088	10.8076	0.005098
Residual	7	0.000705	0.000101		
Total	10	0.003968			

Table 10: Inference in linear regression.

	Coefficient	Standard error	t-test	P value
Intercept	0.085467	0.2280	0.3748	0.7189
Density @ 15°C (ρ)	0.033305	0.2740	0.1215	0.9067
Kinematic viscosity @ 40°C (ν_{40})	0.000241	0.0001	1.7749	0.1192
TAN	−0.02154	0.0064	−3.3717	0.01189

in each observation, is given in Table 10. The P value of inference provides the probability value associated with the two-sided test.

After determining the coefficient of intercepts and independent variables the regression equation is written in the linear form as

$$\text{Coefficient of friction: } \mu = 0.085467 + 0.033305\rho + 0.000241\nu_{40} - 0.02154\text{TAN}, \quad (4)$$

where ρ is density at 15°C, ν_{40} is kinematic viscosity @ 40°C, and TAN is total acid number.

Significance F (Table 10) for relation (4) is 0.005098 which is much less than 0.1, signifying that the formula is more reliable.

4. Conclusion

In the present study, experimental investigations have been carried out to study the performance variability and establish a correlation between the characteristic properties of engine oils. The experiments have been performed to investigate the physicochemical, rheological, and tribological properties of mono- and multigrade engine oils of different API performance standards. Thus, on the basis of the investigations made, the following broad conclusions are drawn:

(i) The commercial engine oils are nearly similar in their physicochemical characteristics. However, the synthetic lubricants possess high VI and TBN and higher concentrations of additives as compared to mineral based oils.

(ii) The rheological behavior of lubricants reveals that the variation of viscosity with temperature for the tested engine oils obeys Reynolds' equation. The lubricants describe non-Newtonian shear thinning behavior with the power law index values close to 0.99.

(iii) The tribological performance of lubricants reveals that the synthetic base lubricant possesses superior antifriction and antiwear properties than the mineral base lubricants. The coefficient of friction varies from 0.0881 to 0.1429 for the tested lubricants. Similarly the wear scar diameter varies from 0.391 mm to 0.746 mm for the tested lubricants The tribo performance of the lubricants is predominantly influenced by the viscosity and the additives present.

(iv) The worn out surfaces reveal that the synthetic base lubricants result in less surface distress while the contemporary mineral base lubricants show rigorous scuffing. All the lubricants are capable of forming thin boundary film on the steel surfaces.

(v) The correlation analysis reveals that the friction and wear behavior of lubricants is influenced by their viscosity. The viscosity in turn is influenced by density, TAN, and TBN values. Moreover, the TAN and TBN are influenced by the concentrations of the trace metals present in the additives used.

(vi) An empirical relation correlating friction, viscosity, density, and TAN values of the lubricants is given by $\mu = 0.085467 + 0.033305\rho + 0.000241\nu_{40} - 0.02154\mathrm{TAN}$ which gives fair degree of reliability with maximum deviation of 14% from the experimental results.

Nomenclature

TAN: Total acid number
TBN: Total base number
ICPAES: Inductively coupled plasma emission spectrometer
FBT: Four-ball tribotester
WSD: Wear scar diameter
EP: Extreme pressure
SEM: Scanning electron microscope
VI: Viscosity index
μ_0: Dynamic viscosity
T_A: Absolute temperature
n: Power law index
μ: Coefficient of friction
ρ: Density at 15°C
ν_{40}: Kinematic viscosity at 40°C
ν_{100}: Kinematic viscosity at 100°C
Multiple R: Coefficient of multiple correlations
R^2: Coefficient of determination
SS: Sum of squares
df: Degree of freedom
MS: Mean squared
Residual MS: Mean squared error
SS Residual: Residual sum of squares
SS Total: Total sum of squares.

Conflicts of Interest

The authors declare that there are no conflicts of interest regarding the publication of this paper.

References

[1] S. Q. A. Rizvi, *A Comprehensive Review of Lubricant Chemistry, Technology, Selection and Design*, ASTM International, Conshocken, PA, USA, 2009.

[2] C. Meunier, D. Mazuyer, P. Vergne, M. E. Fassi, and J. Obiols, "orrelation between the film forming ability and rheological properties of new and aged low sulfated ash, phosphorus and sulfur (low SAPS) automotive lubricants," *Tribology Transactions*, vol. 52, no. 4, pp. 501–510, 2009.

[3] R. B. Bird, R. C. Armsrong, and O. Hassager, *Dynamics of Polymeric Liquids Vol. 1: Fluid Mechanics*, Wiley, New York, NY, USA, 2nd edition, 1987.

[4] N. Sharma, R. K. Sharma, N. Sunil, and S. Kango, "A comparative study for lubrication of surface textured porous journal bearing with two different non-Newtonian fluid models," *International Journal of Surface Science and Engineering*, vol. 10, no. 5, p. 485, 2016.

[5] P. Schümmer, "Mechanics of non-newtonian fluids," *Chemie Ingenieur Technik*, vol. 51, no. 7, p. 766, 1979.

[6] J. Frene, M. Arghir, and H. Zaidi, "Regimes of lubrication in lubricated contacts," *International Journal of Surface Science and Engineering*, vol. 4, no. 2, pp. 175–190, 2010.

[7] J. Abdo, "Design of experiments technique for characterisation of friction in dry contact," *International Journal of Surface Science and Engineering*, vol. 2, no. 1-2, pp. 120–138, 2008.

[8] M. Trzos, M. Szczerek, and W. Tuszynski, "A study on the possibility of the Brugger test application for differentiation between the API GL performance levels of gear oils," *Archives of Civil and Mechanical Engineering*, vol. 13, no. 1, pp. 14–20, 2013.

[9] G. D. Thakre and M. R. Tyagi, "Study on variation of tribo-performance of commercial engine oils," *Industrial Lubrication and Tribology*, vol. 65, no. 5, pp. 341–347, 2013.

[10] R. D. Gaikwad and P. Swamy, "Physico-chemical and viscosity studies in some seed oils from wild & cultivated plants," *Acta Chimica Slovenica*, vol. 55, no. 3, pp. 683–687, 2008.

[11] G. S. Cholakov, K. G. Stanulov, P. Devenski, and H. A. Iontchev, "Quantitative estimation and prediction of tribological performance of pure additive compounds through computer modelling," *Wear*, vol. 216, no. 2, pp. 194–201, 1998.

[12] M. Roegiers and B. Zhmud, "Property blending relationships for binary mixtures of mineral oil & elektrionised vegetable oil: viscosity solvent power & seal compatibility index," *Lubrication Science*, vol. 23, no. 6, pp. 263–278, 2011.

[13] B. K. Sharma and A. J. Stipanovic, "Pressure viscosity coefficient of lubricant base oils as estimated by nuclear magnetic resonance spectroscopy," *Industrial and Engineering Chemistry Research*, vol. 41, no. 19, pp. 4889–4898, 2002.

[14] C. Barus, "Isothermals, isopiestics and isometrics relative to viscosity," *American Journal of Science*, vol. 45, no. 266, pp. 87–96, 1893.

[15] P. W. Michael, J. M. Garcia, S. S. Bair, M. T. Devlin, and A. Martini, "Lubricant Chemistry and Rheology Effects on Hydraulic Motor Starting Efficiency," *Tribology Transactions*, vol. 55, no. 5, pp. 549–557, 2012.

[16] S. Bair and F. Qureshi, "Time-temperature-pressure superposition in polymer thickened liquid lubricants," *Journal of Tribology*, vol. 136, no. 2, Article ID 021505, 6 pages, 2014.

[17] A. K. Yadav, G. Mishra, and R. Yadav, "Synthesis and ultrasonic characterization of nanofluid containing silver nanoparticles," *Material Science An Indian Journal*, vol. 5, no. 3, pp. 237–242, 2009.

[18] D. Knezevic and D. Savic, "Mathematical modelling of changing of dynamic viscosity, as a function of temperature and pressure, of mineral oils for hydraulic systems," *6 FactaUniversitatisSeries: Mechanical Engineering*, vol. 4, no. 1, pp. 27–34, 2006.

[19] S. P. Washington, M. G. Karlaftis, and F. L. Mannering, *Statistical and Econometric Methods for Transportation Data Analysis*, CRC Press & Francis Group, Boca Raton, Fla, USA, 2nd edition, 2010.

[20] D. H. Buckley, *Surface Effects in Adhesion, Friction, Wear and Lubrication*, Elsevier Scientific publishing Company Amsterdam, 1981.

2

Experimental and Theoretical Analysis of Zircaloy-4 and AISI 304 Stainless Steel Material Pair in Water Sliding Conditions

Neelima Khare,[1] Praveen Kumar Limaye,[1] Kulwant Singh,[2] Dhananjay Tatyasaheb Jadhav,[1] Arundhati Bute,[3] and Navnath Kalel[4]

[1]*Refueling Technology Division, Bhabha Atomic Research Centre, Mumbai 400085, India*
[2]*Materials Science Division, Bhabha Atomic Research Centre, Mumbai 400085, India*
[3]*Laser & Plasma Technology Division, Bhabha Atomic Research Centre, Mumbai 400085, India*
[4]*Department of Mechanical Engineering, Rajarambapu Institute of Technology, Sangli 415414, India*

Correspondence should be addressed to Neelima Khare; neelima.khare@gmail.com

Academic Editor: Huseyin Çimenoğlu

Current work was simulated for sliding wear interaction of materials of fuel bundle bearing pad (zircaloy-4) and magazine rotor tube (AISI 304 stainless steel) of Indian Pressurised Heavy Water Reactors (PHWRs). A plan of experiments, based on the techniques of Taguchi, was performed. The objective was to establish a correlation between load and sliding speed with the volume loss and coefficient of friction (COF). These correlations were obtained by multiple linear regressions. The treatment of the experimental results is based on the analysis average and the analysis of variance (ANOVA). Worn surface analyses carried out using SEM and wear mechanisms were identified. ANOVA analysis indicated that load factor has a great influence on the coefficient of friction (~73%). COF suddenly increases to high value after a particular contact pressure due to absence of lubricating film and increase in metal to metal contact. Volume loss of AISI 304 stainless steel and zircaloy-4 is highly affected due to load (~90%) and speed (~65%), respectively. Worn surfaces exhibited deformation, adherence, and compaction of material at all PV conditions. Contact pressures above 475 MPa indicated formation of ratcheting mechanisms and formation of fatigue striation marks. Due to low yield strength of AISI 304 SS, volume loss was on higher side than that of Zr-4.

1. Introduction

Due to their low thermal neutron absorption cross section, good corrosion resistance in water, and good mechanical properties, zirconium and its alloys are preferred materials for nuclear reactor components [1]. Evaluation of tribological properties of zircaloy material is important from viewpoint of its sliding contacts with different materials. Some common applications of zircaloy for Indian Pressurised Heavy Water Reactors (PHWRs) are fuel cladding, coolant tube, and bearing pads for fuel element. Tribological properties of zircaloy material have been evaluated by many researchers; however, considering its application for fuel cladding tube, most of the research work is done for fretting wear analysis [2–5]. Sliding wear studies are rarely performed and only a few cases are available in literature [6, 7]. G. Sharma et al. [6] studied under water sliding wear and friction performance of zircaloy-4 material against 316 stainless steel counterface. Major wear mechanisms identified by them are micropitting, microcutting under low load condition, and delamination of deformed subsurface zones at higher load. P. J. Blau [7] in his work studied dry and oil lubricated performance of few amorphous metals including zircaloy (Zr–Cu–Ni–Ti–Al). He has suggested that more research is needed in order to establish suitability of such materials for specific wear-critical applications. R. Priya et al. [8] studied tribological performance of zirconium-702 and zircaloy-4 in dry and electrolytic environment. They confirmed that synergistic effect of wear and corrosion is more pronounced in Ti and Zr based materials than that in AISI 304L SS.

Tribology research is focusing on statistical analysis of volume loss and friction data in last few decades. R. Sinha

Table 1: Properties of zircaloy-4 and AISI 304 SS [9].

Property	Values	
Material	Zircaloy-4	AISI 304 SS
Composition (wt %)	Sn=1.3, Fe+Cr =0.30, Fe =0.2, Zr =remaining	C= 0.08, Cr=19,, Mn= 2, Ni=9, P= 0.035,S= 0.03,Si=1, Fe=remaining
Density (gm/cc)	6.56	8
Tensile Strength (MPa)	514	505
Yield Strength(MPa)	381	215
Hardness (Rockwell B)	89	70
Hardness (Vickers)	180	129
Elastic Modulus, E (GPa)	99.3	193
Poisson's Ratio, γ	0.37	0.265

et al. [10] performed abrasive wear experiment for Mn/Steel pair using full factorial design on pin-on-disc tribometer. The influence of each input factor was examined to understand its effect for weight loss. The regression equation obtained from analysis of variance table was used for the prediction of weight loss. V. V. Monikandan et al. [11] performed statistical analysis of tribological properties of aluminium matrix composites using full factorial design. ANOVA analysis revealed the statistically and physically significant factors which influence the wear loss and friction coefficient.

One of primary applications of zircaloy-4 in sliding contact is the bearing pads of fuel bundle of Indian PHWRs (Pressurised Heavy Water Reactors). During its lifetime, fuel bundle travels through various locations inside and outside the reactors. To prevent the damage, outer elements of fuel bundle are provided with the bearing pads. During loading to fueling machine, fuel bundle bearing pads come in the sliding interaction with the magazine rotor tube. Considering materials of bearing pad and magazine rotor tube as zircaloy-4 and AISI 304 SS, respectively, similar material pair was taken for experimental evaluation. Operating parameters like sliding speeds and contact pressures were selected considering some extreme operating conditions. The treatment of the experimental results is based on the analysis average and ANOVA.

2. Experimental

2.1. Materials. Sample plates of zircaloy-4 (40 mm × 20 mm × 3 mm dimension) were procured from reputed supplier. Samples were metallographically polished using a polishing machine (Struers Make: LaboPol-5, Denmark) and an average surface roughness (Ra) value of 0.4 μm was achieved. AISI 304 SS balls (12.7 mm dia. AFBMA Grade 25) were selected as a counterface material. Detailed material properties are reported in Table 1.

2.2. Tribological Testing. Tests were performed on wear and friction machine (TE-70, Phoneix Tribology Ltd, UK) with reciprocating ball-on-plate configuration. Tests were carried out in demineralised water (10.5 pH) at room temperature.

Table 2: Assignment of the levels to the factors.

Run	Speed (m/s)	Load (N)
1	0.015	5
2	0.020	8
3	0.025	11

In order to retain uniform test conditions, new ball was used for each test. Specimens were cleaned before the test by immersing in acetone in an ultrasonic bath. Sliding speed (V) and contact pressure (P) were in the range of 0.015, 0.020, 0.025 m/s (7.5, 10, 12.5 Hz frequency) and 475, 555, 616 MPa (5, 8, 11 N load), respectively. Ball was made to slide on the plate sample with selected sliding speeds (keeping sliding amplitude of 0.001 m constant) for different time durations. All tests were performed at constant sliding distance of ~100 m. Each test was repeated three times to verify repeatability of results. Average values of result are reported. Wear track profiles were formed on the plate due to ball sliding. 3D profilometry (Taylor Hobson CCI Optics, UK) was carried out to calculate dimensions of wear scar. Wear mechanisms after sliding were studied in detail by SEM (Carl Zeiss EVO 40, Germany).

2.3. Plan of Experiments (Taguchi's Techniques). For the elaboration of experiments plan, method of Taguchi was used for two factors at three levels. By levels we mean the values taken by the factors. In Table 2 the factors to be studied and the assignment of the corresponding levels are indicated.

The chosen orthogonal array was the L9 (3^2) which has 9 rows corresponding to the number of tests (8 degrees of freedom) with 4 columns at three levels, as shown in Table 3. The factors and the interactions are assigned to the columns. The plan of experiments is made of 9 tests (array rows) in which the first column was assigned to the load (F) and the second column to the velocity (v) and the remaining were assigned to the interactions.

The plan of tests was developed with the aim of relating the influence of the load, velocity in contact with the coefficient of friction, and volume loss. Table 3 presents the results analyzed.

TABLE 3: Design layout and results for AISI 304 SS and Zr-4.

Run	Load (N)	Speed (m/s)	Coefficient of Friction (μ)	Volume loss (mm^3)	
				For Ball (AISI 304 SS)	For Plate (Zr-4)
1	5	0.015	0.234	0.00782	0.00418
2	5	0.020	0.247	0.00947	0.00822
3	5	0.025	0.252	0.00552	0.00671
4	8	0.015	0.244	0.00879	0.00537
5	8	0.020	0.257	0.01058	0.01050
6	8	0.025	0.265	0.01002	0.00785
7	11	0.015	0.272	0.01860	0.00727
8	11	0.020	0.278	0.01760	0.01210
9	11	0.025	0.304	0.01530	0.01010

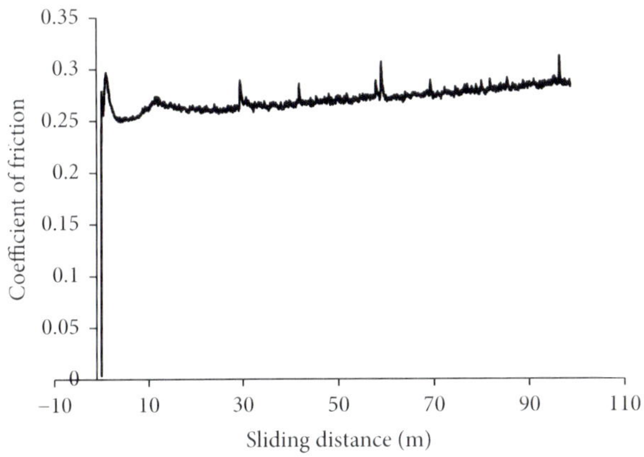

FIGURE 1: Typical variation of COF with time at 616 MPa contact pressure and 0.015 m/s sliding speed.

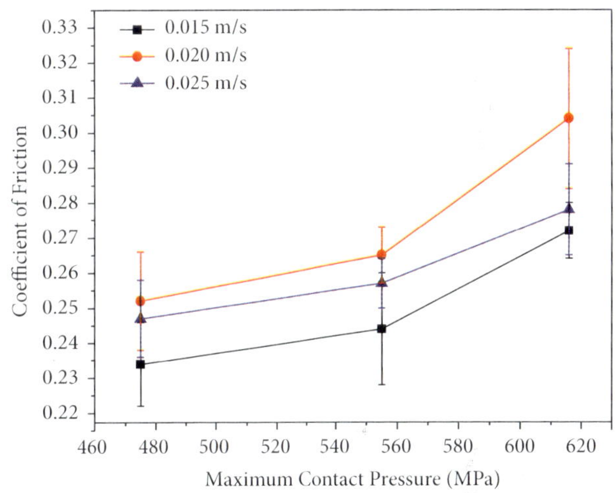

FIGURE 2: Variation of coefficient of friction with contact pressure and sliding speed.

3. Results and Discussion

3.1. Coefficient of Friction. Figure 1 presents typical variation of coefficient of friction (COF) with sliding distance. Once the graph is stabilized, average value of COF was calculated and reported. Figure 2 presents variation of coefficient of friction with sliding speed and contact pressure. Typical trend is COF increasing with increase in contact pressure. General range of COF is 0.23-0.30.

In general, COF decreases with increase in contact pressure; however other parameters like properties of oxide, generation of wear particles, type of lubrication, etc. may affect the trend. Coefficient of friction is the result of transformation and dissipation of input energy in various forms. Hence coefficient of friction can be presented by the following mathematical formula:

$$\mu = \mu_{adh} + \mu_{def} + \mu_{els\_hys}, \quad (1)$$

where μ_{adh} is due to adhesion of asperities, μ_{def} is due to abrasion or plowing, and μ_{els_hys} is due to elastic hysteresis processes.

Figure 2 is indicating increased rate of COF variation with increase in contact pressure. The reason behind this could be explained by the fact that after a particular high value of contact pressure, rate of oxide layer formation is lower than breakage of oxide layer. This results into interaction of clean metals and hence increased adhesion component of friction.

It is observed from Table 4 that the load factor (P= 72.77%) has great influence on the coefficient of friction. It should also be noticed that the error associated with the table ANOVA for the coefficient of friction was approximately 3.71 %, which is rather low. Regression equation for coefficient of friction is presented by the following formula:

$$\mu = 0.1603 + 0.006722 \text{ F} + 2.367 \text{ v} \quad (2)$$

3.2. Wear Analysis

3.2.1. Zircaloy-4. Figure 3 presents image of wear scar taken by 3D profilometer. Volume of wear was determined from scar measurement. Figure 4 presents variation of volume loss with change in contact pressure and sliding speed of zircaloy-4. It shows decreasing trend with increase in contact pressure.

Table 4: ANOVA for coefficient of friction.

Source	DF	Adj SS	Adj MS	F-Value	P Value	% P
Load, F (N)	2	0.002608	0.001304	39.19	0.002	72.77
Speed, v (m/s)	2	0.000843	0.000421	12.66	0.019	23.52
Error	4	0.000133	0.000033			3.71
Total	8	0.003584				

Table 5: ANOVA for volume loss of Zr-4.

Source	DF	Adj SS	Adj MS	F-Value	P Value	% P
Load, F (N)	2	0.000018	0.000009	79.25	0.001	35.29
Speed, v (m/s)	2	0.000033	0.000016	144.83	0.001	64.71
Error	4	0.0000001	0.0000001			0.2
Total	8	0.000051				

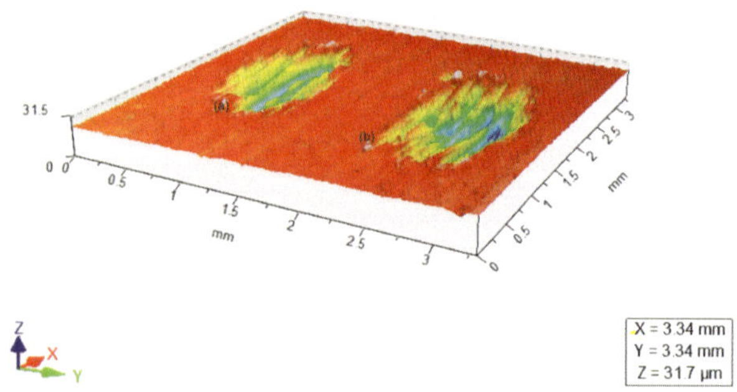

Figure 3: 3D Profilometry at 616 MPa for (a) 0.020 m/s and (b) 0.025 m/s sliding speed.

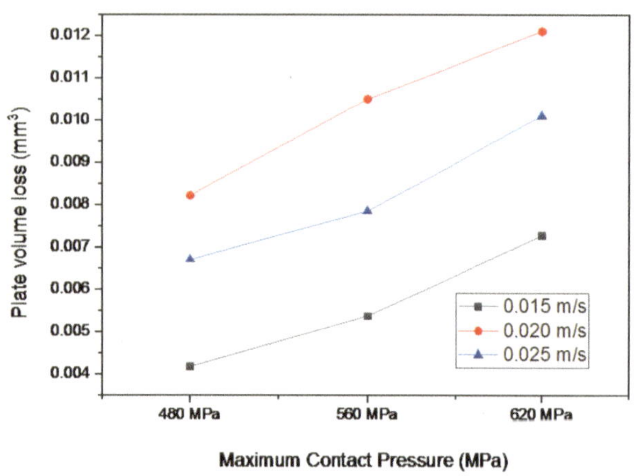

Figure 4: Variation of volume loss of zircaloy-4 plate with contact pressure and sliding speed.

Due to low viscosity, water is ineffective to act as lubricant; hence material properties play important role for components working in water medium. Formation and fracture of brittle oxides caused metal to metal interaction. Generated wear particle may further dictate the wear behavior of material pair. In current case, due to flow of water, oxide particles might have washed away and hence could not take part further after coming out from metal surface. In general adhesion increases with increase in contact pressure; however

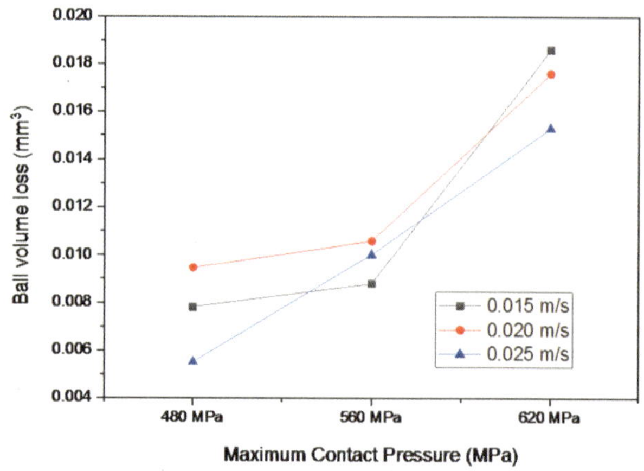

Figure 5: Variation of volume loss of AISI 304 SS with contact pressure and sliding speed.

properties of wear particles also govern performance. Regression equation for volume loss of Zr-4 is presented by the following formula:

$$(\text{Vol. loss})\ \text{Zr-4} = -0.00180 + 0.000576\,F + 0.261\,v \quad (3)$$

It is observed from Table 5 that the speed factor (P= 64.71%) has great influence on the volume loss. It should be noticed that the error associated with the table ANOVA or the volume loss was 0.2 %.

3.2.2. AISI 304 Stainless Steel. Figure 5 presents variation of volume loss of AISI 304 SS with change in contact pressure and sliding speed. Volume loss is decreasing with increase in sliding speed. Considering ball-on-plate contact geometry (Figure 6), the following formula was used to calculate Hertzian contact pressure or maximum contact pressure along the centre line of circular contact area:

$$p_{\max} = \frac{3F}{2\pi a^2}$$

$$a = \sqrt[3]{\frac{3F\left[\left(1-\nu_1^2\right)/E_1 + \left(1-\nu_2^2\right)/E_2\right]}{4\left(1/R_1 + 1/R_2\right)}}, \quad (4)$$

TABLE 6: Ratio of contact pressure and yield strength for zircaloy-4 and AISI 304 SS.

Applied Load, F (N)	Contact pressure (MPa)	$\left(\frac{\text{Contact Pressure}}{\text{Yield Strength}}\right)_{Zr-4}$	$\left(\frac{\text{Contact Pressure}}{\text{Yield Strength}}\right)_{\text{AISI 304SS}}$
5	475	1.24	2.20
8	555	1.45	2.58
11	616	1.61	2.68

TABLE 7: ANOVA for volume loss for AISI 304 SS.

Source	DF	Adj SS	Adj MS	F-Value	P Value	% P
Load, F (N)	2	0.000151	0.000075	41.04	0.002	90.96
Speed, v (m/s)	2	0.000008	0.000004	2.16	0.231	4.82
Error	4	0.000007				4.22
Total	8	0.000166				

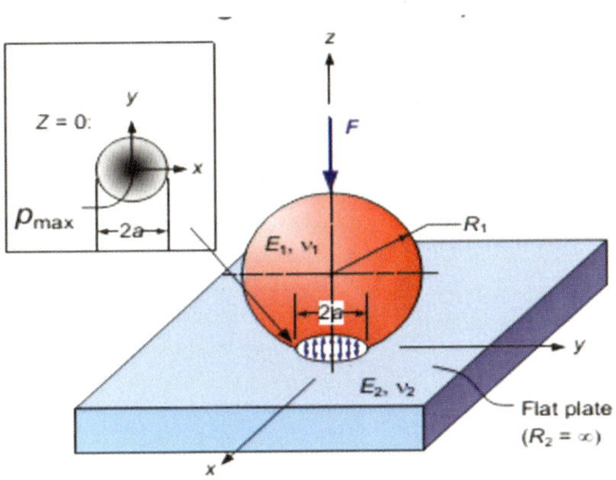

FIGURE 6: Ball-on-plate contact geometry.

where F is applied load in N, a is radius of circular contact area in mm^2, E_1,γ_1, and E_2,γ_2 are elastic modulus and poisons ratio of ball and plate materials, respectively, and R_1 and R_2 are radius of ball and plate. Ratio of contact pressure and yield strength is calculated and reported in Table 6.

Due to availability of very low contact area for AISI 304 SS ball material, oxide layer, once removed, caused increase in volume loss at certain contact pressure. Properties of wear particle are not much influencing this. Relation between yield strength of material and contact pressure dictates severity of wear. Generally if ratio of contact pressure and yield strength is less than 0.5, mild wear occurs; else severe wear may be observed [12]. Considering above calculation, severe wear may be observed; worn surface analysis was carried out to identify wear mechanisms.

It is observed from Table 7 that the load factors (P= 90.96%) have great influence on the volume loss. It should be noticed that the error associated with the table ANOVA for the volume loss was 4.22%, i.e., rather low. Regression equation for volume loss of AISI 304 SS is presented by the following formula:

$$\begin{aligned}&(\text{Vol. loss})\ \text{AISI 304 SS}\\&= 0.0203 - 0.00162\,F - 0.818\,v + 0.118\,F * v\end{aligned} \quad (5)$$

Analysis of variance (ANOVA) Table 7 includes the F and p values. These are used to determine whether the factors are significantly related to the output (response) or not. Generally, in ANOVA, P value is used to determine whether a factor is significant or not by comparing with an alpha value of 0.05 (used in hypothesis testing). If the p value < 0.05, then the factor is significant. In Table 7, P value is 0.002 which is less than 0.05; thus here input factor load (F) is more significant compared to speed (v) factor. Moreover, percentage of contribution (i.e., % P) is evaluated for both the input factors which showed 90.96 % of contribution for load (F), 4.82 % of contribution for speed (v), and 4.22 % of contribution for error (i.e., for interaction effects and other factors (may be environmental)).

4. Worn Surface Analysis

Generally, sliding under water lubricated environment causes adhesive wear at high spots. At junctions where lubrication is not failed, wear may occur by deformation mechanisms. Similarly individual junction conditions based on type of material in contact can determine whether an oxide layer flakes of or if fracture occurs below the oxide.

Worn surfaces analysis was carried out at various PV conditions. Figures 7(a)–7(i) present the SEM images of worn surfaces. Due to combined effects of repeated cycles and absence of lubricating film, most of the worn surfaces are indicating simultaneous occurrence of several wear mechanisms. Deformation, adherence, and compaction of material were observed at most of the PV conditions. Apart from above, other mechanisms were also observed; some mechanisms are dominating at low and others at high PV conditions.

Microcracks were observed at low PV conditions [Figure 7(a)]. Ratcheting mechanism due to repeated cycles sliding and transferred material was seen at 555 MPa and 0.015 m/s conditions [Figure 7(b)]. With further increase in contact pressure (at same sliding speed), ratcheting mechanisms and striation marks due to fatigue were observed [Figure 7(c)]. Formation of microgrooves due to

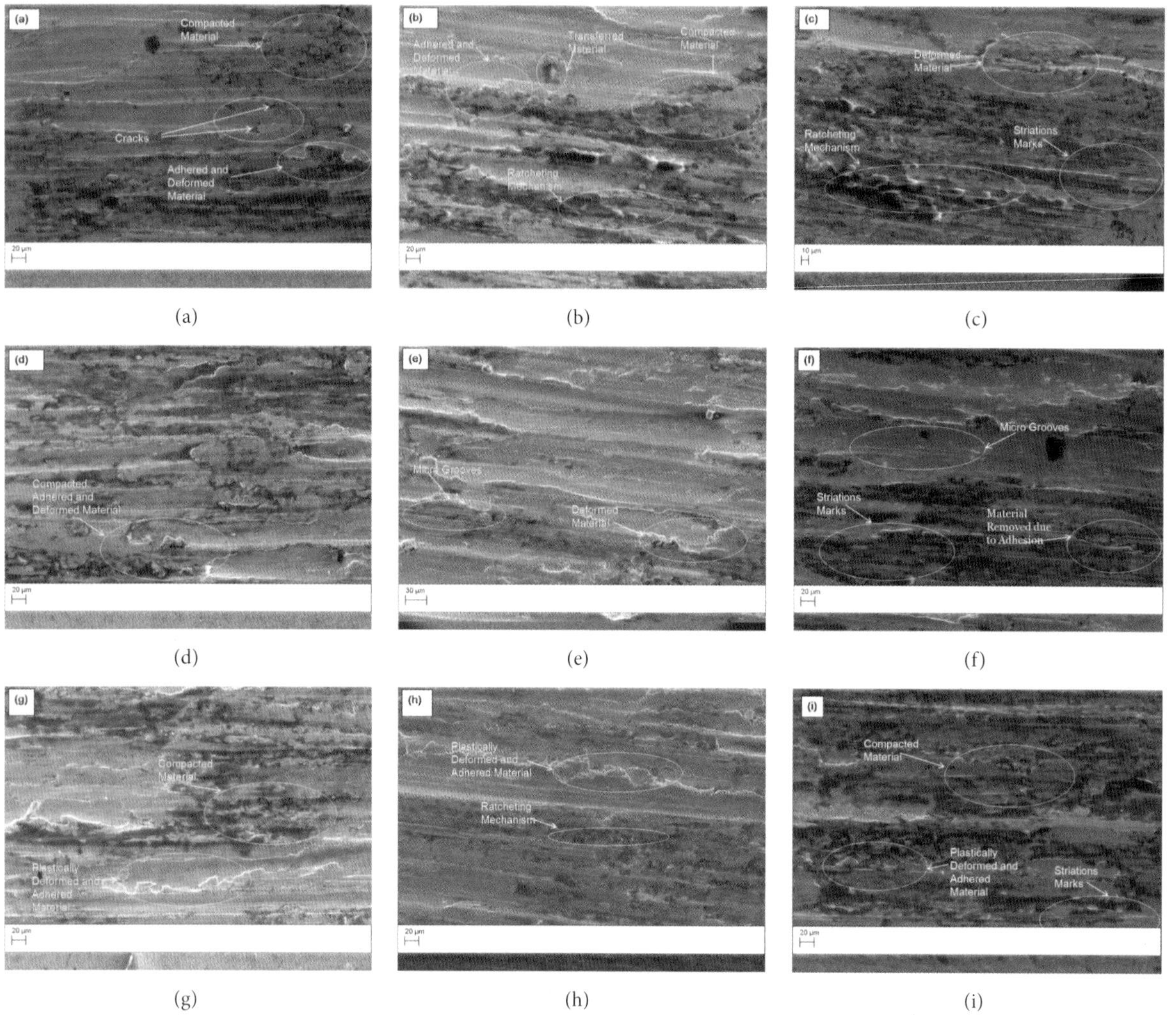

FIGURE 7: SEM images of worn surfaces of Zircaloy-4 at (a) 0.015 m/s, 475 MPa; (b) 0.015 m/s, 555 MPa; (c) 0.015 m/s, 616 MPa; (d) 0.020 m/s, 475 MPa; (e) 0.020 m/s, 555 MPa; (f) 0.020 m/s, 616 MPa; (g) 0.025 m/s, 475 MPa; (h) 0.025 m/s, 555 MPa; (i) 0.025 m/s, 616 MPa.

absence of lubricating film was observed at 0.02 m/s and 555 MPa/616 MPa contact pressure [Figures 7(e) and 7(f)]. At 0.02 m/s and 616 MPa contact pressure, material pulls out due to absence of lubricating film/ oxide layer observed [Figure 7(f)]. Ratcheting mechanisms and fatigue striations marks were observed at higher PV conditions [Figures 7(h) and 7(i)].

Conclusively, it can be stated that, due to high ratio of contact pressure to yield strength and presence of ineffective water lubricant, deformation, adherence, and compaction of material were observed at all the PV conditions. It can also be added that stress and mechanical properties are primary factors affecting wear mechanisms and both are responsible to increase the severity of mechanisms. High contact pressures above 475 MPa are indicating ratcheting mechanisms and formation of fatigue striation marks. Due to severe plastic deformation, crack generated due to repeated cycle fatigue could not be observed much.

5. Conclusions

Study was carried out for determination of underwater tribological properties of pairing materials of bearing pad (Zr-4) and magazine rotor tube (AISI 304 SS) of Indian Pressurised Heavy Water Reactors (PHWRs). Sliding speeds and contact pressures range were selected considering some extreme operating conditions. Coefficient of friction, volume loss, and wear mechanisms were evaluated. The treatment of the experimental results is based on the analysis average and the analysis of variance (ANOVA). Selected conclusions are as follows:

(i) ANOVA analysis indicated that load factor has a great influence on the coefficient of friction (~72.77%) and volume loss of AISI 304 SS (~90.96%). Speed factor has high influence volume loss of Zr-4 (~64.71%).

(ii) The errors associated with the table ANOVA for the coefficient of friction, volume loss of AISI 304 SS, and volume loss of Zr-4 are only 3.71%, 4.22%, and 0.2%, respectively.

(iii) Decreasing volume loss with increase in contact pressure of Zr-4 was observed. It was explained that generated wear particles might have washed away from interface at the interface and led to decreased volume loss of clean metals. Due to high ratio of contact pressure to yield strength of AISI 304 SS, at any PV value volume loss of AISI 304 SS was higher than that of Zr-4.

(iv) SEM examination indicated deformation, adherence, and compaction of material at all the PV conditions. High contact pressures above 475 MPa indicated formation of microgrooves, ratcheting mechanisms, and formation of fatigue striation marks.

(v) Conclusively, it can be stated that, to keep the wear in mild zone, contact pressure should not exceed 0.5 times the yield strength of material under water environment.

Conflicts of Interest

The authors declare that they have no conflicts of interest.

References

[1] R. Krishnan and M. K. Asundi, "Zirconium alloys in nuclear technology," *Advances in Tribology*, vol. 4, no. 1, pp. 41–56, 1981.

[2] K. H. Cho, T. H. Kim, and S. S. Kim, "Fretting wear characteristics of Zircaloy-4 tube," *Wear*, vol. 219, no. 1, pp. 3–7, 1998.

[3] O. Vingsbo, A. R. Massih, and S. Nilsson, "Evaluation of fretting damage of zircaloy cladding tubes," *Journal of Tribology*, vol. 118, no. 4, pp. 705–710, 1996.

[4] J. H. Sung, T. H. Kim, and S. S. Kim, "Fretting damage of TiN coated zircaloy-4 tube," *Wear*, vol. 250-251, pp. 658–664, 2001.

[5] H.-K. Kim, S.-J. Kim, K.-H. Yoon, H.-S. Kang, and K.-N. Song, "Fretting wear of laterally supported tube," *Wear*, vol. 250-251, no. 1, pp. 535–543, 2001.

[6] G. Sharma, P. K. Limaye, and D. T. Jadhav, "Sliding wear and friction behaviour of zircaloy-4 in water," *Journal of Nuclear Materials*, vol. 394, no. 2-3, pp. 151–154, 2009.

[7] P. J. Blau, "Friction and wear of a Zr-based amorphous metal alloy under dry and lubricated conditions," *Wear*, vol. 250-251, no. 1, pp. 431–434, 2001.

[8] R. Priya, C. Mallika, and U. K. Mudali, "Wear and tribocorrosion behaviour of 304L SS, Zr-702, Zircaloy-4 and Ti-grade2," *Wear*, vol. 310, no. 1-2, pp. 90–100, 2014.

[9] http://www.matweb.com.

[10] R. Sinha and A. Mukhopadhyay, "Wear characterization and modelling of Mn–steel liners used in rock crushers," *Perspectives in Science*, vol. 8, pp. 374–376, 2016.

[11] V. V. Monikandan, J. C. Jacob, M. A. Joseph, and P. K. Rajendrakumar, "Statistical Analysis of Tribological Properties of Aluminum Matrix Composites Using Full Factorial Design," *Transactions of the Indian Institute of Metals*, vol. 68, no. S1, pp. 53–57, 2015.

[12] R. G. Bayer, *Wear Analysis for Engineers*, HNB Publishing, New York, NY, USA, 2002.

3

Numerical Investigation of Turbulence Models for a Superlaminar Journal Bearing

Aoshuang Ding, Xiaodong Ren, Xuesong Li, and Chunwei Gu

Department of Thermal Engineering, Tsinghua University, Beijing 100084, China

Correspondence should be addressed to Xuesong Li; xs-li@mail.tsinghua.edu.cn

Academic Editor: Michel Fillon

With rotating machineries working at high speeds, oil flow in bearings becomes superlaminar. Under superlaminar conditions, flow exhibits between laminar and fully developed turbulence. In this study, superlaminar oil flow in an oil-lubricated tilting-pad journal bearing is analyzed through computational fluid dynamics (CFD). A three-dimensional bearing model is established. CFD results from the laminar model and 14 turbulence models are compared with experimental findings. The laminar simulation results of pad-side pressure are inconsistent with the experimental data. Thus, the turbulence effects on superlaminar flow should be considered. The simulated temperature and pressure distributions from the classical fully developed turbulence models cannot correctly fit the experimental data. As such, turbulence models should be corrected for superlaminar flow. However, several corrections, such as transition correction, are unsuitable. Among all the flow models, the SST model with low-Re correction exhibits the best pressure distribution and turbulence viscosity ratio. Velocity profile analysis confirms that a buffer layer plays an important role in the superlaminar boundary layer. Classical fully developed turbulence models cannot accurately predict the buffer layer, but this problem can be resolved by initiating an appropriate low-Re correction. Therefore, the SST model with low-Re correction yields suitable results for superlaminar flows in bearings.

1. Introduction

Lubricant flow in bearings has been considered laminar and lubricant cavitation has been extensively investigated. Tensile stress has also been explored to determine the mechanisms of vaporous and gaseous cavitation [1–3]. Two cavitation mechanisms have been compared under static and dynamic loads to identify the dominant mechanism [4–6]. A new model based on air solubility has also been proposed for gaseous cavitation [7, 8] and improved by considering nonequilibrium dissolution for oil film in bearings [9]. A tapered land thrust bearing has also been utilized to verify the cavitation effect [10, 11]. The relationship between temperature and cavitation has been evaluated [12–15].

With rotating machineries working at high rotating speeds on a large scale, the performances of journal bearings weaken as friction loss increases and the lubricant flow regime of journal bearings changes from laminar to superlaminar or turbulent. Considering the significantly high Reynolds number of a fully developed turbulent flow, oil flow regime in high-speed journal bearings is superlaminar [16]. Thus, superlaminar journal bearings should be accurately predicted to improve the performance of these bearings.

Superlaminar oil films in bearings have been explored by considering turbulence effect. The turbulence effect on bearing lubricant flows was initially examined on the basis of mixing length theory. Compared with the laminar simulation results and experimental data, turbulence simulation results confirm that turbulence effect should be considered for bearing lubricant flows [17, 18]. Ng and Pan [19] further investigated the theories based on law of wall and mixing length theory. Considering the verified theoretical results based on law of wall, Hirs [20] proposed a bulk-flow theory for bearing lubricant simulation. Suganami and Szeri [21] examined the thermal effect and confirmed that this effect is essential for bearing performance. Detailed thermohydrodynamic (THD) analysis has also been conducted to confirm the important role of inertia [22] and superlaminar flow regimes [23] in flow simulation. Considering the surface roughness effect, the

Reynolds equation has been modified for an accurate THD analysis of the flow [24].

Taniguchi et al. [25] established an experimental apparatus for a tilting-pad journal bearing to obtain the bearing surface pressure and temperature data under variable bearing loads and rotating speeds. In this study, the simulation results of a THD model with the modified Reynolds equations can catch the general characteristics of experimental data, but not accurately. The accurate predictions of superlaminar bearings need precise models for turbulence effects in oil flows. Meanwhile, compared with the Reynolds model, the computational fluid dynamic techniques have enough available turbulence models for oil flow simulations. Therefore, CFD is a feasible method for simulations of superlaminar bearing oil flows.

CFD techniques have been utilized to simulate superlaminar oil flows in bearings. Maneshian and Gandjalikhan Nassab [26] reported the CFD results of a journal bearing with the AKN low-Re k-ε turbulence model. On the basis of this work, Solghar and Gandjalikhan Nassab [27] numerically analyzed plain full journal bearings with the same turbulence model.

The turbulence models play an important role in CFD simulations and influence the simulation results significantly. While Bouard et al. have compared three turbulence models using Reynolds equation in tilting-pad journal bearings [28], the turbulence models in CFD techniques are different from Reynolds equation and current CFD studies of superlaminar bearings have not provided a detailed comparison of different turbulence models. Thereby, this paper compares temperature and pressure simulation results of fourteen turbulence models with experimental data from Taniguchi et al. [25] to determine the most suitable turbulence model for superlaminar simulation and explain the underlying mechanisms and what is superlaminar flow.

The remaining parts of this paper are organized as follows. Section 2 presents the geometry and meshing of the test superlaminar journal bearing of Taniguchi et al. [25]. Section 3 describes fourteen turbulence models to simulate superlaminar flow. Section 4 discusses the temperature, pressure, and turbulent viscosity ratio and demonstrates the most suitable turbulence model for the THD simulation of superlaminar oil flow. Section 5 compares the simulated velocity distributions and theoretical results based on law of wall. Section 5 further illustrates the mechanism of superlaminar flows and factors contributing to the different performance levels of the turbulence models. Section 6 shows the conclusions on the selections of the turbulence models for superlaminar flows and reveals the physical meaning of superlaminar flows.

2. Geometry and Meshing

The simulation object is the test tilting-pad journal bearing of Taniguchi et al. [25]. Figure 1 depicts the cross section of the entire experimental apparatus. Table 1 lists the specifications of this journal bearing.

The simulated working condition for this tilting-pad journal bearing is selected as the standard working condition at 3000 r/min and under 180 kN loading. Given the experimental film pressure distributions shown in Figure 2, numbers 2 and 3 bearing pads are loaded, whereas numbers 1 and 4 bearing pads are unloaded. The oil flow in number 3 pad has similar characteristics to number 2 pad because of the load angle between the pads. Thus, the entire journal bearing is replaced by one loaded pad in the simulations to simplify the bearing geometry. The main characteristics of the oil flow can also be obtained from the simulation results of the loaded pad.

Figure 3 shows the measured oil film thickness of number 2 pad at 3000 r/min and under 180 kN loading [25]. The three-dimensional geometry of the structured mesh for oil is fixed based on the measured film thickness and the mesh independence of the structured mesh has been validated [29]. The structured mesh is shown in Figure 4. Along the lubricant flowing direction, the radial thickness is set as the measured film thickness. In the axial direction, the radial thickness is constant and the axial length equals to the bearing length, 300 mm. The side faces are perpendicular to the length direction (z-axis). The rotor-side and pad-side walls are normal to the radial direction.

TABLE 1: Specifications of tested bearing [25].

Bearing type	Tilting pad
Bearing diameter (mm)	479
Bearing length (mm)	300
Radial clearance (mm)	0.612
Pad thickness (mm)	121
Number of pads	4
Pad arc (dleg)	80
Pivot offset	0%
Preload factor	0
Load angle	Between pads

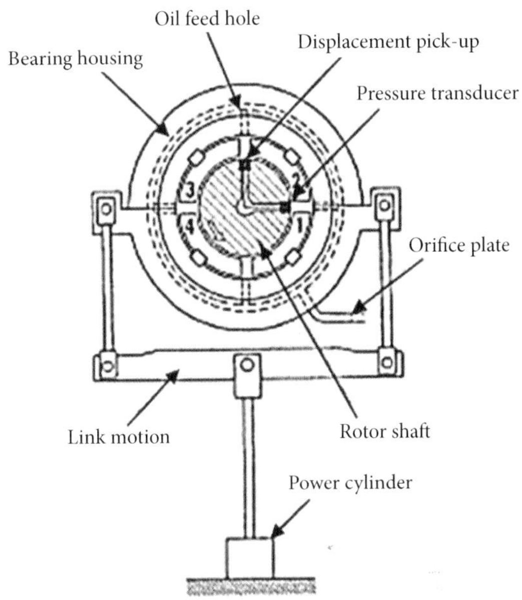

FIGURE 1: Geometry of the stream surface [25].

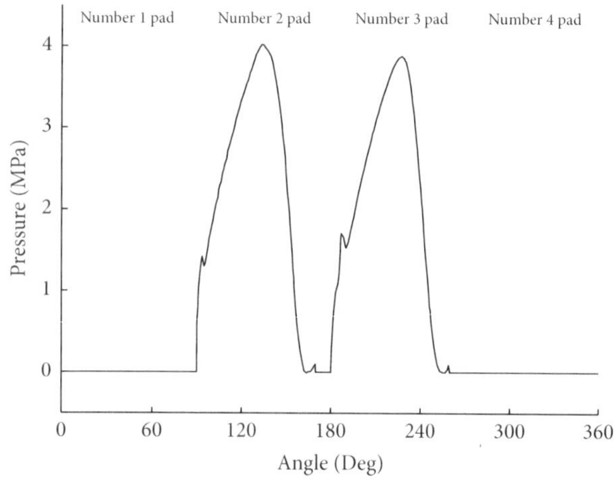

FIGURE 2: Experimental film pressure at 3000 r/min and under 180 kN loading [25].

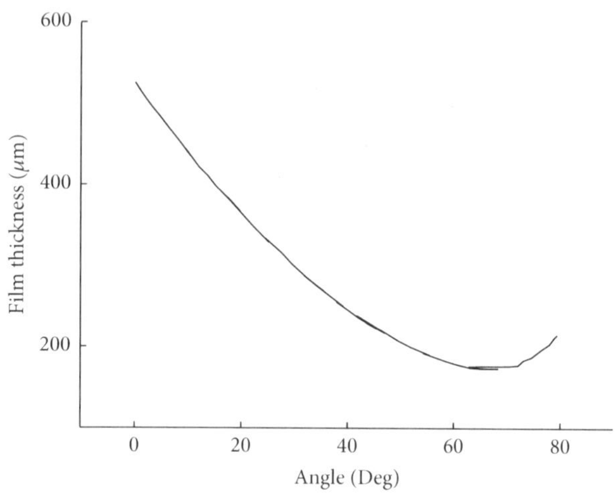

FIGURE 3: Measured film thickness at 3000 r/min and under 180 kN loading [25].

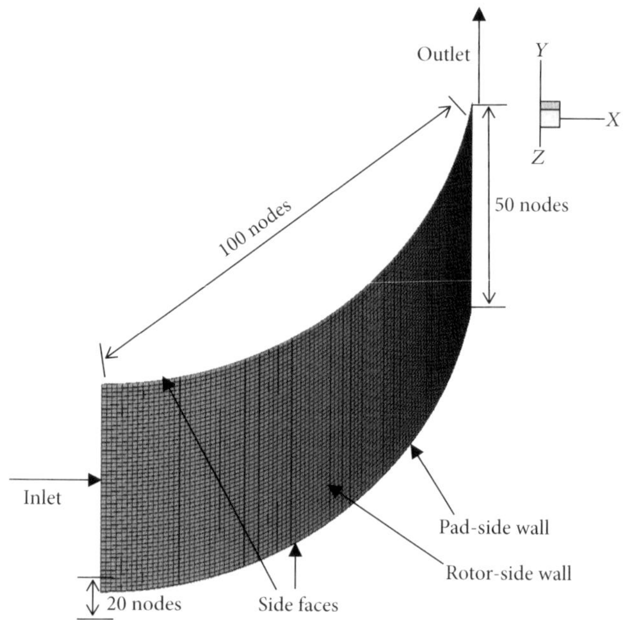

FIGURE 4: Mesh [29].

TABLE 2: Viscosity properties of ISO VG32.

Temperature (K)	Viscosity (kg/(m·s))
313.15	0.0275
318.15	0.0224
323.15	0.0185
328.15	0.0155
333.15	0.0131
338.15	0.0112
343.15	0.0096
348.15	0.0084
353.15	0.0073
373.15	0.0046

TABLE 3: Boundaries.

Boundary	Boundary type	Boundary details
Inlet	Pressure inlet	0.1 MPa 315.15 K
Outlet	Pressure outlet	0.1 MPa
Side faces	Pressure outlet	0 MPa
Pad-side wall	No slip	115 W/(m^2·K)
Rotor-side wall	No slip	115 W/(m^2·K) 314.159 rad/s

The minimum orthogonal quality of number 2 pad mesh is equal to 0.757, where the values near 1 correspond to high quality. The maximum ortho skew of number 2 pad mesh is equal to 0.243, where the values near 0 correspond to high quality. Therefore, this structured mesh has high quality.

The lubricant for this journal bearing is ISO VG32. Considering the thermal effect on material viscosity, the piecewise linear interpolation method is employed to determine the relationship between the temperature and viscosity with the data listed in Table 2. The supply oil pressure and temperature in the oil feed hole are 0.1 MPa and 315.15 K, respectively, based on the data of the experimental apparatus.

Table 3 lists the boundaries in detail with the reference pressure set as 1 bar. Considering the outlet of number 2 pad is at the position of another oil inlet hole and the measured outlet pressure is 0.1 MPa, the relative pressures of the outlet of number 2 pad is set at 0.1 MPa. The inlet mixture temperature is assumed to be uniformly distributed, and its value is set at the measured temperature of 315.35 K in the inlet of number 2 pad. Furthermore, the heat transfer coefficient for the pad-side and rotor-side walls is set at 115 W/(m^2·K) with the free stream temperature equal to the inlet temperature based on the work of Taniguchi et al. [25]. The lubricant velocity on the rotor-side wall is set as the rotational speed of the rotor, 314.159 rad/s. The lubricant velocity on the pad-side wall is set as zero.

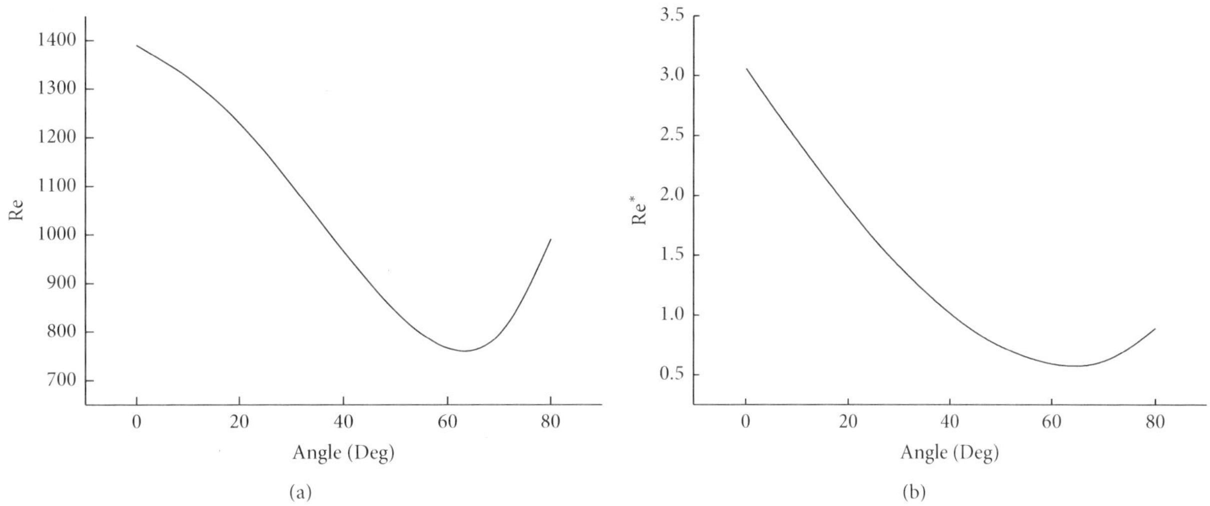

FIGURE 5: Reynolds number of oil film: (a) Re; (b) Re*.

3. Turbulence Models

The oil flow regime in the journal bearing should be discussed clearly before suitable models are selected to facilitate the CFD simulations. This regime can be characterized by the Reynolds number, which can be defined as follows [30]:

$$\mathrm{Re} = \frac{\rho u d}{\mu}, \quad (1)$$

where d is set as the measured oil film thickness, u is set as the rotational speed of the rotor-side wall, and the value of μ is decided by the rotor-side temperature to calculate the Reynolds number of the oil film.

Apart from the noted definition for the Reynolds number, an alternative reduced Reynolds number is also typically defined for considering the effect of the shaft diameter properly in bearing analyses [30]. Its definition is

$$\mathrm{Re}^* = \frac{\rho u R}{\mu}\left(\frac{d}{R}\right)^2, \quad (2)$$

where R is set as the radius of the rotor.

Both the definitions of the Reynolds number are the accepted methods of nondimensionalizing the flow turbulence characteristics in modern bearing treatments [30]. The calculated Reynolds numbers initially decrease and then increase in outlet (Figure 5). The Reynolds numbers change indicates that the flow regime changes from superlaminar to laminar and turbulence effect on the oil film changes with the Reynolds numbers. Thus, the turbulence effect should be considered in the simulations.

This study uses ANSYS Fluent with finite volumes for the bearing simulations. In CFD, the basis of the incompressible turbulence and laminar models is the Navier-Stokes equation and continuous equation. Navier-Stokes equation is expressed as follows:

$$\rho\left(\frac{\partial u}{\partial t} + u\cdot\nabla u\right) = -\nabla p + \rho f + \nabla(\lambda\nabla\cdot u) + \nabla \cdot \left(2\left(\mu+\mu_t\right)E\right). \quad (3)$$

The continuous equation is

$$\nabla\cdot u = 0. \quad (4)$$

Considering the thermal effect, the energy equation is included:

$$\rho\frac{Dh}{Dt} = \frac{Dp}{Dt} + \nabla\cdot(k\nabla T) + \phi, \quad (5)$$

where the viscous dissipation function is given by the following:

$$\phi = 2\left(\mu+\mu_t\right)\left[\left(\frac{\partial u_i}{\partial x_i}\right)\left(\frac{\partial u_i}{\partial x_i}\right) + \frac{1}{2}\left(\frac{\partial u_j}{\partial x_l}\right)\left(\frac{\partial u_l}{\partial u_j}\right)\right]. \quad (6)$$

The values of turbulent viscosity μ_t and kinetic viscosity μ are required to solve the aforementioned equations. While the laminar model considers turbulent viscosity as zero, the turbulence models utilize variable methods to obtain the value of turbulent viscosity.

A total of fourteen turbulence models are employed to simulate the oil flow. Table 4 shows that these models are classified into three groups: the classical fully developed turbulence models, transition turbulence models, and turbulence models with low-Re correction. Each group includes different corrections for the standard turbulence models, which forms the new modified turbulence models.

Different corrections are added to these turbulence models.

The equations of the low-Re corrections for different turbulence models are variable. For example, the purpose

Table 4: Turbulence models.

Classical full-developed turbulence models	Transition turbulence models	Turbulence models with low-Re correction
(1) SA model (2) Standard k-ε model with scalable wall treatment (3) RNG k-ε model with scalable wall treatment (4) Realizable k-ε model with scalable wall treatment (5) Standard k-ω model (6) SST model (7) Quadratic pressure–strain RSM model	(8) SST model with intermittency transition (9) Transition SST model (Menter–Langtry γ-θ Laminar–Turbulent Transition) (10) Transition k-kl-ω model	(11) AKN low-Re k-ε model (12) Abid low-Re k-ε model (13) Lam–Bremhorst low-Re k-ε model (14) SST model with low-Re correction

Table 5: Parameters of k-ε models.

k-ε models	C_μ	$C_{1\varepsilon}$	$C_{2\varepsilon}$	C_2	σ_k	σ_ε
Standard k-ε (scalable wall treatment)	0.09	1.44	1.92		1	1.3
AKN k-ε (low-Re correction)						
Abid k-ε (low-Re correction)						
Yang–Shih k-ε (low-Re correction)						
RNG k-ε (scalable wall treatment)	0.0845	1.42	1.68			
Realizable k-ε (scalable wall treatment)		1.44		1.9	1	1.2

of the low-Re correction for the SST model is to dampen the turbulent viscosity with the coefficient α^* and modify the transport equations of the turbulent kinetic energy and specific dissipation rate. The α^* equation is expressed as follows:

$$\alpha^* = \frac{0.144 + \mathrm{Re}_t}{6 + \mathrm{Re}_t}, \quad \mathrm{Re}_t = \frac{\rho k}{\mu\omega}. \tag{7}$$

The intermittency transition correction is based on the Menter-Langtry γ-θ laminar-turbulent transition model, which is also called the transition SST model. The modifications in the production and destruction terms from the turbulent kinetic energy equation are implemented to ensure that the kinetic energy production is sufficiently low at the transition point.

All of the turbulence models in Table 4 have adjustable parameters. These parameters are set to their default values, which are the same as those in the FLUENT software, to set the turbulence models reasonably. For example, the parameters of the k-ε models are listed in Table 5.

All turbulence models use the half-Sommerfeld cavitation model.

4. Simulation Result Comparisons

4.1. Temperature Distribution Comparisons. An accurate simulation of temperature distributions is necessary to determine the THD characteristics. Therefore, the simulated temperature distributions in the pad-side center line are selected for comparison with the experimental data of number 3 pad from the results of Taniguchi et al. [25] to determine the most suitable turbulence model for superlaminar journal bearing simulation.

Compared with the experimental data, Figure 6 shows the temperature simulation results in the pad-side center line.

Figure 6 shows that the simulated pad-side temperature of the classical fully developed turbulence models is much lower than the experimental data. The temperature difference around the outlet is more than 5 K. By contrast, the laminar pad-side temperature is 5 K higher than the experimental temperature data around the outlet. Regardless of the high or low temperature simulation results, remarkable differences from the experimental data indicate that the classical fully developed turbulence models and laminar model cannot accurately determine the thermodynamic characteristics of the superlaminar oil flow.

Apart from the classical fully developed turbulence models, the performance of transition turbulence models is poor. Among them, the transition k-kl-ω model and SST model with intermittency transition model have almost the same temperature simulation results as the laminar model. The least error of the transition turbulence models is still more than 5 K around the outlet. Therefore, the transition turbulence models are unsuitable for the THD simulation of the superlaminar bearing.

Although the temperature distributions of the k-ε models with low-Re correction are approximately 5 K higher than the experimental data, the SST model with low-Re correction has almost the same temperature simulation results as the experimental data. Its largest error is approximately 2 K, which is lower than the errors of the other thirteen turbulence models. Thus, considering that temperature distribution is a vital thermal characteristic, the SST model with low-Re correction is the best turbulence model for the superlaminar oil flow simulation.

4.2. Load Comparisons. In the numerical simulations, the film geometry of this journal bearing is given by the experimental film thickness distribution under 180 kN and at 3000 r/min. Considering the bearing load is a vital characteristic, the simulated bearing loads are compared with the experimental load.

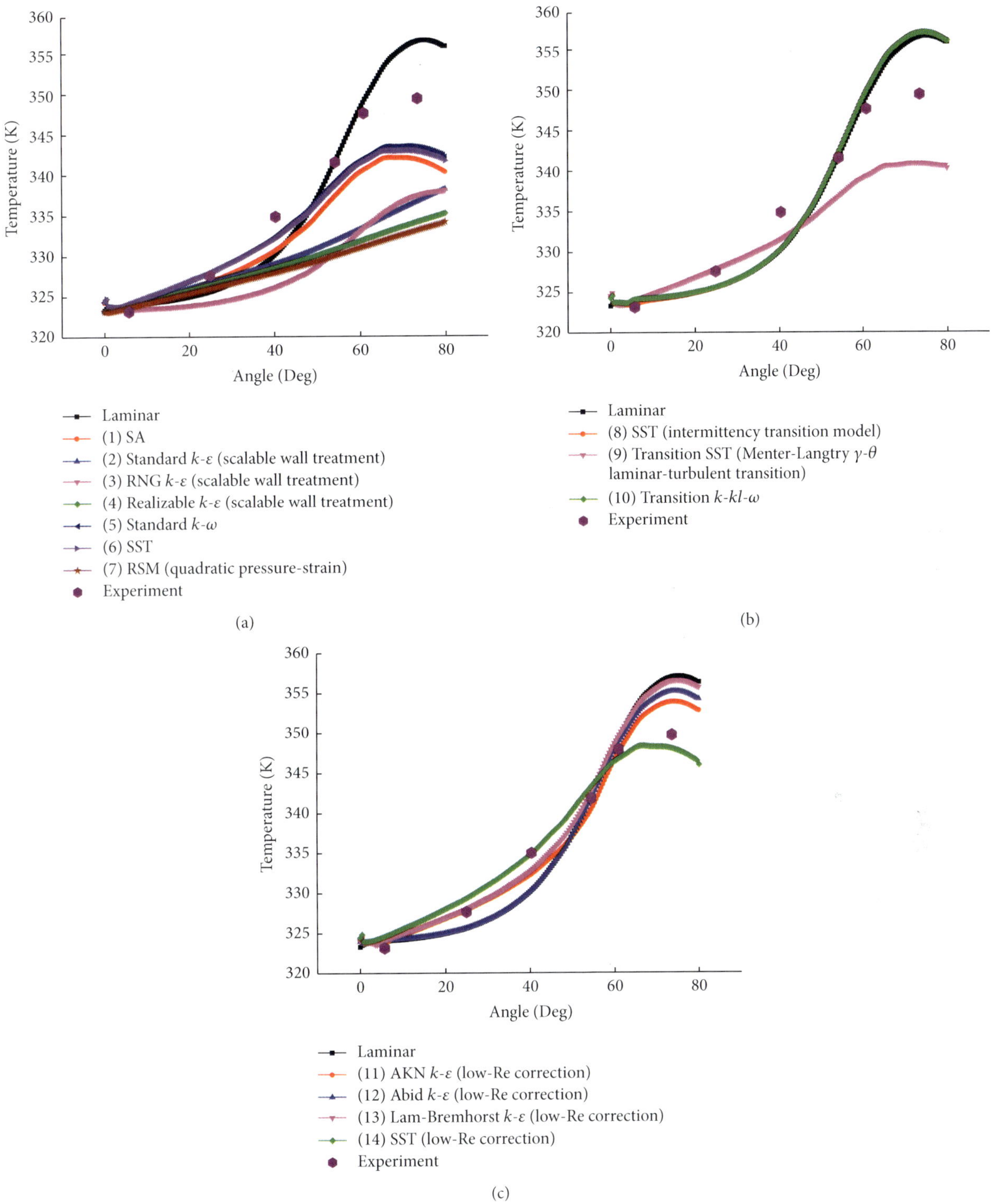

Figure 6: Comparison between the experimental temperature profile at 3000 rpm and under 180 kN and simulated temperature profile with the fixed film geometry in pad-side center line for (a) the group of the classical turbulence models; (b) the group of the transition turbulence models; and (c) the group of the turbulence models with low-Re correction.

As noted in Section 2 and Figure 2, the loaded pads provide load-bearing and the loaded pads, numbers 2 and 3 pads, have almost the same pressure distribution. Thus, one loaded pad provides the half bearing load, equaling to 90 kN. The pad loads simulated by different models are listed in Table 6.

As shown in Table 6, the pad load simulated by the SST model with low-Re correction is closest to the experimental

Table 6: Pad loads.

Group	Model	Corrections	Pad loads (kN)
Classical turbulence models	laminar		54.2
Classical turbulence models	SA	–	79.8
	Standard k-ω		48.2
	SST		58.5
Classical turbulence models	Standard k-ε	Scalable wall treatment	34.2
	RNG k-ε		22.3
	Realizable k-ε		37.6
Classical turbulence models	Quadratic pressure-strain RSM	–	35.2
Transition turbulence models	Transition SST	–	58.8
	Transition k-kl-ω		54.2
Transition turbulence models	SST	Intermittency transition	54.5
Turbulence models with low-Re correction	AKN k-ε	Low-Re correction	58.4
	Abid k-ε		56.6
	Lam–Bremhorst k-ε		57.5
Turbulence models with low-Re correction	SST	Low-Re correction	88.9
Experiment	–	–	90.0

data and its relative error is 1.27%, accurate enough for the bearing load predictions.

Among the classical turbulence models, the SA model has the best pad load and its error is still larger than 10%, unacceptable for the load predictions. As for the transition turbulence models and other turbulence models with low-Re correction, their errors are all larger than 33% and cannot predict the load precisely. Thus, among all these models, the SST model with low-Re correction is the only suitable one for the bearing load simulations.

4.3. Pressure Distribution Comparisons. In addition to temperature distributions, the simulation of the pressure distributions should be corrected to obtain accurate THD characteristics. The simulated pad-side pressure distributions in the pad-side center line are compared with the experimental data from Taniguchi et al. [25] to determine the most suitable turbulence model for the superlaminar journal bearing simulation.

4.3.1. Classical Turbulence Models. Compared with the experimental data, the pressure simulation results of the classical turbulence models in the center line of the pad-side wall are shown in Figure 7.

The highest simulated pad-side pressure of laminar is 2.43 MPa at 40.0° from the inlet, which is 39.6% lower than the highest measured pad-side pressure of 4.02 MPa at 43.3°. Although the simulation error for the angle of the highest pad-side pressure is acceptable, the error for the highest pressure value cannot be neglected. Thus the turbulence effect on the hydrodynamic characteristics is not negligible.

Table 7 shows that the errors in the highest pressure values for the classical fully developed turbulence models are more than 20%. The differences with experimental data are unacceptable. Thus, the turbulent effects in the superlaminar oil flow should be considered different from the fully developed turbulent flow.

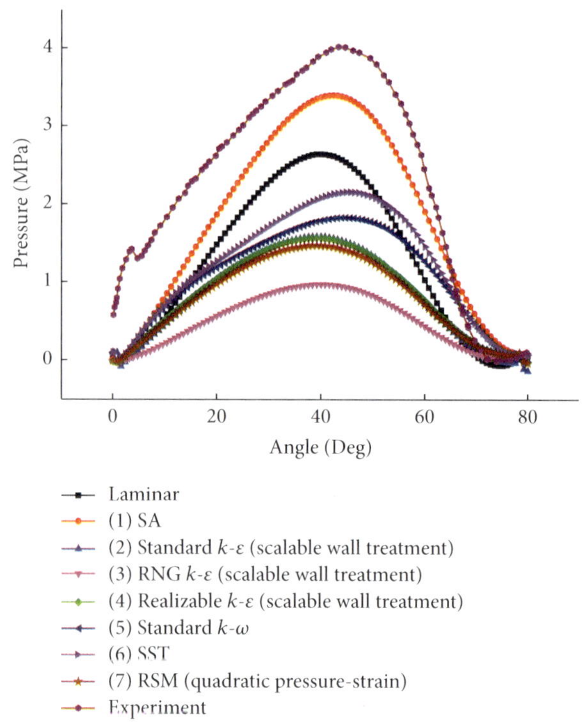

Figure 7: Comparison between the experimental pressure profile at 3000 rpm and under 180 kN and the simulated pressure profiles with the fixed film geometry in pad-side center line for the group of the classical turbulence models.

A small difference is observed among the simulation results of the different classical k-ε models. Compared with the RNG k-ε model with scalable wall treatment, the simulation results of the realizable k-ε and standard k-ε models with scalable wall treatment are more consistent with the experimental pressure data, which are still much worse than the laminar results.

Table 7: Highest pressure of the classical full-developed turbulence models group.

Model	Corrections	Highest pressure	
		Value (MPa)	Position (Degree)
SA	–	3.09	41.6
Standard k-ω		1.83	44.8
SST		2.15	45.6
Standard k-ε	Scalable wall treatment	1.58	39.2
RNG k-ε		0.97	40.0
Realizable k-ε		1.57	39.2
Quadratic pressure-strain RSM		1.47	39.2

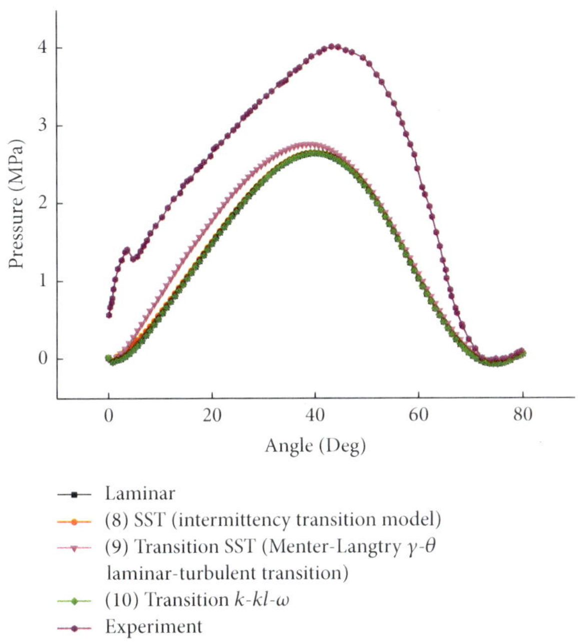

Figure 8: Comparison between the experimental pressure profile at 3000 rpm and under 180 kN and the simulated pressure profiles with the fixed film geometry in pad-side center line for the group of the transition turbulence models.

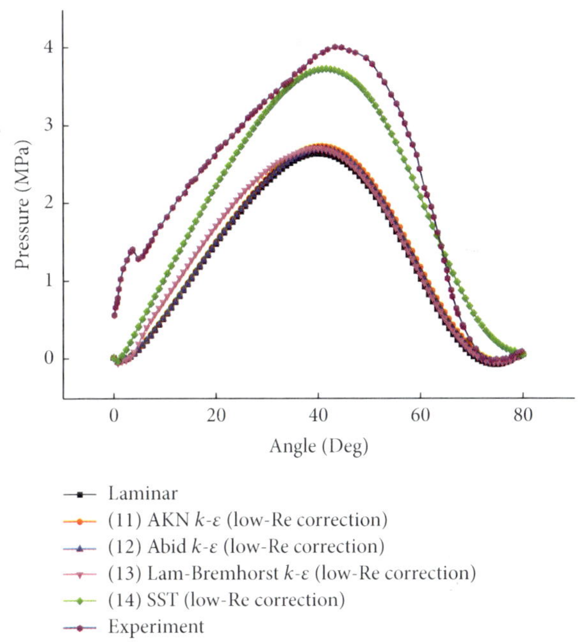

Figure 9: Comparison between the experimental pressure profile at 3000 rpm and under 180 kN and the simulated pressure profiles with the fixed film geometry in pad-side center line for the group of turbulence models with low-Re correction.

4.3.2. Transition Turbulence Models. Compared with the experimental data, the pressure simulation results of the transitional turbulence models in the center line of the pad-side wall are shown in Figure 8.

Table 8 shows that the highest pressure value and angle of the transition SST model are 31.09% lower and 9.47% earlier than the experimental data, which are the best results in the transition turbulence models. However, these values are still unacceptable for the precious bearing simulation.

The transition k-kl-ω model and the SST model with intermittency transition model perform worse than several classical fully developed turbulence models, such as SA model. Thus, the transition model is unsuitable for the simulations of superlaminar oil flow and other corrections should be added to the turbulence models for this simulation.

4.3.3. Turbulence Models with Low-Re Correction. Compared with the experimental data, the pressure simulation results of the turbulence models with low-Re correction in the center line of the pad-side wall are shown in Figure 9.

Table 9 shows that the simulation results vary for the different turbulence models, and the best model in this group is the SST model with low-Re correction. The error of the highest pad-side pressure value for this model is 7.21%, which is also the best among the fourteen turbulence models.

Compared with the theoretical data [25], the highest pressure value of the SST model with low-Re correction is higher and has a smaller difference with the experimental data. The pressure simulated by SST model with low-Re correction around the outlet decreases faster and is nearer to the change tendency of the experimental data than the theoretical data. Furthermore, the theoretical data is obtained by solving the modified Reynolds equations, which is different from the

TABLE 8: Highest pressure of the transition turbulence models group.

Model	Corrections	Highest pressure Value (MPa)	Position (Degree)
Transition SST	–	2.77	39.2
Transition k-kl-ω		2.65	40.0
SST	Intermittency transition	2.65	40.0

TABLE 9: Highest pressure of the turbulence models with low-Re correction group.

Model	Corrections	Highest pressure Value (MPa)	Position (Degree)
AKN k-ε	Low-Re correction	2.74	40.8
Abid k-ε		2.71	40.0
Lam–Bremhorst k-ε		2.71	19.2
SST	Low-Re correction	3.73	41.6

CFD methods. Hence, the CFD simulation with the SST model with low-Re correction provides a relevant method to simulate the characteristics of the bearing oil flow.

Meanwhile, the low-Re AKN k-ε model, utilized by Solghar and Gandjalikhan Nassab [27] and Maneshian and Gandjalikhan Nassab [26], simulates the pressure distribution similar to laminar in this case. This scenario implies that the turbulence effect is not considered suitably by the low-Re AKN k-ε effect, and the corrections for the transport equations and turbulence viscosity should be modified for the superlaminar flow. Compared with the SST model with low-Re correction, the low-Re k-ε model generally performs even worse than the simulation results of the standard k-ε model with scalable wall treatment. This condition indicates that the low-Re correction for k-ε turbulence models may be unsuitable for the bearing oil flow simulation.

The pad-side pressure contours of the laminar and typical turbulence models in each group are shown in Figure 10. The left border is the oil inlet, whereas the right border is the oil outlet.

The aforementioned contours show that the highest pressure positions are almost the same, which are consistent with the pad-side pressure results in center line and reflects the necessity to analyze the simulated center line pressure.

Therefore, the aforementioned pressure analysis reveals that, among these fourteen turbulence models and laminar model, the SST model with low-Re correction performs the best pressure simulation.

The SST model with low-Re correction can simulate the THD characteristics of superlaminar oil flow accurately based on the pressure and temperature analysis.

4.4. Turbulent Viscosity Ratio Distribution Comparison. The turbulence strength should also be discussed for a detailed comparison. Figure 11 shows the turbulent viscosity ratio profile in center line between the pad-side wall and rotor-side wall. The definition of the turbulent viscosity ratio is the ratio of the turbulent viscosity to laminar viscosity (μ_t/μ), which is directly proportional to the turbulent Reynolds number and represents the turbulence strength.

Figure 11 shows a significant increase in the turbulent viscosity ratio for almost all turbulence models around the inlet. This scenario is due to the automatic adjustment to the turbulent kinetic energy, turbulent dissipation rate, and specific dissipation rate of the inlet boundary in the turbulence models.

By ignoring the inlet increase, compared with the Reynolds number of oil flow in Figure 5, the turbulent viscosity ratios of the SST model with low-Re correction and SST model show a similar change tendency of decreasing first and then increasing. This condition means that both of them can catch the change tendency of turbulence effect properly.

The low-Re AKN k-ε model and SST model with intermittency transition simulate the oil flow as pure laminar with the turbulent viscosity ratio equal to zero. The Menter-Langtry γ-θ laminar-turbulent transition model also simulates the oil flow as laminar around the outlet, which explains why transition models exhibit similar temperature distributions to the laminar. Therefore, transition models are unsuitable for the superlaminar flow.

Figure 12 depicts the contours of the turbulent viscosity ratio simulated by the SST model with low-Re correction. The upper boundary is the rotor-side, whereas the lower boundary is the pad-side. The angle of number 2 pad is 80°, and the three selected locations are 20°, 40°, and 60° from the inlet.

Figure 12 shows that the turbulent viscosity ratio simulated by the SST model with low-Re correction decreases slowly from 20° to 60°, which is consistent with the change tendency in the turbulence viscosity ratio in Figure 11. The contours are symmetric with two low turbulent viscosity ratio regions, which reflect the influence of the two wall boundaries.

These results generally infer that the SST model with low-Re correction produces better results for superlaminar flows

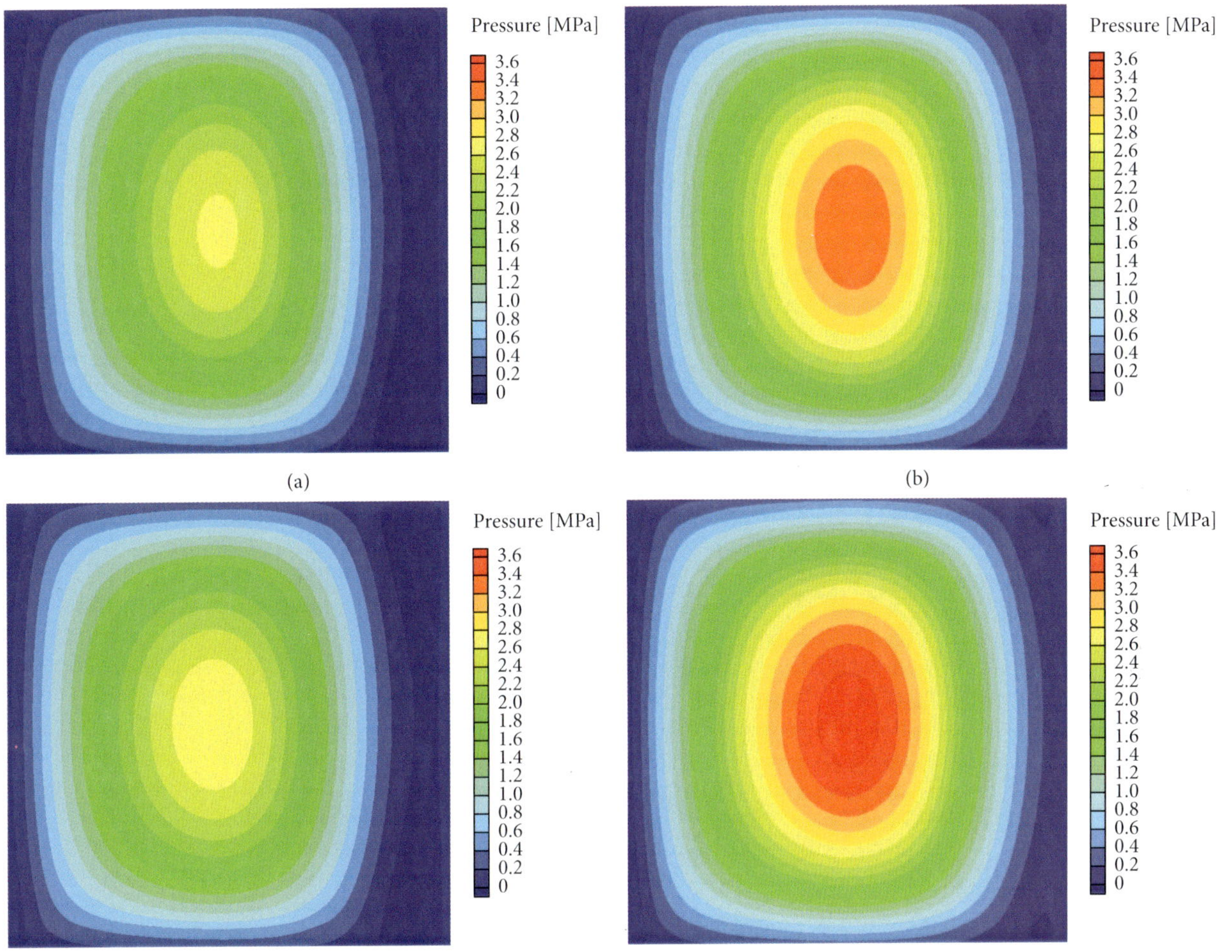

FIGURE 10: Contours of pad-side pressure. (a) Laminar model; (b) SA model; (c) transition SST model; and (d) SST model with low-Re correction.

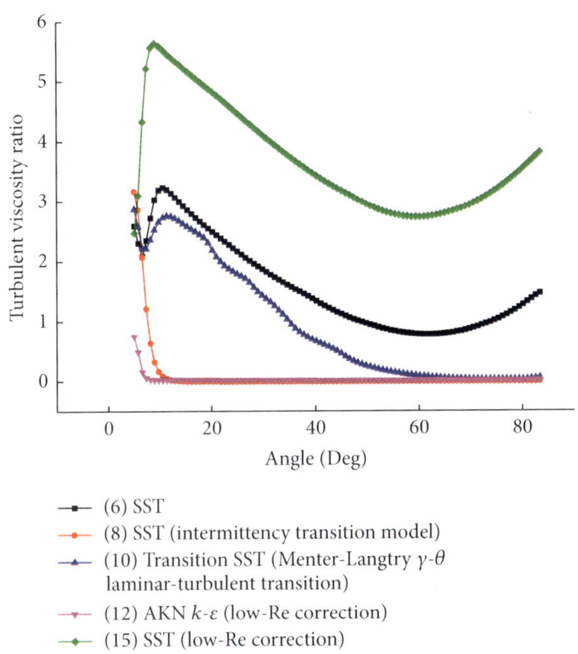

FIGURE 11: Turbulent viscosity ratio in center line.

in the bearing compared with the other thirteen turbulence models and laminar model.

5. Velocity Profile Analysis

The oil flow field should be examined in detail to analyze the SST model with low-Re correction. First, a middle plane, which is perpendicular to the length direction and located in the middle of the pad length, is drawn. Five profiles at different degrees from the inlet are then located in the middle plane. The five velocity profiles are shown in Figure 13.

The difference among the velocity profiles is most evident at 40° from the inlet, which also shows the influence of the boundary layers. The laminar velocity profile is a straight line from the pad-side wall to the rotor-side wall, whereas the velocity profiles of the SST model and SST model with low-Re correction are S-shaped curves. The S-shaped curves show that the two SST models consider the viscous sublayer and buffer layer. The rotor-side boundary layer also interacts with the pad-side boundary layer in the middle buffer layer region.

The velocity profiles of the three models for 0°, 20°, 60°, and 80° from the inlet do not have such a significant

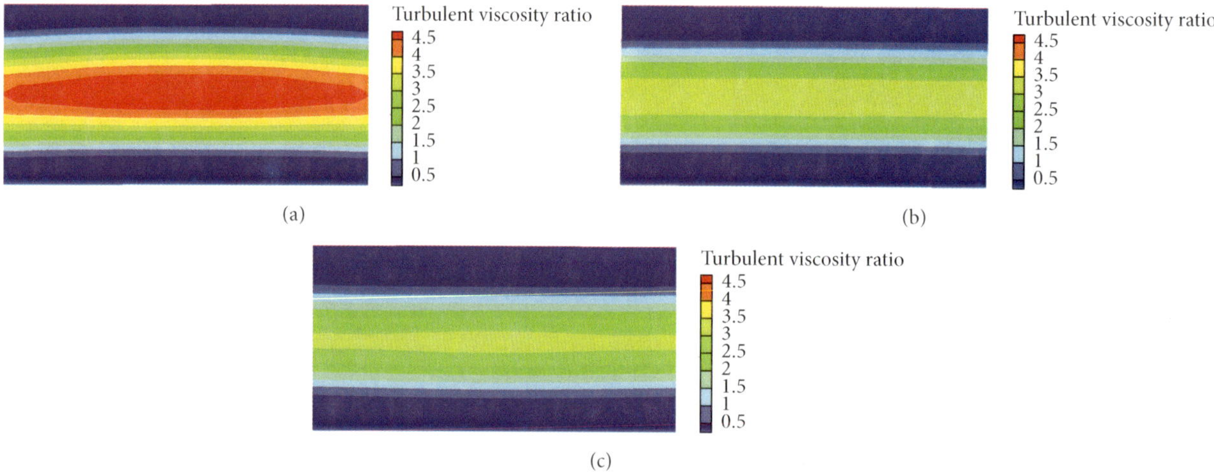

FIGURE 12: Contours of turbulent viscosity ratio in the sections simulated by SST model with low-Re correction. (a) 20° from the inlet; (b) 40° from the inlet; and (c) 60° from the inlet.

difference from the velocity as the 40 degrees from the inlet, which may be caused by the influence of the inlet and the outlet boundary and cannot show the difference among the velocity profiles of the three models.

However, Figure 13 shows that the velocity profile difference between the SST model with low-Re correction and SST model is not sufficiently clear to analyze the two turbulence models. Thus, the dimensionless parameters u^+ and y^+ are adopted to present the velocity distribution.

Considering that the difference between the velocity profiles is most evident at 40° from the inlet, then the locations for u^+ and y^+ are set at 40° from the inlet.

Hence, y^+ can be calculated as follows:

$$y^+ = \frac{\rho y u_\tau}{\mu}, \tag{8}$$

where u_τ can be calculated as follows:

$$u_\tau = \sqrt{\frac{\tau_w}{\rho}} \tag{9}$$

and u^+ can be calculated as follows:

$$u^+ = \frac{u}{u_\tau}. \tag{10}$$

Given that y^+ requires the distance value to the nearest wall (y), then the oil flow region is divided into two parts: the pad-side and the rotor-side parts. The dividing line is in the middle of the flow region. The rotor-side and the pad-side parts utilize the rotor-side and the pad-side wall shear stresses (u_τ), respectively, to address the two boundary layers individually.

Figure 14 shows that the change tendency of u^+ with y^+ increasing at 40° from the inlet. The viscous sublayer and the log-law lines are drawn for comparison with the CFD results based on the u^+ and y^+ functions in the viscous sublayer and the log-law region.

The function of u^+ and y^+ in the viscous sublayer is expressed as follows:

$$u^+ = y^+. \tag{11}$$

The function of u^+ and y^+ in the log-law region for a smooth wall is expressed as follows [31]:

$$u^+ = \frac{1}{0.41} \ln y^+ + 5. \tag{12}$$

The pad-side and rotor-side parts of the SST model with low-Re correction obtain similar results. The simulated u^+ is initially almost the same as the viscous sublayer, which means that this flow region is actually the viscous sublayer. The simulated u^+ is then lower than the viscous sublayer and becomes near the log-law in the end. This scenario means that the boundary layer changes from the viscous sublayer to the buffer layer at the latter part. If sufficient space is available for the boundary layer to be fully developed, then the simulated u^+ can be the same as the log-law at the latter part.

A relatively large difference exists between the pad-side and rotor-side parts for the SST model in terms of value. However, the change tendencies of the simulated u^+ in two parts are almost the same. The simulated u^+ change is initially similar to the viscous sublayer, which means that the viscous sublayer is correctly considered in this turbulence model. However, the simulated u^+ then becomes higher and increases faster than the log-law, which are inconsistent with the law of the wall. This scenario means that the buffer layer is incorrectly considered in the SST model.

Therefore, the buffer layer plays the most important role for the turbulence boundary layer of the superlaminar bearing. This condition indicates that the mechanism of superlaminar flows is near-all turbulence. The classical fully developed turbulence models cannot predict the buffer layer accurately, but an appropriate low-Re correction can address this problem. This scenario explains why the SST model with low-Re correction produces suitable results for superlaminar flows in the bearing.

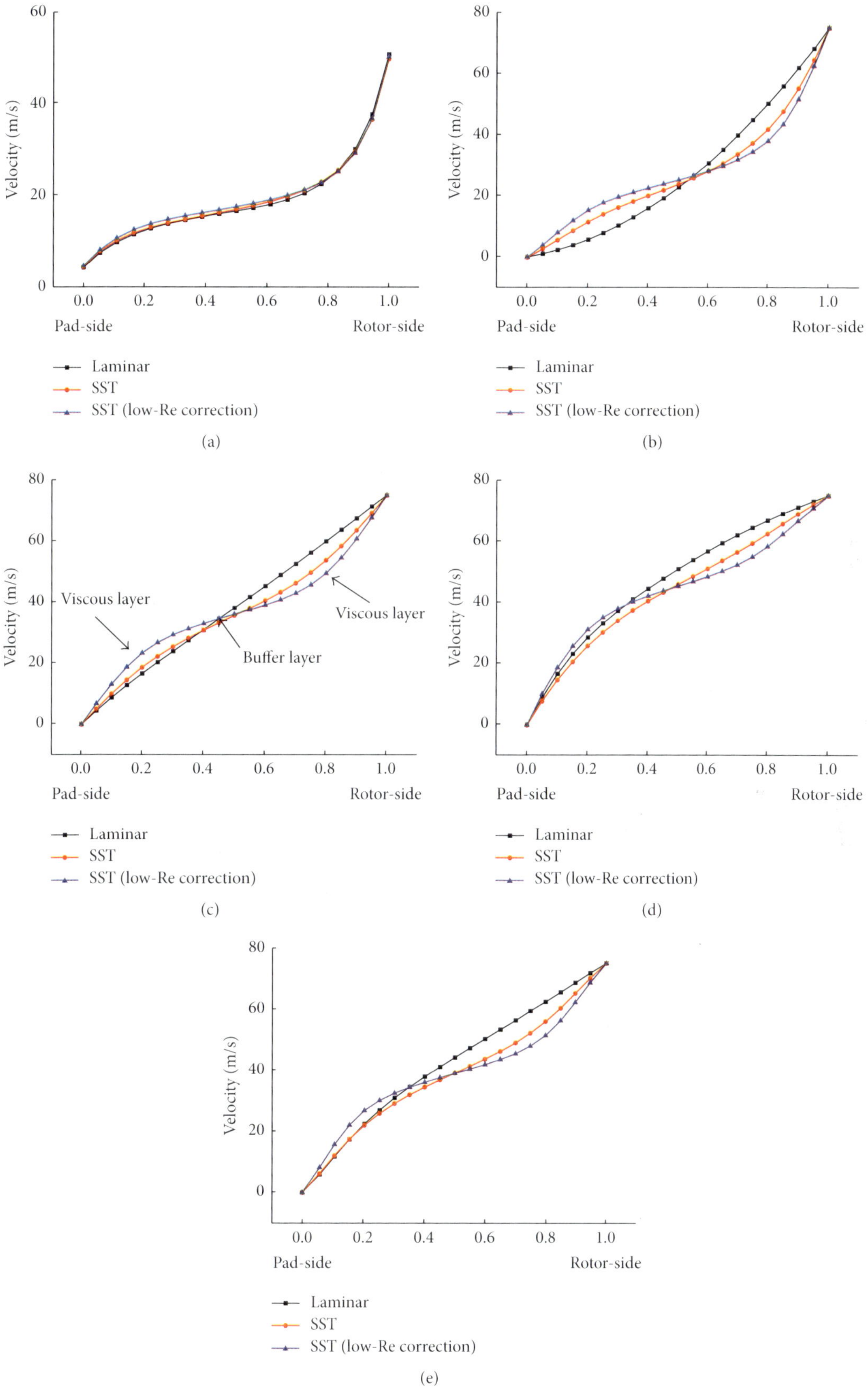

FIGURE 13: Velocity profiles in the middle plane. (a) In 0° from the inlet; (b) in 20° from the inlet; (c) in 40° from the inlet; (d) in 60° from the inlet; and (e) in 80° from the inlet.

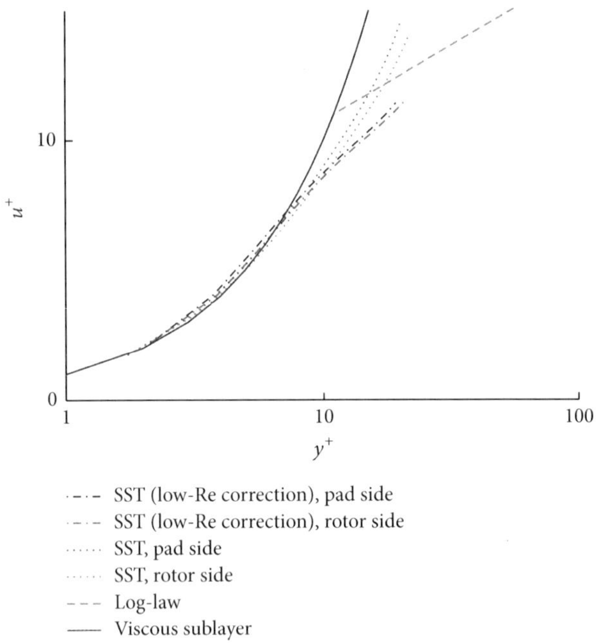

FIGURE 14: u^+ and y^+ in 40° from the inlet.

6. Conclusions

A three-dimensional model with a structured mesh for a tilting-pad journal bearing is established in this study. A total of fourteen turbulence models and laminar model are utilized to simulate the superlaminar oil flow at 3000 r/min and under 180 kN loading. The simulation results are then compared with the experimental data obtained by Taniguchi et al. [25]. The conclusions of the current study are summarized as follows:

(i) Compared with experimental data, the laminar and classical turbulence models cannot accurately simulate the temperature and pressure distributions. Therefore, the laminar and classical fully developed turbulent models fail to adapt to superlaminar flow.

(ii) The transition turbulence models simulate the rapidly decreasing turbulent viscosity ratio and consider the oil flow around the outlet as laminar. This pattern is different from the change tendency of the experimental Reynolds number. Thus, the transition models generally obtain similar laminar simulation results and cannot properly simulate oil flow properly with an evident difference between the simulated THD characteristics and experimental results.

(iii) The performances of the low-Re correction in different turbulence models vary. The k-ε models with low-Re correction simulate the oil flow as laminar, which results in unacceptable differences in the temperature and pressure results and experimental data. However, the results of the SST model with low-Re correction are consistent with the experimental data. The low-Re correction for SST is the best correction to simulate the superlaminar oil flow in the tilting-pad journal bearing among the compared models.

(iv) The velocity profiles indicate that the buffer layer plays the most important role for the turbulence boundary layer of the bearing. This finding suggests that the superlaminar flow mechanism is near-wall turbulence. Although the classical fully developed turbulence models cannot accurately predict the buffer layer, an appropriate low-Re correction can address this problem and help the SST model with low-Re correction yield suitable results for the superlaminar flow in the bearing.

In summary, this study reveals the physical meaning of the superlaminar oil flow in bearings and the significant importance of the buffer layer in the superlaminar flow. A suitable turbulence model, the SST model with low-Re correction, is pointed out for simulating the superlaminar flow accurately.

Nomenclature

C_μ: Constant
$C_{1\varepsilon}$: Constant
$C_{2\varepsilon}$: Constant
C_2: Constant
d: Film thickness
f: Body force
i, j, l: Three dimensions
k: Turbulence kinetic energy
P: Stress tensor
R: Radius of the rotor
Re: Reynolds number
Re*: Reduced Reynolds number
Re$_t$: Turbulent Reynolds number
t: Time
u: Flow velocity
u^+: Dimensionless velocity
u_τ: Friction velocity
y: Distance to the nearest wall
y^+: Wall coordinate
α^*: Turbulent viscosity coefficient
ε: Turbulent dissipation rate
μ: Kinetic viscosity
ρ: Density
σ_k: Constant
σ_ε: Constant
τ_w: Wall shear stress
ω: Specific dissipation rate.

Conflicts of Interest

The authors declare that they have no conflicts of interest.

References

[1] J. C. Fisher, "The fracture of liquids," *Journal of Applied Physics*, vol. 19, no. 11, pp. 1062–1067, 1948.

[2] H. Temperley, "Two-dimensional ising models," in *Phase Transitions and Critical Phenomena*, C. Domb and M. S. Green, Eds., Academic Press, Cambridge, Mass, USA, 1974.

[3] L. Floberg, "Lund Technical University," *Journal of Tribology*, vol. 6, no. 4, pp. 145-146, 1973.

[4] D. Dowson and C. M. Taylor, "Cavitation in bearings," *Annual Review of Fluid Mechanics*, vol. 11, pp. 35–66, 1979.

[5] C. M. Woods and D. E. Brewe, "Solution of the Elrod algorithm for a dynamically loaded journal bearing using multigrid techniques," *Journal of Tribology*, vol. 111, no. 2, pp. 302–308, 1989.

[6] D. E. Brewe and B. O. Jacobson, "The effect of vibration amplitude on vapour cavitation in journal bearings," *Wear*, vol. 115, no. 1-2, pp. 63–73, 1987.

[7] X.-S. Li, Y. Song, Z.-R. Hao, and C.-W. Gu, "Cavitation mechanism of oil-film bearing and development of a new gaseous cavitation model based on air solubility," *Journal of Tribology*, vol. 134, no. 3, Article ID 031701, 2012.

[8] X. Li, Y. Song, Z. Hao, and C. Gu, "Cavitation Mechanism of Oil-Film Bearing and Development of a New Gaseous Cavitation Model Based on Air Solubility," *Journal of Tribology*, vol. 134, no. 3, pp. 1047–1052, 2012.

[9] Z.-R. Hao and C. W. Gu, "Numerical modeling for gaseous cavitation of oil film and non-equilibrium dissolution effects in thrust bearings," *Tribology International*, vol. 78, pp. 14–26, 2014.

[10] Y. Song, C.-W. Gu, and X. Ren, "Development and validation of a gaseous cavitation model for hydrodynamic lubrication," *Proceedings of the Institution of Mechanical Engineers, Part J: Journal of Engineering Tribology*, vol. 229, no. 10, pp. 1227–1238, 2015.

[11] Y. Song, X. Ren, C.-W. Gu, and X.-S. Li, "Experimental and numerical studies of cavitation effects in a tapered land thrust bearing," *Journal of Tribology*, vol. 137, no. 1, Article ID 011701, 2015.

[12] D. G. Huang and Y. Q. Zhuang, "Temperature and cavitation," *Proceedings of the Institution of Mechanical Engineers, Part C: Journal of Mechanical Engineering Science*, vol. 222, no. 2, pp. 207–211, 2008.

[13] D. Huang, Y. Zhuang, and R. Cai, "A computational method for cavitational flows based on energy conservation," *Proceedings of the Institution of Mechanical Engineers, Part C: Journal of Mechanical Engineering Science*, vol. 221, no. 11, pp. 1333–1338, 2007.

[14] Y. Y. Zhang, X. J. Sun, and D. G. Huang, "A numerical study on cavitation suppression using local cooling," *International Journal of Fluid Machinery & Systems*, vol. 3, no. 4, pp. 292–300, 2010.

[15] Y. Wang, X. J. Sun, Y. J. Dai, G. Q. Wu, Y. Cao, and D. G. Huang, "Numerical investigation of drag reduction by heat-enhanced cavitation," *Applied Thermal Engineering*, vol. 75, pp. 193–202, 2015.

[16] D. F. Wilcock, "Super Laminar Flow in Bearings: A Review of the Second Leeds-Lyon Symposium, Lyons, France, September 17–19, 1975," *Journal of Lubrication Technology*, vol. 98, no. 1, pp. 5–7, 1976.

[17] D. F. Wilcock, "Turbulence in High Speed Journal Bearing," *Trans. ASME*, vol. 173, no. 1, pp. 825–834, 1950.

[18] V. N. Constantinescu, "On turbulent lubrication," *Archive Proceedings of the Institution of Mechanical Engineers*, vol. 173, pp. 881–900, 1959.

[19] C. W. Ng and C. H. T. Pan, "A linearized turbulent lubrication theory," *Journal of Fluids Engineering*, vol. 87, no. 3, pp. 675–682, 1965.

[20] G. G. Hirs, "A bulk-flow theory for turbulence in lubricant films," *Journal of Lubrication Technology*, vol. 95, no. 2, pp. 137–146, 1973.

[21] T. Suganami and A. Z. Szeri, "Parametric study of journal bearing performance: the 80 deg partial arc bearing," *Journal of Lubrication Technology*, vol. 101, no. 4, pp. 486–491, 1979.

[22] S. J. Chowdhury and G. Ahmadi, "Thermohydrodynamic analysis of wide thrust bearings operating in turbulent inertial flow regimes," *Journal of Tribology*, vol. 110, no. 2, pp. 327–334, 1988.

[23] D. Ives and W. B. Rowe, "The performance of hybrid journal bearings in the superlaminar flow regimes," *Tribology Transactions*, vol. 35, no. 4, pp. 627–634, 1992.

[24] H. Hashimoto, "Surface roughness effects in high-speed hydrodynamic journal bearings," *Journal of Tribology*, vol. 119, no. 4, pp. 776–780, 1997.

[25] S. Taniguchi, T. Makino, K. Takeshita, and T. Ichimura, "Thermohydrodynamic analysis of large tilting-pad journal bearing in laminar and turbulent flow regimes with mixing," *Journal of Tribology*, vol. 112, no. 3, pp. 542–550, 1990.

[26] B. Maneshian and S. A. Gandjalikhan Nassab, "Thermohydrodynamic characteristics of journal bearings running under turbulent condition," *International Journal of Engineering, Transactions A: Basics*, vol. 22, no. 2, pp. 181–194, 2009.

[27] A. A. Solghar and S. A. Gandjalikhan Nassab, "Numerical analysis of turbulent lubrication in plain full journal bearings," *Industrial Lubrication and Tribology*, vol. 65, no. 2, pp. 91–99, 2013.

[28] L. Bouard, M. Fillon, and J. Frêne, "Comparison between three turbulent models - Application to thermohydrodynamic performances of tilting-pad journal bearings," *Tribology International*, vol. 29, no. 1, pp. 11–18, 1996.

[29] A. Ding and X. Ren, "Validation of Turbulence Models for the Superlaminar Flows in Journal Bearings," in *Proceedings of the ASME Turbo Expo 2017: Turbomachinery Technical Conference and Exposition*, Charlotte, NC, USA.

[30] T. Dimond, A. Younan, and P. Allaire, "A review of tilting pad bearing theory," *International Journal of Rotating Machinery*, vol. 2011, Article ID 908469, 23 pages, 2011.

[31] H. Schlichting, *Boundary-Layer Theory*, 2000.

4

A Study of Hydromagnetic Longitudinal Rough Circular Step Bearing

Jatinkumar V. Adeshara,[1] M. B. Prajapati,[2] G. M. Deheri,[3] and R. M. Patel[4]

[1] *Research Scholar, Department of Mathematics, H. N. G. University, Patan, 384 265, Gujarat State, India*
[2] *Head, Department of Mathematics, HNG University, Patan 384 265, Gujarat, India*
[3] *Department of Mathematics, SP University, Vallabh Vidyanagar 388 120, Gujarat, India*
[4] *Department of Mathematics, Gujarat Arts and Science College, Ahmedabad 380 006 Gujarat, India*

Correspondence should be addressed to R. M. Patel; rmpatel2711@gmail.com

Academic Editor: Huseyin Çimenoğlu

This article discusses the effect of longitudinal roughness on the performance of hydromagnetic squeeze film in circular step bearing. To characterize the random roughness of the bearing surfaces the stochastic model of Christensen and Tonder has been employed. The stochastically averaged Reynolds' type equation is solved using suitable boundary conditions to obtain the pressure distribution and then the load bearing capacity is computed. The results are presented in graphical form. The graphical results presented here establish that the hydromagnetic lubrication offers significant help to the longitudinal roughness pattern to enhance the performance of the bearing system. Of course, conductivities of the plates, standard deviation, and the supply pressure contribute towards reducing the negative effect induced by variance (+ve) and skewness (+ve).

1. Introduction

It is documented that gears, braking units, hydraulic dampers, skeletal bearings, and synovial joints make use of squeeze film mechanism. Generally an electrically conducting fluid with high thermal and electrical conductivity is applied as a lubricant for squeeze film to work under such extreme circumstances. Also, a use of external magnetic field then advances the performance of lubrication.

The performance of an oil lubricated circular step bearing is analyzed by Majumdar [1]; this study underlines the importance of radii ratio. Lin [2] discussed the couple stress effect on the performance of an externally pressurized circular step bearing. It was established that the couple stress fluid modified the performance of the bearing system. Deheri et al. [3] reformed and developed the study of Majumdar [1] to use the magnetization effect by considering a magnetic fluid as the lubricant, the flow being regulated by Neuringer–Rosensweig's model [4].

It is well known that the effect of surface roughness is quite important in different types of bearing systems. Most of the discussions have considered the stochastic averaging method of Christensen and Tonder [5–7] who presented a stochastic model to evaluate the effect of rough surfaces. (Ting [8], Prakash and Tiwari [9], Prajapati [10], and Andharia et al. [11]).

The hydromagnetic squeeze films between two conducting rough circular plates are studied by Vadher et al. [12]. It was investigated that the negative effect of roughness got minimized up to some extent by applying magnetization. Patel and Deheri [13] evaluated the performance of ferrofluid squeeze film in rotating curved circular plates by using Shliomis model. Here, the outcomes of this article suggested that Shliomis model based ferrofluid lubrication was relatively better than the model of Neuringer–Rosensweig [4]. Patel et al. [14] extended the study of Deheri et al. [3] to study the effect of transverse surface roughness. It was indicated that the bearing performance was adversely affected by the roughness of the bearing surfaces. But the situation was found to be a little enhanced in case of variance (-ve) was considered. A modified lubrication equation for hydromagnetic

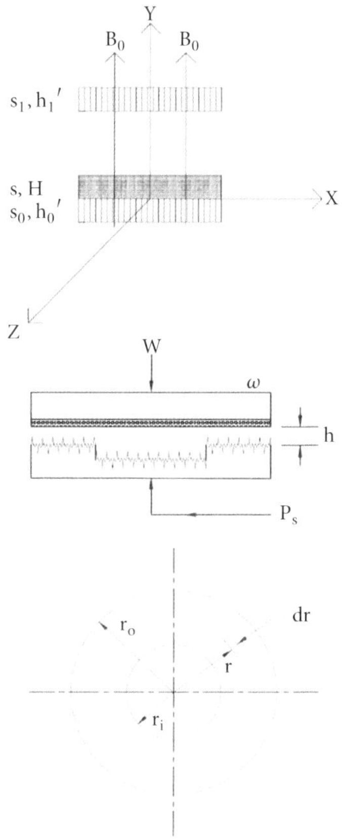

FIGURE 1: Configuration of bearing system.

non-Newtonian cylindrical squeeze films was presented by Lin et al. [15]. The hydromagnetic non-Newtonian effects offered quite better performance for circular squeeze films in comparison with the case of a nonconducting Newtonian lubricant.

Patel and Deheri [16] studied the magnetic fluid lubrication of a squeeze film in transversely rough porous circular plates with concentric circular pocket. Longitudinal roughness effect has been a matter of discussions in Andharia and Deheri [17], Andharia and Deheri [18], Andharia and Deheri [19], and Shimpi and Deheri [20]. Recently, Lin [21] examined the ferrofluid lubrication of a longitudinally rough journal bearing.

Patel et al. [22] discussed the hydromagnetic lubrication of a rough porous circular step bearing. Hence, it has been sought to analyze the effect of longitudinal roughness pattern on behavior of a hydromagnetic squeeze film in circular step bearings.

2. Analysis

The configuration of the bearing system is presented in Figure 1.

For the computation of different performance characteristics, generally the following assumptions are considered:

(1) The recess is deep enough for the pressure in it to be uniform.

(2) The bearing has very low rotational velocity and its effect is neglected for the pressure generation.

As shown in Figure 1 a thrust load w is applied and the bearing supports the load without metal to metal contact. The load w is supported by the fluid within the pocket and land. The fluid escapes in the radial direction through the restrictions by a land or sill around the recess.

In view of the discussions of Christensen and Tonder [5–7] the surfaces of bearing are taken to be longitudinally rough. The thickness h(x) of the lubricant film is given by

$$h(x) = \bar{h}(x) + h_s \tag{1}$$

where $\bar{h}(x)$ is the mean film thickness and h_s is the deviation from the mean film thickness characterizing the random roughness surface. h_s is considered in nature and governed by the probability density function $f(h_s)$, $-c \le f(h_s) \le c$, where c is the maximum deviation from the mean film thickness. The mean α and standard deviation σ and the measure of symmetry ε of the random variable h_s are defined by the relationships.

$$\alpha = E(h_s)$$
$$\sigma^2 = E\left[(h_s - \alpha)^2\right] \tag{2}$$

and

$$\varepsilon = E\left[(h_s - \alpha)^3\right] \tag{3}$$

where E denotes the expected value defined by

$$E(R) = \int_{-c}^{c} Rf(h_s)\,dh_s \tag{4}$$

The concerned Reynolds' type equation gives the pressure induced flow for a circular step bearing as (Majumdar (1985), Patel, Deheri, and Vadher (2015))

$$Q = -\frac{2\pi r\,(dp/dr)\left[\left(2/M^3 m(h)\right)[M/2 - \tanh(M/2)]\right]\left[(\phi_0 + \phi_1 + 1)/(\phi_0 + \phi_1 + (\tanh(M/2))/(M/2))\right]}{12\mu} \tag{5}$$

where

$$m(h) = h^{-3}\left[1 - 3\alpha h^{-1} + 3h^{-2}\left(\sigma^2 + \alpha^2\right) - \frac{3}{40}h^{-3}\left(\varepsilon + 3\sigma^2\alpha + \alpha^3\right)\right] \tag{6}$$

Using Reynolds' boundary conditions

$$r = r_o; \quad p = 0;$$

$$r = r_i; \quad p = p_s; \tag{7}$$

the governing equation for the film pressure p is given by

$$p = p_s\frac{\ln(r/r_o)}{\ln(r_i/r_o)} \tag{8}$$

where in

$$p_s = \frac{6}{\pi\left[(2/M^3 m(h))[M/2 - \tanh(M/2)]\right]\left[(\phi_0 + \phi_1 + 1)/(\phi_0 + \phi_1 + (\tanh(M/2))/(M/2))\right]}\ln\left(\frac{r_o}{r_i}\right) \tag{9}$$

Introduction of the dimensionless terms

$$\begin{aligned}
P_s^* &= \frac{6\ln(1/k)}{\pi\left((2/M^3)/(M*(h))\right)[M/2 - \tanh(M/2)]\left[(\phi_0 + \phi_1 + 1)/(\phi_0 + \phi_1 + (\tanh(M/2))/(M/2))\right]}\\
M*(h) &= m(h)\,h^3 = \left(1 - 3\alpha^* + 3\left(\alpha^{*2} + \sigma^{*2}\right) - \frac{3}{40}(\varepsilon^* + 3\sigma^{*2}\alpha^* + \alpha^{*3}\right)\\
\alpha^* &= \left(\frac{\alpha}{h}\right)\\
\sigma^* &= \left(\frac{\sigma}{h}\right)\\
\varepsilon^* &= \left(\frac{\varepsilon}{h^3}\right)\\
k &= \left(\frac{r_i}{r_o}\right)
\end{aligned} \tag{10}$$

paves the way for the expression of pressure distribution in nondimensional form as

$$P = P_s^*\frac{\ln(r/r_o)}{\ln(r_i/r_o)} \tag{11}$$

The load bearing capacity w is calculated by integrating the pressure which takes the dimensionless form

$$W = \frac{P_s^*\ln\left(1 - k^2\right)}{\ln(1/k)} \tag{12}$$

3. Results and Discussions

The representation for dimensionless pressure profile is given in (11) while the nondimensional load bearing capacity is governed by (12). From these equations it is clearly observed that the load carrying capacity

$$W \propto P_s^* \tag{13}$$

and

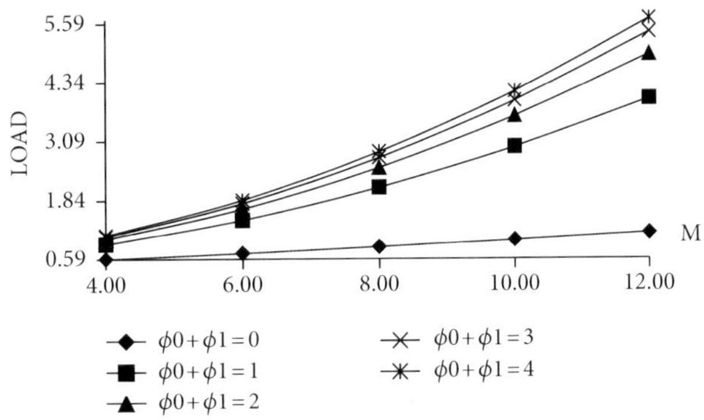

FIGURE 2: Variation of load carrying capacity with respect to M and $\phi_0 + \phi_1$.

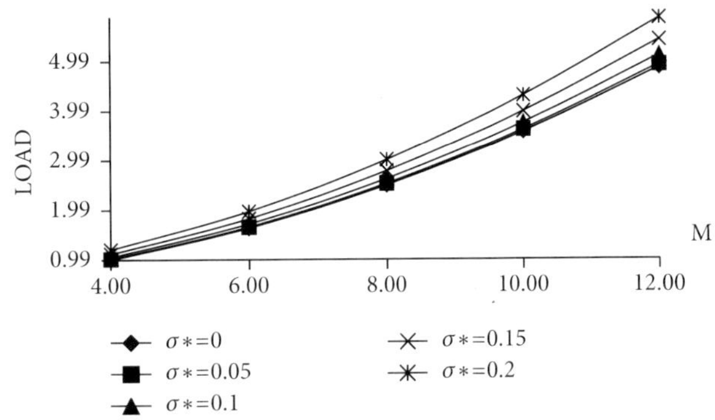

FIGURE 3: Distribution of load bearing capacity with respect to M and σ^*.

$$p_s = \frac{Q}{((2/M^3)/m(h))[M/2 - \tanh(M/2)][(\phi_0 + \phi_1 + 1)/(\phi_0 + \phi_1 + (\tanh(M/2))/(M/2))]} \tag{14}$$

This suggests that for a constant flow rate the load increases as stochastically averaged squeeze film thickness decreases. Therefore, the bearing turns out to be self-compensating provided the flow rate is considered as constant. It is depicted from (11) and (12) that the effect of conductivity parameters on the distribution of pressure and load bearing capacity is determined by

$$\frac{\phi_0 + \phi_1 + (\tanh(M/2))/(M/2)}{\phi_0 + \phi_1 + 1} \tag{15}$$

which turns to

$$\frac{\phi_0 + \phi_1}{\phi_0 + \phi_1 + 1} \tag{16}$$

for large values of M, because $\tanh(M) \cong 1$ and $(2/M) \cong 0$. Further, it is noticed that the pressure and load carrying capacity increases with increases in $\phi_0 + \phi_1$ because both the functions are increasing functions of $\phi_0 + \phi_1$.

It is clearly depicted from Figures 2–6 that the hydromagnetization parameter induces a sharp increase in the load bearing capacity while Figures 7–10 gives the profile of load with respect to $\phi_0 + \phi_1$ for different values of standard deviation, variance, skewness, and radii ratio, respectively. Here, also the load carrying capacity increases with an increase in conductivity parameter. Figures 7–10 also tells that the load increases sharply up to some extent ($\phi_0 + \phi_1 \cong 1.32$) and then this rate of increase gets slower.

It is seen from Figures 11–13 that the standard deviation associated with longitudinal roughness causes increased load bearing capacity with respect to various parameters such as variance, skewness, and radii ratio. Besides, variation of load carrying capacity for different values of variance is described by Figures 14 and 15. From these figures it is noticed that the load decreases with increase in positive variance while the variance (– ve) increases it. The trends of load bearing capacity with respect to skewness are similar to that of variance which is observed from Figure 16.

Form Figures 5 and 9 it is clearly observed that the effect of ε^* on the variation of load carrying capacity with respect to magnetization and conductivity, respectively, remains negligible while this effect of ε^* with regards to variance is registered to be nominal as suggested by Figure 14.

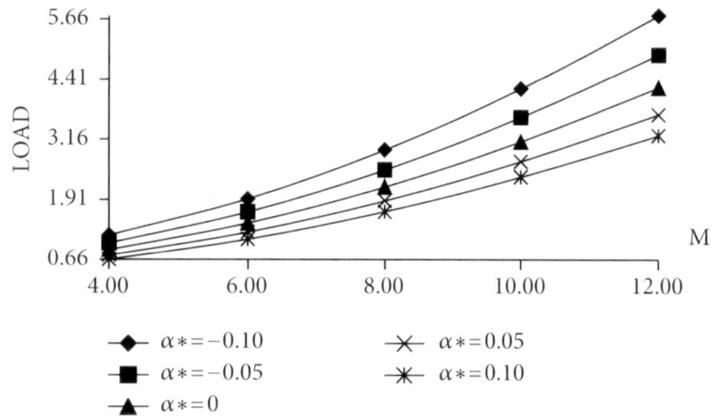

Figure 4: Profile of load with respect to M and α^*.

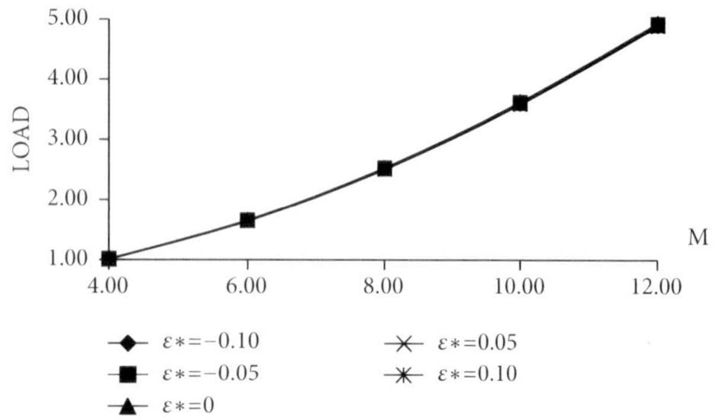

Figure 5: Variation of load carrying capacity with respect to M and ε^*.

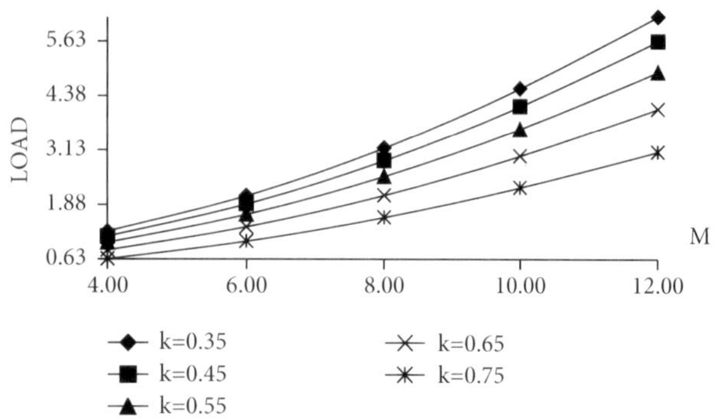

Figure 6: Distribution of load bearing capacity with respect to M and k.

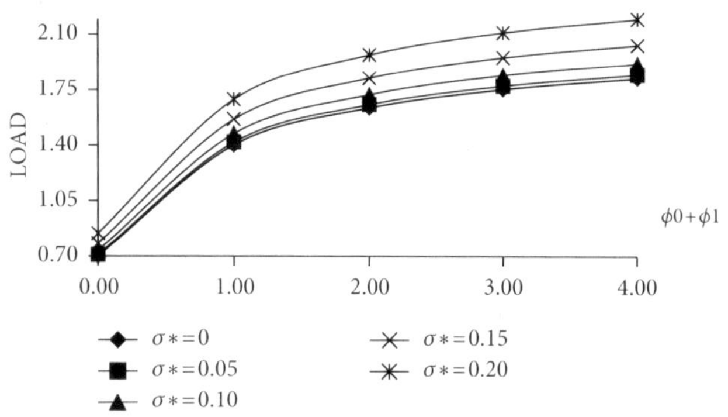

Figure 7: Profile of load with respect to $\phi_0 + \phi_1$ and σ^*.

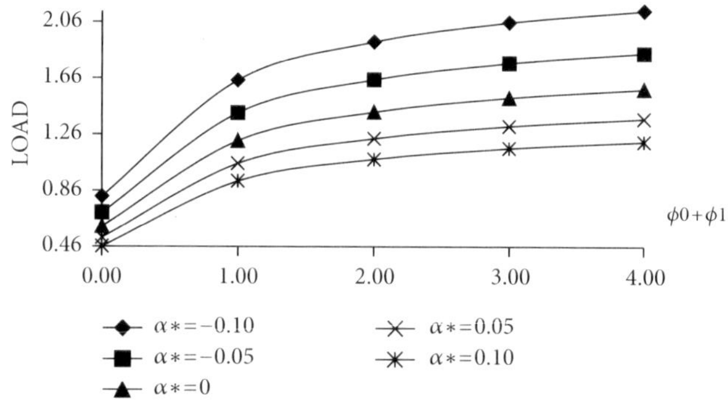

Figure 8: Variation of load carrying capacity with respect to $\phi_0 + \phi_1$ and α^*.

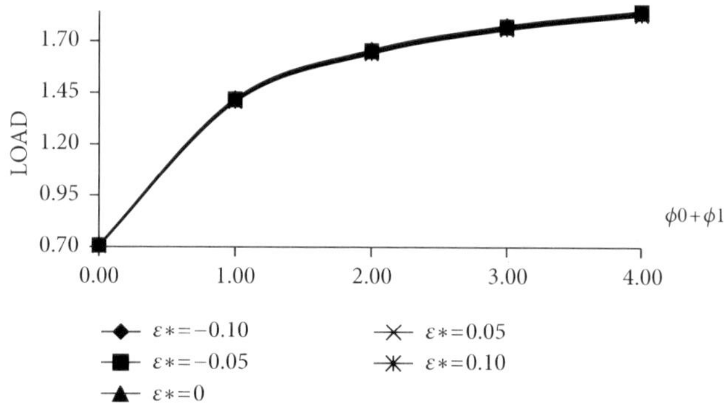

Figure 9: Distribution of load bearing capacity with respect to $\phi_0 + \phi_1$ and ε^*.

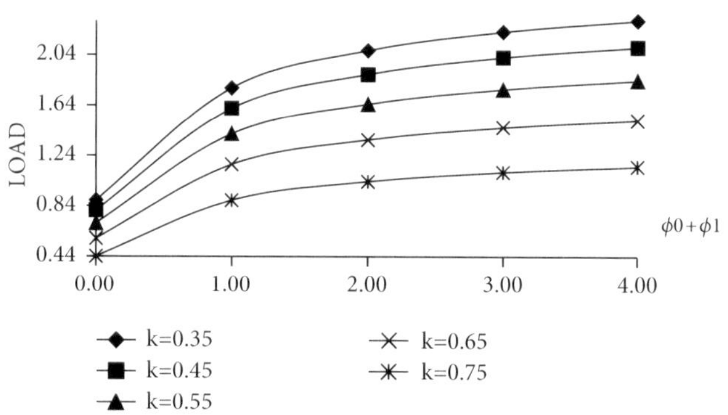

Figure 10: Profile of load with respect to $\phi_0 + \phi_1$ and k.

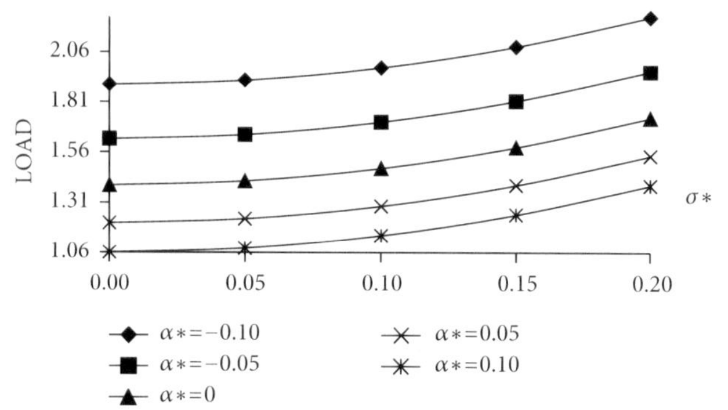

Figure 11: Variation of load carrying capacity with respect to σ^* and α^*.

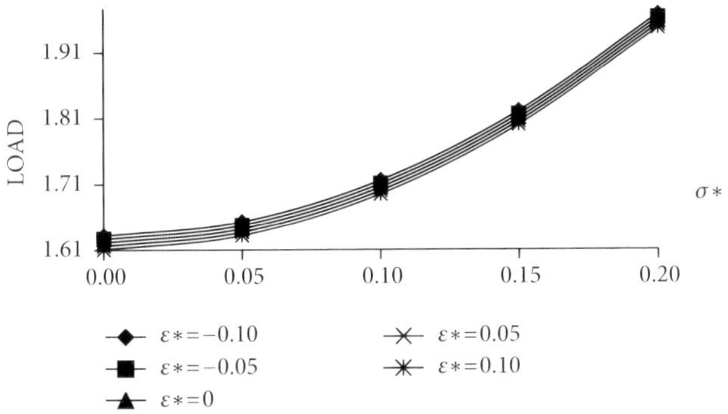

FIGURE 12: Distribution of load bearing capacity with respect to σ^* and ε^*.

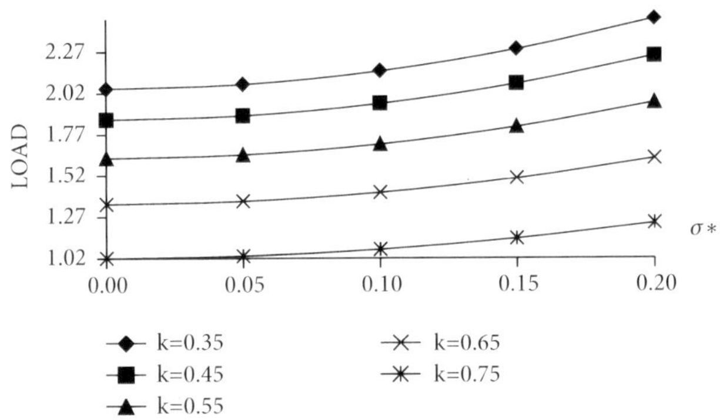

FIGURE 13: Profile of load with respect to σ^* and k.

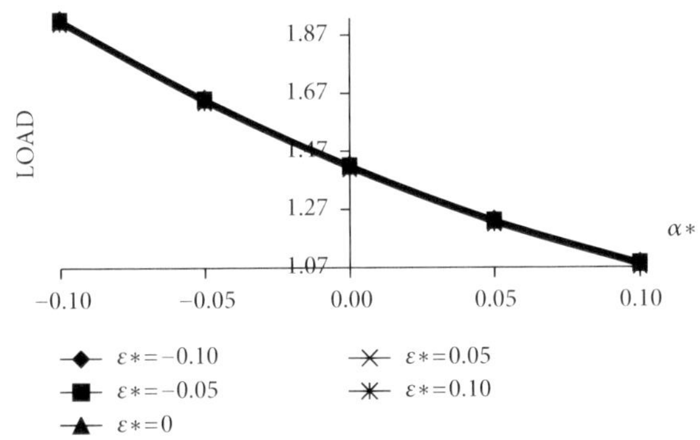

FIGURE 14: Variation of load carrying capacity with respect to α^* and ε^*.

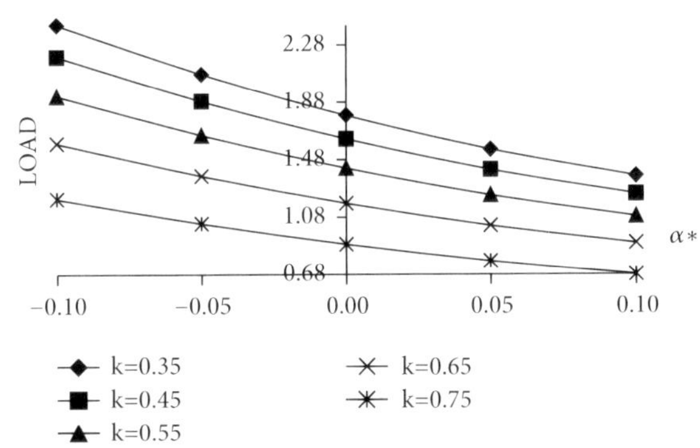

FIGURE 15: Distribution of load bearing capacity with respect to α^* and k.

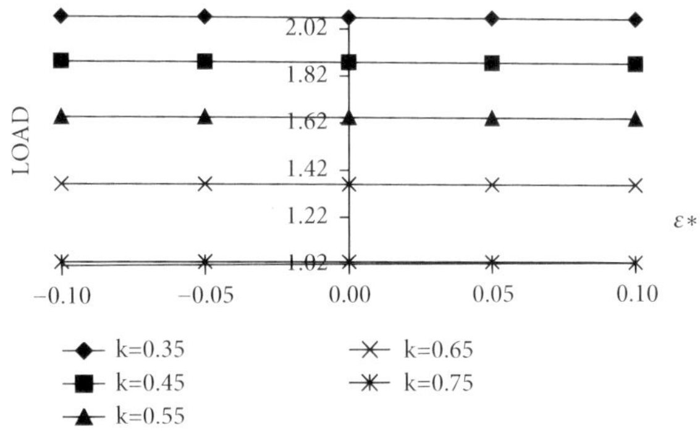

FIGURE 16: Profile of load with respect to ε^* and k.

This means that this type of bearing system the suitable combination of negatively skewed roughness and negative variance may go a long way for better performance of the bearing system.

It is seen that for small as well as large values of M the performance of the bearing suffers when the plates are considered to be electrically conducting, in matching with the hydromagnetic case, when the plates are taken to be nonconducting. Further, this can physically be clarified by fringing phenomena which happens when the plates are electrically conducting. Finally, the load is acquired to be more in comparison with the Neuringer Rosensweig model based ferrofluid squeeze film.

4. Conclusions

The trio of hydromagnetization, supply pressure, and standard deviation counters fruitfully the negative effect induced by positively skewed roughness and variance (+ve) in the case of negatively skewed roughness when variance comes out to be negative. Of course, in bettering the bearing performance the plate conductivities and radii ratio may play an important role. Needless to say, the longitudinal roughness parameter behaves in a little better way as compared to the transverse roughness. Therefore, it becomes very much essential that the longitudinal roughness aspect must be given due consideration from longevity point of view while designing this type of bearing system.

Nomenclature

- r: Radial coordinate
- r_o: Outer radius
- r_i: Inner radius
- k: Radii ratio (r_i/r_o)
- h: Lubricant film thickness
- H: Thickness of solid housing
- s: Electrical conductivity of the lubricant
- μ: Viscosity of the lubricant
- B_o: Uniform transverse magnetic field applied between the plates
- M: Hartmann Number $[B_0 h(s/\mu)^{1/2}]$
- p_s: Supply Pressure

Q: Flow rate
P_s^*: Dimensionless supply pressure
p: Lubricant pressure
P: Nondimensional pressure
w: Load carrying capacity
W: Dimensionless load carrying capacity
h_0': Surface width of lower plate
h_1': Surface width of upper plate
s_0: Electrical conductivity of lower surface
s_1: Electrical conductivity of upper surface
$\phi_0(h)$: Electrical permeability of lower surface $(s_0 h_0'/sh)$
$\phi_1(h)$: Electrical permeability of upper surface $(s_1 h_1'/sh)$
σ^*: Nondimensional standard deviation
α^*: Dimensionless variance
ε^*: Nondimensional skewness.

Conflicts of Interest

The authors declare that they have no conflicts of interest.

References

[1] B. C. Majumdar, *Introduction to Tribology of Bearing*, Wheeler publishers, Wheeler Co .Ltd, 1985.

[2] J. Lin, "Static and dynamic characteristics of externally pressurized circular step thrust bearings lubricated with couple stress fluids," *Tribology International*, vol. 32, no. 4, pp. 207–216, 1999.

[3] G. M. Deheri, H. C. Patel, and R. M. Patel, "Performance of magnetic fluid based circular step bearings," *Mechanika*, vol. 57, no. 1, pp. 22–27, 2006.

[4] J. L. Neuringer and R. E. Rosensweig, "Magnetic fluids," *Physics of Fluids*, vol. 7, no. 12, pp. 1927–1937, 1964.

[5] H. Christensen and K. Tonder, "Tribology of rough surfaces: Parametric study and comparison of lubrication models," *SINTEF Report*, vol. 93, no. 3, pp. 324–329, 1969.

[6] H. Christensen and K. Tonder, "Tribology of rough surfaces: Stochastic models of hydrodynamic lubrications," *SINTEF Report*, vol. 93, no. 3, pp. 324–329, 1969.

[7] H. Christensen and K. Tonder, "Tribology of rough surfaces: A stochastic model of mixed lubrication," *SINTEF Report*, vol. 93, no. 3, pp. 324–329, 1970.

[8] L. L. Ting, "Engagement behavior of lubricated porous annular disks. Part I: squeeze film phase—surface roughness and elastic deformation effects," *Wear*, vol. 34, no. 2, pp. 159–172, 1975.

[9] J. Prakash and K. Tiwari, "Roughness effects in porous circular squeeze-plates with arbitrary wall thickness," *Journal of Lubrication Technology*, vol. 105, no. 1, pp. 90–95, 1983.

[10] B. L. Prajapati, *On Certain Theoretical Studies in Hydrodynamics And Elastohydrodynamics Lubrication [PhD. thesis]*, S. P. University, V. V. Nagar, India, 1995.

[11] P. I. Andharia, J. L. Gupta, and G. M. Deheri, "Effect of transverse surface roughness on the behavior of squeeze film in a spherical bearing," in *Proceedings of the International Conference: Problem of Non-Conventional Bearing Systems (NCBS '99)*, 1999.

[12] P. A. Vadher, P. C. Vinodkumar, G. M. Deheri, and R. M. Patel, "Behaviour of hydromagnetic squeeze films between two conducting rough porous circular plates," *Journal of Engineering Tribology*, vol. 222, no. 4, pp. 569–579, 2008.

[13] J. R. Patel and G. M. Deheri, "Shliomis model based magnetic fluid lubrication of a squeeze film in rotating rough curved circular plates, Cerib," *Journal of Science and Technology*, vol. 1, pp. 138–150, 2013.

[14] R. Patel, G. Deheri, and H. Patel, "Effect of transverse roughness on the performance of a circular step bearing lubricated with a magnetic fluid," *Annals*, vol. 8, no. 2, pp. 23–30, 2010.

[15] J.-R. Lin, L.-M. Chu, L.-J. Liang, and P.-H. Lee, "Derivation of a modified lubrication equation for hydromagnetic non-Newtonian cylindrical squeeze films and its application to circular plates," *Journal of Engineering Mathematics*, vol. 77, pp. 69–75, 2012.

[16] R. M. Patel and G. M. Deheri, "Magnetic fluid based squeeze film behavior between rotating porous circular plates with a concentric circular pocket and surface roughness effects," *International Journal of Applied Mechanics and Engineering*, vol. 8, no. 2, pp. 271–277, 2003.

[17] P. I. Andharia and G. M. Deheri, "Effect of longitudinal surface roughness on the behaviour of squeeze film in a spherical bearing," *International Journal of Applied Mechanics and Engineering*, vol. 6, no. 4, pp. 885–897, 2001.

[18] P. I. Andharia and G. Deheri, "Longitudinal roughness effect on magnetic fluid-based squeeze film between conical plates," *Industrial Lubrication and Tribology*, vol. 62, no. 5, pp. 285–291, 2010.

[19] P. I. Andharia and G. M. Deheri, "Performance of magnetic-fluid-based squeeze film between longitudinally rough elliptical plates," *ISRN Tribology*, vol. 2013, Article ID 482604, 6 pages, 2013.

[20] M. E. Shimpi and G. M. Deheri, "Combine effect of bearing deformation and longitudinal roughness on the performance of a ferrofluid based squeeze film together with velocity slip in truncated conical plates," *Imperial journal of Interdisciplinary research*, vol. 2, no. 8, pp. 1423–1430, 2016.

[21] J.-R. Lin, "Longitudinal surface roughness effects in magnetic fluid lubricated journal bearings," *Journal of Marine Science and Technology (Taiwan)*, vol. 24, no. 4, pp. 711–716, 2016.

[22] R. M. Patel, G. M. Deheri, and P. A. Vahder, "Hydromagnetic rough porous circular step bearing," *Eastern Academic Journal*, vol. 3, pp. 71–87, 2015.

5

Engineering Software Solution for Thermal Elastohydrodynamic Lubrication using Multiphysics Software

Thomas Lohner, Andreas Ziegltrum, Johann-Paul Stemplinger, and Karsten Stahl

Gear Research Centre (FZG), Technical University of Munich (TUM), Boltzmannstraße 15, 85748 Garching, Germany

Correspondence should be addressed to Andreas Ziegltrum; ziegltrum@fzg.mw.tum.de

Academic Editor: Michel Fillon

The complexity of thermal elastohydrodynamic lubrication (TEHL) problems has led to a variety of specialised numerical approaches ranging from finite difference based direct and inverse iterative methods such as Multilevel Multi-Integration solvers, via differential deflection methods, to finite element based full-system approaches. Hence, not only knowledge of the physical and technical relationships but also knowledge of the numerical procedures and solvers is necessary to perform TEHL simulations. Considering the state of the art of multiphysics software, the authors note the absence of a commercial software package for solving TEHL problems embedded in larger multiphysics software. By providing guidelines on how to implement a TEHL simulation model in commercial multiphysics software, the authors want to stimulate the research in computational tribology, so that, hopefully, the research focus can be shifted even more on physical modelling instead of numerical modelling. Validations, as well as result examples of the suggested TEHL model by means of simulated coefficients of friction, coated surfaces, and nonsmooth surfaces, highlight the flexibility and simplicity of the presented approach.

1. Introduction

Measurements, numerical simulations, and derived analytical solutions have, for decades, created detailed insights into the behaviour of thermal elastohydrodynamically lubricated (TEHL) contacts. Due to the large number of numerical investigations, it is astonishing that, so far, no commercial software package embedded in larger multiphysics software has been available to solve TEHL problems.

The simulation of TEHL contacts is, mainly due to elastic deformations and the large pressure-induced increase of lubricant viscosity, highly nonlinear and has led to various specialised numerical simulation approaches. Great computational efforts, and the tendency to instabilities at high loads when using computational fluid dynamics (CFD) [1], make the use of simulations based on the Reynolds equation necessary. Hartinger et al. [2] thereby compare TEHL solutions obtained with CFD with solutions based on the Reynolds equation which show very good agreement. Lubrecht [3] and Venner [4] develop a direct iterative finite difference based Multilevel Multi-Integration (MLMI) solver for isothermal EHL problems. Habchi [1] introduces the finite element method (FEM) based full-system approach to solve the fully coupled system of the highly nonlinear elastohydrodynamic equations, which are coupled with the energy equations by an iterative procedure. Raisin et al. [5] extend the approach based on line contacts to transient operating conditions. Although both authors [5, 6] mention the use of the commercial software, COMSOL Multiphysics [7], to solve the TEHL problem, explicit implementation guidelines are not given. For the isothermal EHL line and point contact with Newtonian fluid behaviour under transient and steady-state operating conditions, Tan et al. [8] describe some information for implementation in COMSOL Multiphysics in more detail.

Considering the large number of self-developed approaches and solvers, the modelling of TEHL contacts is hardly transparent. Hence, the promotion of using commercial software packages for solving TEHL problems is necessary to make simulation techniques available to a broader audience of the tribology society and to focus research more on physical relationships rather than on numerical procedures. Accordingly, the authors provide guidelines for the implementation and solving of TEHL problems in the commercial

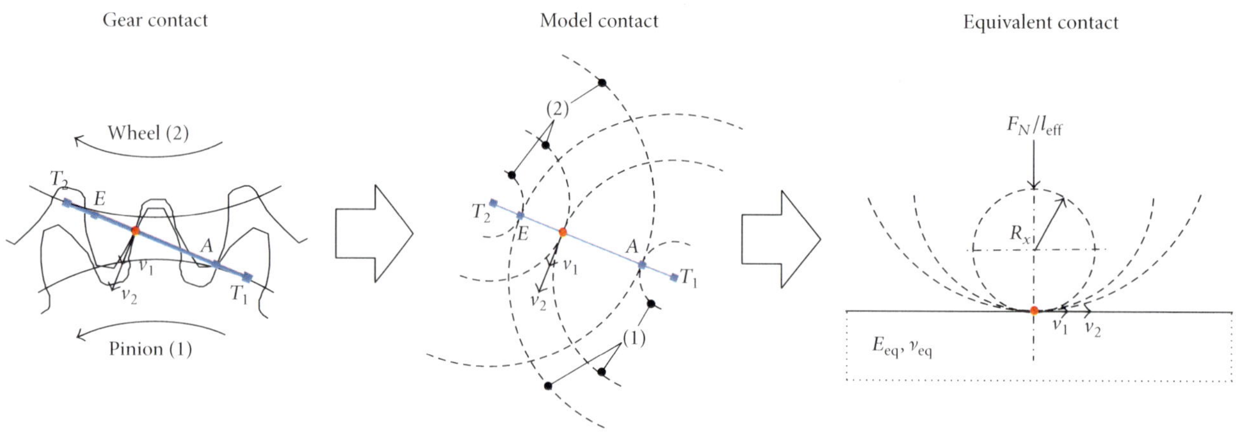

FIGURE 1: Line contact in a spur gear and its simplification to a model (disc) contact and to an equivalent contact.

software, COMSOL Multiphysics, to stimulate and accelerate research in computational tribology. The possibility of developing a complex TEHL model with moderate effort is demonstrated in this paper.

2. Theory

In this section, the theory and governing equations for simulation of TEHL contacts are addressed. For simplification, a two-dimensional line contact is considered.

2.1. General. Contact conditions of, for example, gears are characterised by varying load, motion, and geometry along the path of contact. This transient contact can be interpreted as a transient model contact of two rolling elements with varying load, motion, and geometry for each meshing point. In TEHL simulations, the transient contact is usually further simplified to an equivalent contact between a single (inelastic) roller and an elastic flat body. This is adopted in this study and exemplarily shown for a line contact of a spur gear in Figure 1.

Characteristic quantities of the equivalent contact are described by the Hertzian contact parameters, for example, p_H, b_H, and R_x [9]. Poisson's ratio ν_{eq} and Young's Modulus E_{eq} of the equivalent elastic flat body are calculated according to Habchi [1]. The kinematics are characterised by the sum velocity v_Σ and the siding velocity v_g, whereby, for all analyses in this study, the velocity of the lower body is higher than the velocity of the upper body $v_1 > v_2$:

$$\begin{aligned} v_g &= v_1 - v_2, \\ v_\Sigma &= v_1 + v_2. \end{aligned} \tag{1}$$

The coefficient of friction μ of the TEHL contact is evaluated by integrating the shear stress in the middle of the lubricant film (see also Habchi [1]):

$$\mu = \frac{\int_{x_{in}}^{x_{ex}} \tau_{zx}\Big|_{z=h/2} dx}{F_N/l_{\text{eff}}}. \tag{2}$$

2.2. Generalised Reynolds Equation. Derived from the two-dimensional compressible Navier-Stokes equations by considering reasonable assumptions for TEHL contacts [1, 10] and no-slip boundary conditions, the shear rate $\dot{\gamma}_x$ and velocity distribution v_x in the lubricant film can be expressed by

$$\begin{aligned} \dot{\gamma}_x(x,z,t) &= \frac{\partial v_x}{\partial z} \\ &= \frac{1}{\eta}\frac{\partial p}{\partial x}\left(z - \frac{\int_0^h (z/\eta)\,dz}{\int_0^h (1/\eta)\,dz}\right) \\ &\quad + \frac{v_2 - v_1}{\eta \int_0^h (1/\eta)\,dz}, \end{aligned} \tag{3}$$

$$\begin{aligned} v_x(x,z,t) &= \frac{\partial p}{\partial x}\left(\int_0^z \frac{\tilde{z}}{\eta}d\tilde{z} - \int_0^z \frac{1}{\eta}d\tilde{z}\frac{\int_0^h (z/\eta)\,dz}{\int_0^h (1/\eta)\,dz}\right) \\ &\quad + \frac{v_2 - v_1}{\int_0^h (1/\eta)\,dz}\int_0^z \frac{1}{\eta}d\tilde{z} + v_1. \end{aligned} \tag{4}$$

Inserting the velocity equation in the integrated transient compressible continuity equation results in the generalised Reynolds equation according to Yang and Wen [11]:

$$\begin{aligned} &\frac{\partial}{\partial x}\left[\frac{\partial p}{\partial x}\int_0^h \rho\left(\int_0^z \frac{\tilde{z}}{\eta}d\tilde{z} - \int_0^z \frac{1}{\eta}d\tilde{z}\frac{\int_0^h (z/\eta)\,dz}{\int_0^h (1/\eta)\,dz}\right)dz\right. \\ &\quad + \frac{\partial}{\partial x}\left[\frac{v_2 - v_1}{\int_0^h (1/\eta)\,dz}\int_0^h \rho\left(\int_0^z \frac{1}{\eta}d\tilde{z}\right)dz\right. \\ &\quad \left. + \int_0^h \rho\,dz \cdot v_1\right] + \frac{\partial}{\partial t}\int_0^h \rho\,dz = 0. \end{aligned} \tag{5}$$

In TEHL line contacts, the generalised Reynolds equation determines the hydrodynamic pressure. Due to the integrations when deriving the generalised Reynolds equation, the pressure solution for line contacts is one-dimensional. The integral terms in (3) to (5) can be understood as integral viscosity and density terms. As a boundary condition, $p(x_{in}) = p(x_{ex}) = 0$ has to be prescribed. In addition, the cavitation model according to Wu [12] fulfils the Reynolds cavitation boundary condition [13]: $p \geq 0$ in Ω_P and $p(x_{\text{cav}}) = \partial p/\partial x(x_{\text{cav}}) = 0$.

2.3. Contact Mechanics. The calculation of the elastic deformation of the equivalent body is based on the finite element method as introduced by Habchi [1]. This has advantages, in terms of accuracy and resolution, over the half-space assumption [10]. The elastic deformation of the equivalent body in gap height direction δ is calculated by applying the linear elasticity equation and neglecting the dynamic response of the solids [5]:

$$\nabla \cdot \sigma = 0 \quad \text{with } \sigma = C \cdot \varepsilon(U),$$
$$\delta(x,t) = |U_v(x,t)| \quad \text{with the displacement vector } U = \begin{pmatrix} U_u \\ U_v \end{pmatrix}. \tag{6}$$

Boundary conditions of the linear elasticity equation are zero displacement on the bottom of the calculation domain Ω_δ and the hydrodynamic pressure p between x_{in} and x_{ex} on the top of the calculation domain Ω_δ, which is applied as a normal stress (Figure 3, top left). A free boundary condition assuming zero normal and tangential stress is applied to the remaining boundaries of the computational domain Ω_δ.

The film thickness equation describes the height of the separating lubricant film and consists of the constant parameter h_0, the undeformed geometry, the elastic deformation of the equivalent body $\delta(x,t)$, and the deviation from the smooth profile $\mathfrak{R}$:

$$h(x,t) = h_0 + \frac{x^2}{2R_x} + \delta - \mathfrak{R}. \tag{7}$$

Hydrodynamic pressure and the applied load have to be in balance. This is ensured by the load balance equation, which determines the constant parameter h_0 in (7):

$$\int_{\Omega_P} p(x,t)\, d\Omega_P = \left(\frac{F_N}{l_{\text{eff}}}\right)(t). \tag{8}$$

2.4. Energy Conservation. Similarly to the generalised Reynolds equation, the transient energy equations are simplified with reasonable assumptions typically used in TEHL calculations [1, 10]. The resulting energy equation of the lubricant can be written as

$$\rho c_p \cdot \left(\frac{\partial T}{\partial t} + v_x \frac{\partial T}{\partial x}\right) - \left(\frac{\partial}{\partial x}\left(\lambda \frac{\partial T}{\partial x}\right) + \frac{\partial}{\partial z}\left(\lambda \frac{\partial T}{\partial z}\right)\right) = -\frac{T}{\rho}\frac{\partial \rho}{\partial T}\left(\frac{\partial p}{\partial t} + v_x \frac{\partial p}{\partial x}\right) + \eta \left(\frac{\partial v_x}{\partial z}\right)^2. \tag{9}$$

The transient energy equations of the solids (1, 2) are shown in (10). The transitions between the solids and the lubricant are described by a temperature and conductive heat flux continuity condition:

$$\rho_i c_{p,i} \cdot \left(\frac{\partial T}{\partial t} + v_i \frac{\partial T}{\partial x}\right) - \left(\frac{\partial}{\partial x}\left(\lambda_i \frac{\partial T}{\partial x}\right) + \frac{\partial}{\partial z}\left(\lambda_i \frac{\partial T}{\partial z}\right)\right) = 0 \quad \text{with } i = 1, 2. \tag{10}$$

Thus, based on heat sources (shearing and compression of the lubricant), heat sinks (expansion of the lubricant), and heat transfer due to convection and conduction, the temperature distribution in lubricant and solids can be calculated. Boundary conditions ensure that lubricant entering the computational domain Ω_T ($v_x(x_{in}, z, t) \geq 0$ and $v_x(x_{ex}, z, t) \leq 0$) has bulk temperature T_M and for lubricant leaving Ω_T ($v_x(x_{in}, z, t) \leq 0$ and $v_x(x_{ex}, z, t) \geq 0$) the conductive heat flux is zero ($-\lambda \cdot \partial T/\partial x = 0$). Similarly, solids entering the computational domains $\Omega_{T,1}$ and $\Omega_{T,2}$ ($v_i(x_{in}, t) \geq 0$ and $v_i(x_{ex}, t) \leq 0$) have bulk temperature T_M and for solids leaving $\Omega_{T,1}$ and $\Omega_{T,2}$ ($v_i(x_{in}, t) \leq 0$ and $v_i(x_{ex}, t) \geq 0$) the conductive heat flux is zero ($-\lambda_i \cdot \partial T/\partial x = 0$). The height of solids $\Omega_{T,1}$ and $\Omega_{T,2}$ is with $d = 3.15 \cdot b_H$ [14] sufficient so that the conductive heat flux into the solids becomes zero and bulk temperature T_M can be assumed at d.

2.5. Lubricant Properties. Pressure, temperature, and shear rate distributions in TEHL contacts have a significant influence on lubricant properties. The accompanying changes in viscosity have the largest influence on the TEHL contact itself. The models for pressure and temperature dependency of the viscosity as suggested in [15] are used. The dynamic viscosity for a given temperature and ambient pressure $\eta(T)$ is described by the Vogel model [16]:

$$\eta(T) = A \cdot \exp\left(\frac{B}{C + (T - 273.15\,\text{K})}\right). \tag{11}$$

The pressure dependency of the viscosity $\eta(T, p)$ is modelled by the Roelands [17] equation

$$\eta(T,p) = \eta(T) \cdot \exp\left\{(\ln(\eta(T)) + 9.67) \cdot \left[-1 + \left(1 + \frac{p}{p_{\eta 0}}\right)^{z(T)}\right]\right\} \tag{12}$$

with a temperature dependent pressure exponent $z(T)$ according to

$$z(T) = \frac{\alpha_p(T) \cdot p_{\eta 0}}{\ln(\eta(T)) + 9.67}. \tag{13}$$

The pressure-viscosity coefficient $\alpha_p(T)$ is temperature dependent:

$$\alpha_p(T) = E_{\alpha_p 1} \cdot \exp\left(E_{\alpha_p 2} \cdot T\right). \tag{14}$$

The rheological behaviour of the lubricant is described by an Eyring model [18]:

$$\eta(T, p, \dot{\gamma}_x) = \frac{\tau_c}{\dot{\gamma}_x}\operatorname{asinh}\left(\frac{\eta(T, p)\cdot\dot{\gamma}_x}{\tau_c}\right) \quad \text{with } \tau(T, p, \dot{\gamma}_x) = \eta(T, p, \dot{\gamma})\cdot\dot{\gamma}_x. \tag{15}$$

The temperature and pressure dependency of lubricant density $\rho(T, p)$ is modelled by the Bode model [19]:

$$\rho(T, p) = \frac{\rho_s\cdot(1-\alpha_s\cdot T)}{1 - D_{\rho 0}\cdot\ln\left(\left(D_{\rho 1} + D_{\rho 2}\cdot T + p\right)/\left(D_{\rho 1} + D_{\rho 2}\cdot T\right)\right)}. \tag{16}$$

Based on measurements, Larsson and Andersson [20] derived a model for the temperature and pressure dependency of the thermal conductivity $\lambda(p)$ and of the specific heat capacity per volume $(c_p\cdot\rho)(T, p)$. Temperature dependency of the thermal conductivity is not considered; as in [20], the influence is assessed as negligible. The equations for $\lambda(p)$ and $(c_p\cdot\rho)(T, p)$ are

$$\lambda(p) = \lambda_0\cdot\left(1 + \frac{d_{\lambda 1}\cdot p}{1 + d_{\lambda 2}\cdot p}\right), \tag{17}$$

$$(c_p\cdot\rho)(T, p) = (c_p\cdot\rho)(295\,\mathrm{K})\cdot\left(1 + \frac{A_{c1}\cdot p}{1 + A_{c2}\cdot p}\right) \cdot\left[1 + A_{c3}\cdot\left(1 + A_{c4}\cdot p + A_{c5}\cdot p^2\right)\cdot(T - 295\,\mathrm{K})\right]. \tag{18}$$

Among others, parameters for paraffinic mineral oil are available for (17) and (18) in [20].

3. Model Description and Implementation

A guideline for the implementation of a TEHL simulation model in COMSOL Multiphysics (abbr. by COMSOL) and MathWorks MATLAB [21] (abbr. by MATLAB) is given in the following sections. All COMSOL-specific notations are written in italics.

3.1. Dimensionless Parameters. For a convenient numerical solution procedure of the nonlinear system of equations with good conditioning, the variables are transferred to a dimensionless form. Following Habchi [1], Tan et al. [8], Venner [4], and Lubrecht [3], the definitions are

$$\begin{aligned}
X &= \frac{x}{b_H}, \\
\overline{T} &= \frac{T}{T_M}, \\
D &= \frac{d}{b_H}, \\
P &= \frac{p}{p_H}, \\
H &= \frac{h\cdot R_x}{b_H^2}, \\
\overline{\delta} &= \frac{\delta\cdot R_x}{b_H^2}, \\
\overline{\rho} &= \frac{\rho}{\rho(T_M)}, \\
\overline{\eta} &= \frac{\eta}{\eta(T_M)}, \\
\overline{t} &= \frac{t\cdot v_\Sigma}{2b_H}, \\
C_v &= \frac{v_\Sigma(t)}{v_\Sigma(t=0)}, \\
C_R &= \frac{R_x(t)}{R_x(t=0)}, \\
C_w &= \frac{(F_N/l_{\mathrm{eff}})(t)}{(F_N/l_{\mathrm{eff}})(t=0)}, \\
Z &= \begin{cases} \dfrac{z}{b_H} & \text{for solids} \\ \dfrac{z}{h} & \text{for lubricant.} \end{cases}
\end{aligned} \tag{19}$$

3.2. Numerical Solution Procedure. The numerical solution scheme shown in Figure 2 is similar to other schemes of, for example, Habchi [1] and Wang et al. [22]. In this work, it is based on two self-developed decoupled COMSOL FEM-models, denoted as FEM-model (P, H) and FEM-model $(\overline{T})$, and a MATLAB sequence controller, which also performs additional calculations.

After reading all the required input parameters, an initial solution based on a simple steady-state isothermal Newtonian approach is calculated in the FEM-model (P, H) for the initial operating conditions. The procedure thereby is the same as described in the following paragraph. Based on the initial solution, the time loop is launched and repeated until the last time-step has been calculated. The coupling of two consecutive time-steps is realised by an implicit backward differentiation formula (BDE) scheme of the first order as used in [23]. It is implemented by passing the required

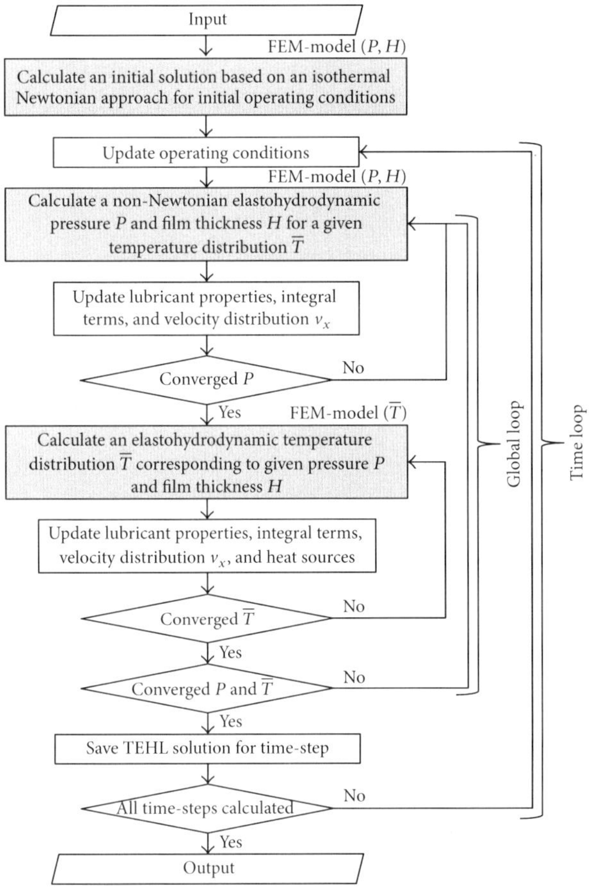

FIGURE 2: Numerical solution scheme.

solution variables of the previous time-step for calculating the time derivatives of the generalised Reynolds equation and energy equations from the sequence controller to the FEM-model (P, H) and FEM-model $(\overline{T})$.

Within the time loop, a global loop is launched for each time-step. Firstly the pressure and film thickness distribution under consideration of non-Newtonian fluid behaviour is calculated for a given temperature distribution $\overline{T}$ in the FEM-model (P, H) described in Section 3.3.1. Then, the lubricant properties acc. to Section 2.5, the integral terms in (5), and the velocity distribution v_x (4) are updated in the sequence controller. Thereby, the shear stress τ is analytically derived (15). After calling up the FEM-model (P, H) again, the new calculated pressure distribution is compared with the previous calculated pressure distribution. The iterative procedure is repeated until the maximum absolute difference of two consecutive pressure distributions is smaller than 10^{-3}.

After the convergence of P, the pressure and film thickness distributions P and H are kept constant and an iterative calculation of the temperature distribution $\overline{T}$ in lubricant and solid bodies, based on the FEM-model $(\overline{T})$ described in Section 3.3.2, is launched. For each new temperature distribution $\overline{T}$ calculated, the lubricant properties acc. to Section 2.5, the integral terms in (5), the heat sources in (9), and the velocity distribution v_x (4) are updated in the sequence controller. Thereby, the shear stress τ is analytically derived (15). The iterative calculation is repeated until the maximum absolute difference of two consecutive temperature distributions in the lubricant is smaller than 10^{-3}.

Finally, the converged pressure distribution and temperature distribution in the lubricant of two consecutive global loops are compared. Again, convergence is assumed when the maximum absolute difference between two consecutive solutions of pressure and temperature distributions is smaller than 10^{-3}. After saving the solution of the considered time-step, the next global loop is calculated with updated operating conditions and time derivatives of the generalised Reynolds equation and energy equations.

3.3. Implementation in COMSOL. In the following, the characteristics of the FEM-model (P, H) and FEM-model $(\overline{T})$, as well as of the solver settings and sequence controller, are described in detail. To a great extent, the structure of the FEM-models follows the outstanding work of Habchi [1], who introduces the FEM based full-system approach to TEHL achieving very high convergence rates without underrelaxation required. Thereby, the FEM approach offers the possibility of using a mesh with nonregular unstructured elements and high order approximation functions. As shown in the following, it is possible to implement the approach of Habchi in commercial multiphysics software.

3.3.1. FEM-Model (P, H). Figure 3 (top left) shows the computational domains Ω_δ and Ω_P on which the pressure and film thickness distribution are calculated fully coupled. Figure 3 (bottom left) illustrates in parts the corresponding model tree. For each new call-up of FEM-model (P, H), the solution of the previous call is used as an initial value.

The elastic deformation of the equivalent body in Z-direction $\overline{\delta}(X, \overline{t})$ is calculated on a quadratic two-dimensional computational domain Ω_δ with the extensions $X \in [-30, 30]$ and $Z \in [-60, 0]$. Hence, an influence of the boundary conditions of Ω_δ on the contact area can be excluded. The *physics Solid Mechanics* is used with the displacement vector $\overline{U}(X, \overline{t})$ as the dependent variable. A *linear elastic* isotropic material specified by Young's Modulus E_{eq} and Poisson's ratio ν_{eq} is assumed. The boundary conditions described in Section 2.3 are implemented as a *fixed constraint* condition on the bottom boundary of Ω_δ, a *boundary load* condition with the *load type pressure* P on Ω_P, and a *free boundary* condition on the remaining boundaries of Ω_δ. Lagrange quadratic elements are used for the linear elasticity equations.

The pressure and film thickness distributions P and H, as well as the constant parameter H_0 based on the load balance in (8), are calculated on the one-dimensional computational domain Ω_P with an extension from $X_{in} = -4.5$ to $X_{ex} = 3$. Thereby, X_{ex} of Ω_P has already been enlarged for simulations with varying load over time, for example, [5]. The dependent variables are the pressure P, the constant parameter H_0, and the displacement vector $\overline{U}$. The film thickness in (7) is implemented as a *variable* and the deviation from the smooth profile $\mathfrak{R}$ is read as the input parameter by an *interpolation* function. The load balance in (8) is implemented as a *Global*

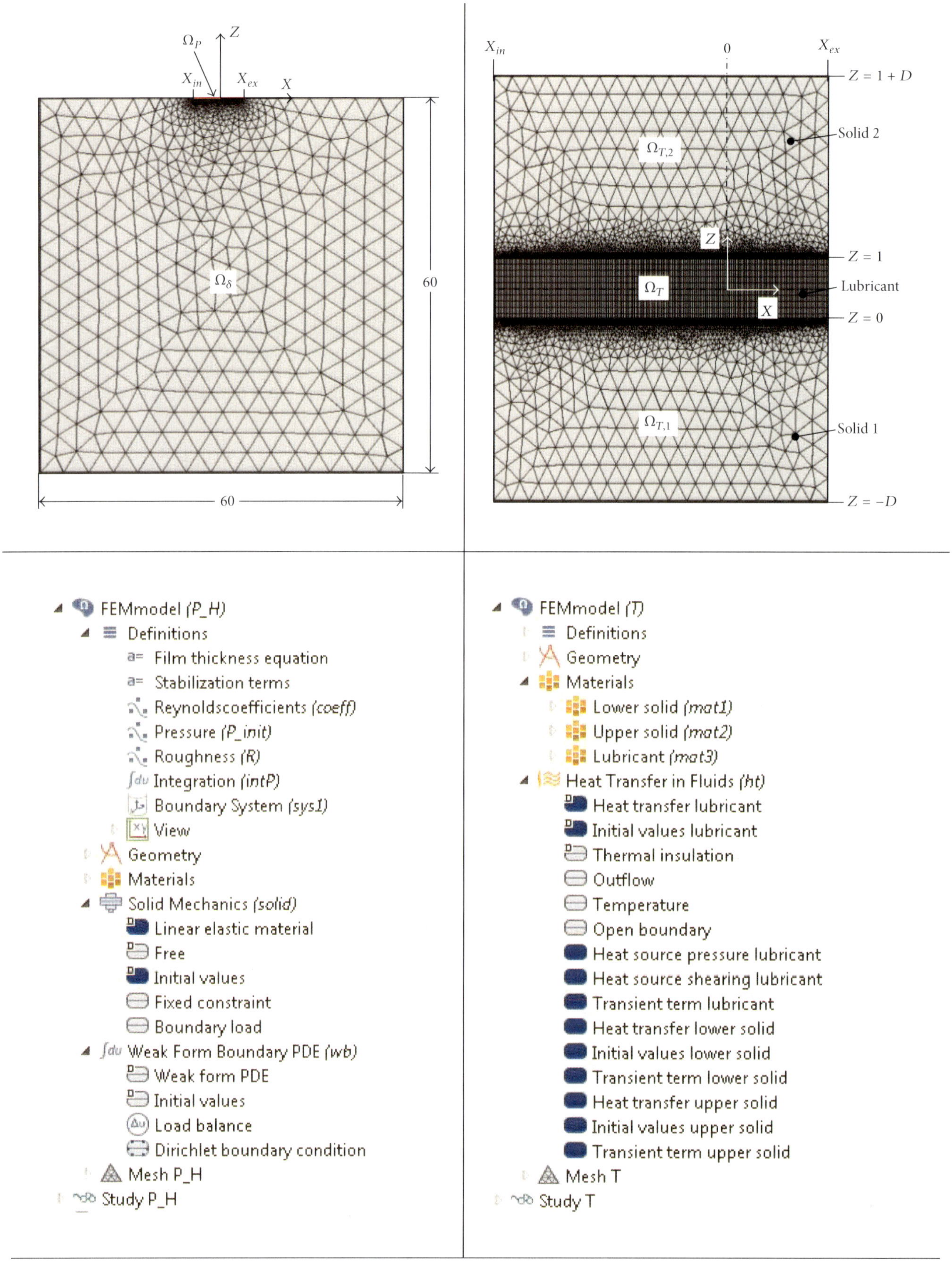

FIGURE 3: FEM-model (P, H) for pressure and film thickness calculation with corresponding COMSOL model tree (left) and FEM-model $(\overline{T})$ for temperature calculation with corresponding COMSOL model tree (right); element numbers are adjusted for presentation purposes.

Equation with a *component coupling integration* on Ω_P for the left-hand side of the load balance equation. Its result is the constant parameter H_0. As no specific module for the generalised Reynolds equation in (5) is available in COMSOL, it is defined in the *physics Weak Form Boundary PDE* by using Lagrange quintic elements. Based on the dimensionless parameters in (19), the weak form of the generalised Reynolds equation in dimensionless form is written as

$$\begin{aligned}
&-\int_{\Omega_P}\left((\overline{\varepsilon}+ID)\frac{\partial P}{\partial X}\right)\frac{\partial W_P}{\partial X}dX \\
&-\int_{\Omega_P}(\overline{\rho}H)_H\frac{\partial W_P}{\partial X}dX+\int_{\Omega_P}\frac{\partial(\overline{\rho}H)_{\overline{t}}}{\partial\overline{t}}W_P d\Omega_P \\
&+\int_{\Omega_P}Q_\xi W_P d\Omega_P+\left\{\begin{matrix}SUPG\\GLS\end{matrix}\right\}=0 \\
&\qquad\text{with}\ \left\{\begin{matrix}SUPG\\GLS\end{matrix}\right\}=SUPG\ \text{or}\ GLS.
\end{aligned}\tag{20}$$

W_P is the test function for the pressure P and is denoted as $test(P)$ in COMSOL. *ID* and *SUPG*/*GLS* represent numerical stabilisation terms, which are implemented as *variables* and given here for completeness. Note that, for moderate cases, the TEHL simulation model can work without stabilisation terms. The numerical stabilisation terms have been applied to the generalised Reynolds equation by Habchi [1, 24] in order to avoid the oscillatory behaviour of the Reynolds equation at high loads. *SUPG* (Streamline-Upwind-Petrov-Galerkin) and *GLS* (Galerkin-Least-Squares) are consistent stabilisation terms whereas *ID* (Isotropic Diffusion) is a nonconsistent stabilisation term. For implementation of *ID* and *SUPG*/*GLS*, the reader is referred to [1, 24]. The penalty term $Q_\xi = P_f \cdot P^-$ as an additional source term with P_f in the order of 10^6 [1] ensures the Reynolds cavitation boundary conditions [13] and represents the cavitation model according to Wu [12]. Zero pressure on the boundaries of Ω_P is ensured by a *Dirichlet boundary condition*. The coefficients $\overline{\varepsilon}$, $(\overline{\rho}H)_H$, and $(\overline{\rho}H)_{\overline{t}}$ represent the first, second, and third term of (5) in dimensionless form:

$$\begin{aligned}
\overline{\varepsilon}&=\frac{H^3}{\overline{\lambda}}\int_0^1\overline{\rho}\left(\int_0^Z\frac{\widetilde{Z}}{\overline{\eta}}d\widetilde{Z}-\int_0^Z\frac{1}{\overline{\eta}}d\widetilde{Z}\frac{\int_0^1(Z/\overline{\eta})\,dZ}{\int_0^1(1/\overline{\eta})\,dZ}\right)dZ \\
&\qquad\text{with}\ \overline{\lambda}=\frac{R_x^2\cdot\eta(T_M)\cdot v_\Sigma}{2\cdot p_H\cdot b_H^3}, \\
(\overline{\rho}H)_H&=\frac{2H}{v_\Sigma}\left(\frac{v_2-v_1}{\int_0^1(1/\overline{\eta})\,dZ}\int_0^1\overline{\rho}\int_0^Z\frac{1}{\overline{\eta}}d\widetilde{Z}\,dZ\right. \\
&\quad\left.+\int_0^1\overline{\rho}v_1 dZ\right), \\
(\overline{\rho}H)_{\overline{t}}&=H\int_0^1\overline{\rho}\,dZ.
\end{aligned}\tag{21}$$

Note thereby that all the parts of the coefficients in (21) and of the stabilisation terms in (20) can, despite the film thickness H, be prepared in the sequence controller and transferred to the FEM-model (P, H) by *interpolation* functions.

The mesh in the FEM-model (P, H) is *free triangular* with a normal element size and a refinement at Ω_P by a *distribution* with typically 1000 elements.

3.3.2. FEM-Model $(\overline{T})$. Figure 3 (top right) shows the computational domains Ω_T, $\Omega_{T,1}$, and $\Omega_{T,2}$ on which the temperature distribution $\overline{T}$ in lubricant and solid bodies is calculated fully coupled. Figure 3 (bottom right) illustrates in parts the corresponding model tree. The lubricant domain Ω_T has the dimensionless height of 1 with $Z \in [0,1]$, whereas the solid domains $\Omega_{T,1}$ and $\Omega_{T,2}$ have the dimensionless heights of $D = 3.15$ with $Z \in [-D, 0]$ and $Z \in [1, D+1]$, respectively. The length of the computational domains extends from X_{in} to X_{ex} (see Figure 3, top left). For each new call-up of the FEM-model $(\overline{T})$, the solution of the previous call-up is used as initial value. The transient energy equations in dimensionless form are implemented and calculated in the *physics Heat Transfer in Fluids* for the lubricant domain Ω_T and in *Heat Transfer in Solids* for the solid domains $\Omega_{T,1}$ and $\Omega_{T,2}$. The dependent variable $\overline{T}$ is solved based on Lagrange quadratic elements. The numerical stabilisation terms *ID*, *SUPG*, and *GLS* are already implemented in the considered *heat transfer physics* in COMSOL and applied when required. The velocity distribution (4), viscosity distribution (15), density distribution (16), thermal conductivity distribution (17), and heat capacity distribution (18) in Ω_T are calculated in the sequence controller and transferred to the FEM-model $(\overline{T})$ by *interpolation* functions. Furthermore, the *heat sources* in the *physics Heat Transfer in Fluids,* due to the shearing and compression of the lubricant, are calculated in the sequence controller and transferred to FEM-model $(\overline{T})$. Note thereby that, similar to the coefficients in (21), $\overline{T}$ as a dependent variable needs to be applied in the FEM-model $(\overline{T})$. The implementation of the energy equations on $\Omega_{T,1}$ and $\Omega_{T,2}$ in the *Heat Transfer in Solids* modules with the translation motions of solids v_1 and v_2 as input parameters is straightforward.

The boundary conditions in Section 2.4 are implemented by defining bulk temperature $\overline{T}_M = 1$ at the left and upper boundary of $\Omega_{T,2}$ and at the left and lower boundary of $\Omega_{T,1}$, an *Outflow* boundary condition at the right-hand boundaries of $\Omega_{T,1}$ and $\Omega_{T,2}$ and an *Open boundary* condition at the left- and right-hand boundary of the lubricant domain Ω_T with lubricant entering Ω_T having a bulk temperature $\overline{T}_M = 1$. As a common coordinate system in the FEM-model $(\overline{T})$ is used, manual adjustments are necessary to respect the different dimensionless definitions of Z in Ω_T and $\Omega_{T,1}$ and $\Omega_{T,2}$, respectively (19). To be precise, the dimensionless scale of the Z axis of the solid domains $\Omega_{T,1}$ and $\Omega_{T,2}$ has to be adjusted according to the dimensionless scale of the Z axis of the lubricant domain Ω_T, so that the continuity condition of the conductive heat flux and temperature across the lubricant-solid boundaries is respected. This is handled in COMSOL by multiplying the dimensionless energy equations of the solids by the conversion factor Z_{cf}. This factor is the ratio of the dimensionless scales of the Z axes of the lubricant domain and the solid domains:

$$Z_{cf} = \frac{Z\,(\text{lubricant})}{Z\,(\text{solid})} = \frac{z/h}{z/b_H} = \frac{R_x}{H \cdot b_H}. \qquad (22)$$

The mesh in the FEM-model ($\overline{T}$) is divided into the lubricant domain Ω_T and the solid domains $\Omega_{T,1}$ and $\Omega_{T,2}$. Whereas Ω_T is meshed by a *mapped* equidistant mesh, $\Omega_{T,1}$ and $\Omega_{T,2}$ are meshed with *free triangular* elements with a refinement on the solid-lubricant boundaries and an *extra-fine* element size. The lubricant domain Ω_T may consist of 40 elements in height and 1000 elements in the length direction.

3.3.3. Sequence Controller. The communication between the MATLAB sequence controller and the FEM-model (P, H) and FEM-model ($\overline{T}$) is achieved by COMSOL's *LiveLink for MATLAB* function. The MATLAB sequence controller basically represents the implementation of the global numerical scheme shown in Figure 2. There, it is very convenient, for example, to implement the models for the lubricant properties, calculate the integrals by adaptive Simpson quadrature rule with typically 40 elements in Z-direction, and perform other calculations with the current values for pressure, film thickness, and temperature. As no other possibility is available, the data exchange of vectors and matrices from MATLAB to COMSOL is handled in a somewhat cumbersome manner: The quantities, calculated in the sequence controller, for example, velocity distribution, are written into a text file and read by the COMSOL *interpolation* functions. For data exchange from COMSOL to MATLAB, the *mpheval* command is used.

3.3.4. Solver. For solving the nonlinear system of equations, two separate COMSOL solvers for the FEM-model (P, H) and for the FEM-model ($\overline{T}$) are required. The solver for the FEM-model (P, H) includes the dependent solution variables P, H_0, and $\overline{U}$, whereas the solver for the FEM-model ($\overline{T}$) includes the temperature $\overline{T}$ as the dependent solution variable. For both FEM-models, a direct COMSOL *MUMPS* (Multifrontal Massively Parallel Direct Solver) [25] is used with a damped Newton-Raphson approach [26] and 10^{-3} as the tolerance factor as defined in [7].

4. Model Validation

In order to show the functionality and the plausibility of the TEHL simulation model described in Section 3, its solutions are compared with steady-state solutions from Hartinger et al. [2] and with transient solutions presented by Venner [4].

4.1. Steady-State Validation. Hartinger et al. [2] investigated the TEHL contact by CFD simulations for various conditions. They show that the assumptions of the generalised Reynolds equation are reasonable and provide representative results for thermal and isothermal contacts. For comparison of their results in [2] with the presented TEHL simulation model, the same lubricant property models and properties for lubricant and solids are used. Note that cavitation effects considered by the CFD approach cannot be depicted by the cavitation model of Wu [12]. Figure 4 shows the calculated pressure and film thickness distributions for a thermal (TEHL) and isothermal (EHL) contact with high sliding proportion ($v_\Sigma = 5\,\text{m/s}$, $v_g = 2.5\,\text{m/s}$) in comparison to the presented TEHL simulation model (a) and CFD model of Hartinger et al. [2] (b). The results show very good agreement. Furthermore, the local distributions of shear rate, temperature, and viscosity (not shown) correlate very well with the results of Hartinger et al. [2]. The calculation time for the TEHL solution in Figure 4 is approximately 6 to 8 minutes on a computer with a 3.5 GHz processor for the presented TEHL simulation model and in the order of 24 hours for the CFD model [2].

4.2. Transient Validation. Venner [4] investigated an EHL contact with a moving surface indentation under high sliding conditions by using a transient isothermal Newtonian approach. For the sake of comparison, the authors also assume isothermal and Newtonian simplifications in their presented TEHL simulation model. Input parameters and the investigated surface indentation are thereby adopted from [4]. Figure 5 shows the comparison of pressure and film thickness distribution for the example time, when the surface indentation is located at the dimensionless position $X_d = -0.25$. It is thereby shown that the moving surface indentation initiates a significant film thickness change, which hurries ahead the actual position of the indentation. This can occur under high sliding conditions. The comparison in Figure 5 shows very good agreement. Small differences in the results may be caused by different elastic deformation calculations (FEM in the TEHL simulation model in this study and elastic half-space assumption in [4]).

The validation of the TEHL simulation model shows its functionality and plausibility.

5. Example Results

In this section, examples of possible application opportunities, such as simulating coated surfaces, coefficients of friction, and nonsmooth surfaces, of the described TEHL simulation model are demonstrated.

5.1. Thin Surface Coatings. Firstly, the possibility of investigating the effect of thin surface coatings on the TEHL contact behaviour is shown. Results based on the TEHL model described in this study with slightly different lubricant properties have already been published by Lohner et al. [27]. There, the influence of thin surface coatings on the steady-state TEHL contact temperature was analysed by adding coating layers with separate thermophysical properties to the FEM-model ($\overline{T}$). The FEM-model (P, H) is unchanged. In the FEM-model ($\overline{T}$), additional geometries with the *physics Heat Transfer in Solids* including adopted boundary and continuity conditions as described in Section 3.3.2, additional interpolation functions in COMSOL, and additional meshes for

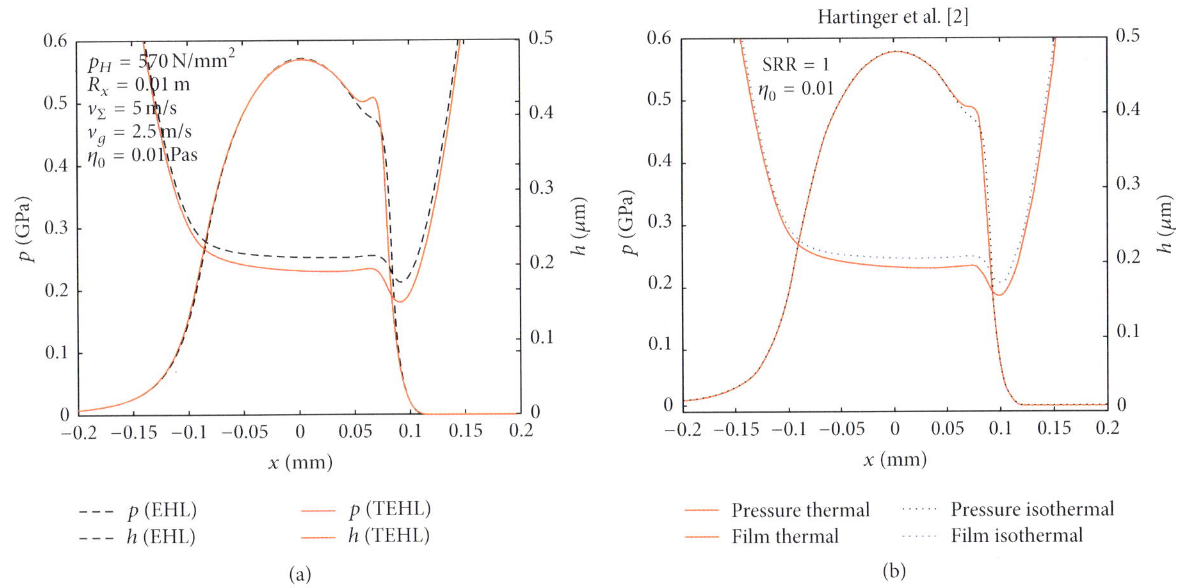

FIGURE 4: Comparison of simulated pressure and film thickness distributions for a thermal (TEHL) and an isothermal (EHL) contact between the TEHL simulation model of this study (a) and the CFD model of Hartinger et al. [2] (b).

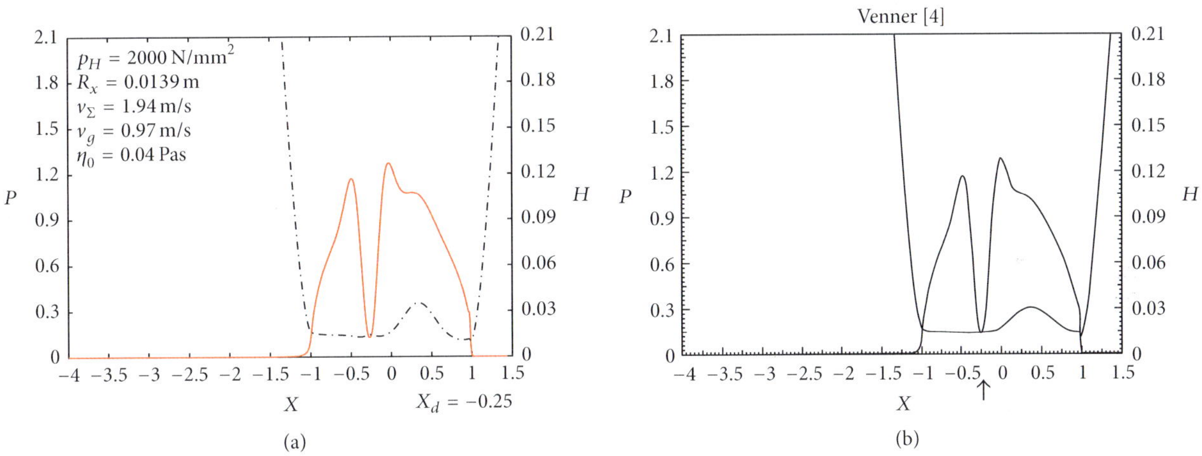

FIGURE 5: Comparison of simulated pressure and film thickness distribution for a transient isothermal Newtonian EHL contact between the TEHL simulation model of this study (a) and the solution of Venner [4] (b).

the thin surface coatings are integrated. The corresponding handling in the sequence controller is the same as for solids $\Omega_{T,1}$ and $\Omega_{T,2}$. This is very convenient to implement and shows the flexibility of the presented TEHL model approach. For simplification, it has been assumed that the deformation of the solid bodies is not affected by the very thin coatings (1.5 μm) [6].

Figure 6 shows an example of the pressure and film thickness (a) and the corresponding temperature distribution of the lubricant and solid bodies (b) for a TEHL contact including a coating with a thickness of 1.5 μm on the upper solid body. The bulk material is case-carburised steel 16MnCr5 and the coating material SiO_2 with a significant different thermal conductivity (λ(16MnCr5) = 44 W/m/K, $\lambda(SiO_2)$ = 1.4 W/m/K). The SiO_2 coating shows an insulating effect and leads to significantly higher temperatures compared to an uncoated TEHL contact. This also affects the coefficient of friction as shown by, for example, Habchi [6] for TEHL point contacts.

5.2. Coefficient of Friction. For comparing measurement and simulation results, the coefficient of friction is a suitable quantity with integral character. Coefficient of friction measurements in fluid film lubrication regime can be performed at the FZG twin disk test rig with, for example, cylindrical discs (r_1 = r_2 = 40 mm, l_{eff} = 5 mm) made from case-carburised steel (16MnCr5) with a polished surface structure. Further information about the test rig is documented in [28], for example. Table 1 shows the main properties of the considered lubricant MIN100 and the physical properties of the bulk material. Model constants of Section 2.5 of the

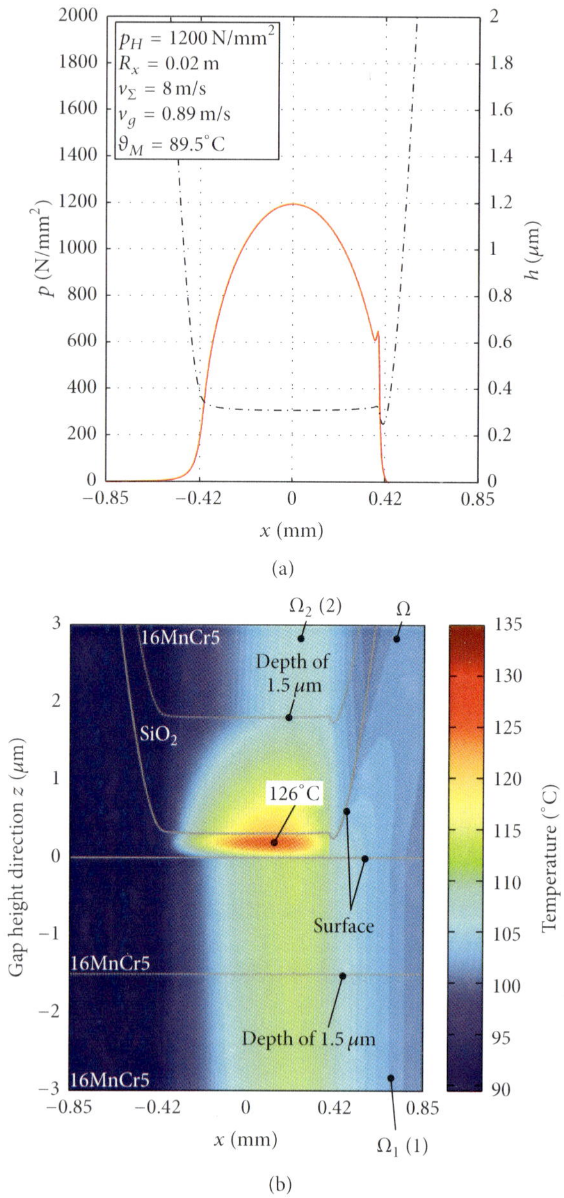

FIGURE 6: Pressure and film thickness (a) and temperature distribution (b) of a TEHL contact with coated (SiO_2) upper solid acc. to Lohner et al. [27].

TABLE 1: Lubricant and bulk material properties ($i = 1, 2$).

	Bulk material properties
	16MnCr5
E_i in N/mm^2	206000
ν_i	0.30
ρ_i in kg/m^3	7760
λ_i in W/m/K	44
$c_{p,i}$ in J/kg/K	431
	Lubricant properties
	MIN100
υ (40°C) in mm^2/s	95
υ (100°C) in mm^2/s	10.5
ρ (15°C) in kg/m^3	885
λ_0 in W/m/K	0.137
c_p (15°C) in J/kg/K	1921
τ_c in N/mm^2	6

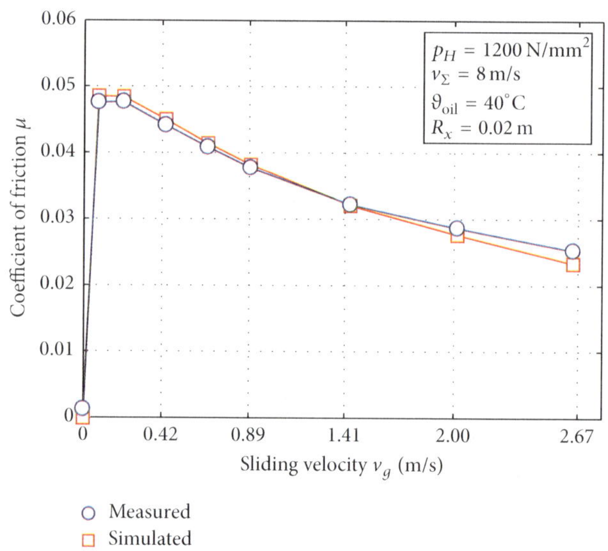

FIGURE 7: Simulated and measured coefficients of friction μ over the sliding velocity v_g.

considered lubricant have been obtained from regression of measurement data. The Eyring shear stress of $\tau_c = 6\,\text{N/mm}^2$ has been adopted from [29] as there a mineral oil of the same viscosity has been considered. Note that based on the Eyring shear model the increase of the shear stress with shear rate is not limited. To achieve a more realistic representation of the rheological behaviour of the lubricant, a shear stress model based on a limiting shear stress is currently being developed by the authors. However, this is not the focus of this study as it can be easy implemented once it is ready.

A comparison of measured and simulated coefficients of the friction of lubricant MIN100 with very good correlation is shown in Figure 7. The considered sliding velocities $v_g = \{0, 0.08, 0.21, 0.42, 0.64, 0.89, 1.41, 2.00, 2.67\}$ m/s are thereby accompanied with corresponding bulk temperatures of the slower disc of $\vartheta_M = \{42, 47, 56, 69, 80, 90, 109, 123, 140\}$°C.

5.3. Nonsmooth Surfaces. Surface features have a significant influence on TEHL contact behaviour [30, 31]. As an example, the contact of a deterministically structured surface (upper body) with repetitive protruding bars with a height of 0.2 μm and a length of 50 μm every 30 μm shown in Figure 8 with a smooth surface (lower body) is investigated. The deterministically structured surface is represented by the term $\mathfrak{R}$ in (7).

The considered lubricant and material parameters are shown in Table 1 and the operation conditions are $p_H = 1200\,\text{N/mm}^2$, $v_g = 0.89$ m/s, $v_\Sigma = 8$ m/s, and $\vartheta_M = 89.5$°C. The model constants of the considered lubricant MIN100, as well as the macrogeometry of the discs, are the same as in Section 5.2. Figure 9 shows the simulated pressure, film

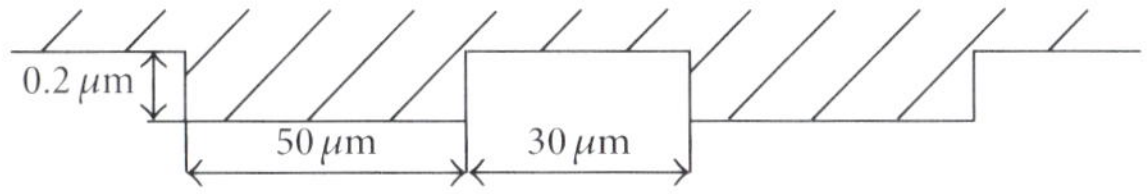

FIGURE 8: Schematic representation of the deterministically structured surface.

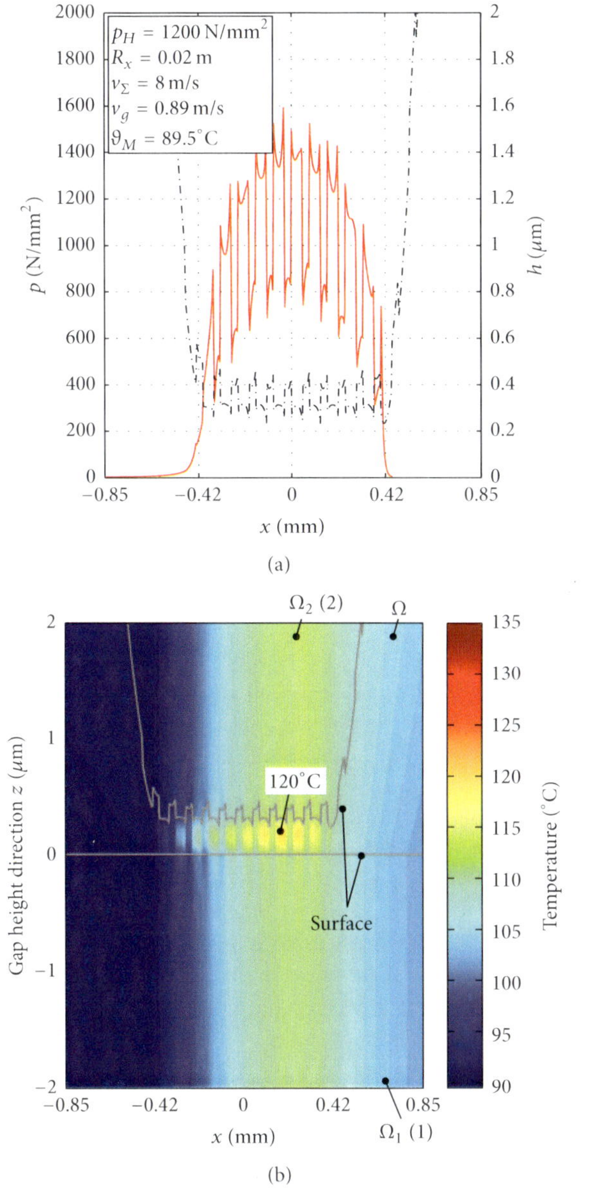

FIGURE 9: Temperature (a) and pressure and film thickness distribution (b) of a transient TEHL contact with a deterministically structured surface shown in Figure 8.

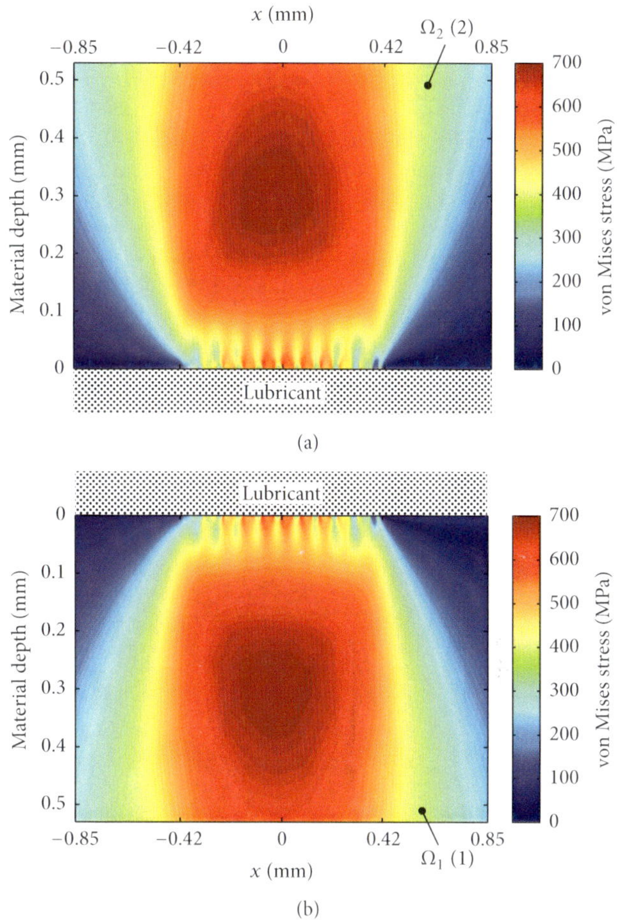

FIGURE 10: von Mises stress in the lower solid body (a) and in the upper solid body (b) correlating to Figure 9.

thickness, and temperature distribution, which are significantly influenced by the deterministically structured surface. Very local changes of the pressure, film thickness, and temperature distributions are observed thereby. The maximum temperature rise in the lubricant is simulated to be 4 K higher for the deterministically structured surface than for a TEHL contact with smooth surfaces. The transient calculation time is approximately 5 to 10 hours on a computer with a 3.5 GHz processor. Postprocessing the simulated pressure and shear stress distribution by applying it to the upper and lower solid body with bulk material properties acc. to Table 1 easily delivers the stress distributions in the solid bodies based on linear elastic isotropic material behaviour. Figure 10 shows an example of the von Mises stress distribution in the lower (a) and upper (b) solid body corresponding to the pressure and shear stress distribution in Figure 9. Stresses due to temperature are omitted. Due to the deterministically structured surface, stress maxima close to the surface occur in addition to the well-known stress maxima in the deeper material depth.

6. Conclusion

In this paper, guidelines for the implementation of a TEHL model in commercial multiphysics software are presented. The introduced approach can be implemented with moderate effort, and the resulting TEHL model is very easy to extend for various applications. The TEHL model divided in a sequence controller in MathWorks MATLAB and in two self-developed decoupled COMSOL Multiphysics FEM-models is also capable of calculating the transient TEHL contact along the path of contact of gears. Even more challenging tasks, too, such as the extension of the described model to mixed lubrication regimes, are possible. By providing implementation guidelines for a TEHL simulation model in commercial multiphysics software, the authors are confident that the research in computational tribology has been stimulated and accelerated.

Nomenclature

$A_{c1,\dots,5}$: Coefficients of lubricant heat capacity model
A, B, and C: Coefficients of lubricant Vogel temperature model
b_H: Hertzian contact half-width in m, $b_H = 4 \cdot R_x \cdot p_H / E'$
$\vec{C}$: Compliance matrix in Pa
c_p: Lubricant specific heat capacity in J/(kg K)
$c_{p,1}, c_{p,2}$: Specific heat capacity of solids 1 and 2 in J/(kg K)
d: Height of solids in the FEM-model $(\overline{T})$ in m
$d_{\lambda 1}, d_{\lambda 2}$: Pressure coefficients of lubricant thermal conductivity model in 1/Pa
r_1, r_2: Radius of solids 1 and 2 in m
$D_{\rho 0}, D_{\rho 1}$, and $D_{\rho 2}$: Coefficients of lubricant Bode density model
E_1, E_2: Young's Modulus of solids 1 and 2 in Pa
$E_{\alpha_p 1}, E_{\alpha_p 2}$: Coefficients of lubricant pressure-viscosity coefficient model in 1/Pa and K
E_{eq}: Equivalent Young's Modulus in Pa, $E_{\text{eq}} = (E_1^2 E_2 (1 + \upsilon_2)^2 + E_2^2 E_1 (1 + \upsilon_1)^2) / [E_1 (1 + \upsilon_2) + E_2 (1 + \upsilon_1)]^2 \cdot b_H / (R_x \cdot p_H)$
E': Reduced Young's Modulus in Pa, $E' = 2 \cdot E_2 \cdot E_1 / (E_1 \cdot (1 - \upsilon_2^2) + E_2 \cdot (1 - \upsilon_1^2))$
F_N: Normal force in N
F_R: Friction force in N
h: Film thickness in m
h_0: Constant parameter of film thickness in m
l_{eff}: Effective contact length in width direction in m
p: Pressure in Pa
p_H: Hertzian pressure in Pa
$p_{\eta 0}$: Coefficient Roelands equation ($p_{0,\text{Roe}} = 1.96 \cdot 10^8$ Pa)
R_x: Reduced radius in m, $R_x = r_1 \cdot r_2 / (r_1 + r_2)$
$\mathfrak{R}$: Deviation from the smooth profile in m
t: Time in s
T: Temperature in K
T_M: Bulk temperature in K
$\vec{U} = \left(\begin{smallmatrix} U_u \\ U_v \end{smallmatrix}\right)$: Displacement vector in m
v_Σ: Sum velocity in m/s
v_1, v_2: Velocity of solids 1 and 2 in m/s
v_g: Sliding velocity in m/s
v_x: Lubricant velocity distribution in m/s
x: Space coordinate in gap length direction in m
x_{in}, x_{ex}: Left and right boundary of Ω_P in m
z: Space coordinate in gap height direction in m
$z(T)$: Roelands pressure-viscosity parameter.

Greek Symbols

α_p: Pressure-viscosity coefficient in 1/Pa
α_s: Coefficient of lubricant Bode density model in 1/K
$\dot{\gamma}_x$: Shear rate in 1/s
δ: Deformation of the equivalent body in m
$\vec{\varepsilon}$: Strain tensor
η: Lubricant viscosity in Pas
ϑ_M: Bulk temperature in °C
ϑ_{oil}: Oil inlet temperature in °C
λ: Lubricant thermal conductivity in W/(m K)
λ_1, λ_2: Thermal conductivity of solids 1 and 2 in W/(m K)
μ: Coefficient of friction
ν_1, ν_2: Poisson's ratios of solids 1 and 2
ν_{eq}: Equivalent Poisson's ratio $\nu_{\text{eq}} = (E_1 \nu_2 (1 + \nu_2) + E_2 \nu_1 (1 + \nu_1)) / (E_1 (1 + \nu_2) + E_2 (1 + \nu_1))$
ρ: Lubricant density in kg/m^3
ρ_1, ρ_2: Density of solids 1 and 2 in kg/m^3
ρ_s: Coefficient of the lubricant Bode density model in kg/m^3
$\vec{\sigma}$: Stress tensor of equivalent body in Pa
τ: Shear stress in Pa, $\tau = \eta \cdot \dot{\gamma}$
τ_c: Eyring shear stress in Pa
υ: Kinematic viscosity in mm^2/s
Ω_T: Lubricant domain in the FEM-model $(\overline{T})$
$\Omega_{T,1}, \Omega_{T,2}$: Solid domains 1 and 2 in the FEM-model $(\overline{T})$
Ω_P: Domain of Reynolds equation in the FEM-model (P, H)
Ω_δ: Equivalent body domain in the FEM-model (P, H).

Disclosure

Upon request, the authors are pleased to offer to provide all the input parameters for recalculation of the result examples in Section 5.

Conflict of Interests

The authors declare that there is no conflict of interests regarding the publication of this paper.

References

[1] W. Habchi, *A full-system finite element approach to elastohydrodynamic lubrication problems: application to ultra-low-viscosity fluids [Dissertation]*, Institut National des Sciences Appliquées de Lyon, Lyon, France, 2008.

[2] M. Hartinger, M.-L. Dumont, S. Ioannides, D. Gosman, and H. Spikes, "CFD modeling of a thermal and shear-thinning elastohydrodynamic line contact," *Journal of Tribology*, vol. 130, no. 4, Article ID 041503, 2008.

[3] A. A. Lubrecht, *The numerical solution of the elastohydrodynamically lubricated line- and point contact problem using multigrid techniques [Ph.D. dissertation]*, University of Twente, Enschede, The Netherlands, 1987.

[4] C. H. Venner, *Multilevel solution of the EHL line and point contact problems [Dissertation]*, University of Twente, Enschede, The Netherlands, 1991.

[5] J. Raisin, N. Fillot, D. Dureisseix, P. Vergne, and V. Lacour, "Characteristic times in transient thermal elastohydrodynamic line contacts," *Tribology International*, vol. 82, pp. 472–483, 2015.

[6] W. Habchi, "A numerical model for the solution of thermal elastohydrodynamic lubrication in coated circular contacts," *Tribology International*, vol. 73, pp. 57–68, 2014.

[7] Comsol: COMSOL Multiphysics Reference Manual version 5.1, 2015.

[8] X. Tan, C. E. Goodyer, P. K. Jimack, R. I. Taylor, and M. A. Walkley, "Computational approaches for modelling elastohydrodynamic lubrication using multiphysics software," *Proceedings of the Institution of Mechanical Engineers, Part J: Journal of Engineering Tribology*, vol. 226, no. 6, pp. 463–480, 2012.

[9] H. Hertz, "Contact theory of rigid and elastic bodies," *Journal Für die Reine und Angewandte. Mathematik*, vol. 92, pp. 156–171, 1881.

[10] D. Bartel, *Simulation von Tribosystemen: Grundlagen und Anwendungen*, Habilitation Treatise, Vieweg+Teubner, Wiesbaden, Germany, 1st edition, 2009.

[11] P. Yang and S. Wen, "A generalized Reynolds equation for non-Newtonian thermal elastohydrodynamic lubrication," *Journal of Tribology*, vol. 112, no. 4, pp. 631–636, 1990.

[12] S. R. Wu, "A penalty formulation and numerical approximation of the Reynolds-Hertz problem of elastohydrodynamic lubrication," *International Journal of Engineering Science*, vol. 24, no. 6, pp. 1001–1013, 1986.

[13] O. Reynolds, "On the theory of the lubrication and its application to Mr. Beauchamps Tower's experiments including an experimental determination of the viscosity of olive oil," *Philosophical Transactions of the Royal Society of London*, vol. 177, pp. 157–234, 1886.

[14] J. Wang and P. Yang, "A numerical analysis for TEHL of eccentric-tappet pair subjected to transient load," *Journal of Tribology*, vol. 125, no. 4, pp. 770–779, 2003.

[15] P. Hepermann and R. Beilicke, *Örtliche Fresstragfähigkeit—Bestimmung der Örtlichen Fresstragfähigkeit: Einfluss von Schräg- und Hochverzahnungen*, Vorhaben-Nr. 598 I, FVA-Heft 1024, Forschungsvereinigung Antriebstechnik, Frankfurt, Germany, 2012.

[16] H. Vogel, "Principle of temperature dependency of viscosity of fluids," *Zeitschrift für Physik*, vol. 22, pp. 645–647, 1921.

[17] C. J. A. Roelands, *Correlation aspects of the viscosity-temperature relationship of lubricating oil [Ph.D. thesis]*, Technische Hogeschool Delft, 1966.

[18] H. Eyring, "Viscosity, plasticity, and diffusion as examples of absolute reaction rates," *The Journal of Chemical Physics*, vol. 4, no. 4, pp. 283–291, 1936.

[19] B. Bode, "Modell zur Beschreibung des Fließverhaltens von Flüssigkeiten unter hohem Druck," *Tribologie und Schmierungstechnik*, vol. 36, pp. 182–189, 1989.

[20] R. Larsson and O. Andersson, "Lubricant thermal conductivity and heat capacity under high pressure," *Journal of Engineering Tribology*, vol. 214, no. 4, pp. 337–342, 2000.

[21] MathWorks, *Matlab and Statistic Toolbox Release 2014*, MathWorks, Natick, Mass, USA, 2014.

[22] Y. Wang, H. Li, J. Tong, and P. Yang, "Transient thermoelastohydrodynamic lubrication analysis of an involute spur gear," *Tribology International*, vol. 37, no. 10, pp. 773–782, 2004.

[23] P. Yang and S. Wen, "Behavior of non-newtonian thermal EHL film in line contacts at dynamic loads," *Journal of Tribology*, vol. 114, no. 1, pp. 81–85, 1992.

[24] W. Habchi, D. Eyheramendy, P. Vergne, and G. Morales-Espejel, "Stabilized fully-coupled finite elements for elastohydrodynamic lubrication problems," *Advances in Engineering Software*, vol. 46, no. 1, pp. 4–18, 2012.

[25] MUMPS Documentation: MUMPS—aMUltifrontal Massively Parallel sparse direct Solver, http://mumps-solver.org/.

[26] P. Deuflhard, "A modified Newton method for the solution of ill-conditioned systems of nonlinear equations with application to multiple shooting," *Numerische Mathematik*, vol. 22, pp. 289–315, 1974.

[27] T. Lohner, J. Mayer, and K. Stahl, "EHL contact temperature—comparison of theoretical and experimental determination," in *Proceedings of the STLE 70th Annual Meeting & Exhibition*, Dallas, Tex, USA, May 2015.

[28] T. Lohner, R. Merz, J. Mayer, K. Michaelis, M. Kopnarski, and K. Stahl, "On the effect of plastic deformation (PD) additives in lubricants," *Tribologie und Schmierungstechnik*, vol. 62, no. 2, pp. 13–24, 2015.

[29] L. Bobach, D. Bartel, R. Beilicke et al., "Reduction in EHL friction by a DLC coating," *Tribology Letters*, vol. 60, article 17, 2015.

[30] T. Almqvist and R. Larsson, "Thermal transient rough EHL line contact simulations by aid of computational fluid dynamics," *Tribology International*, vol. 41, no. 8, pp. 683–693, 2008.

[31] Q. J. Wang, D. Zhu, H. S. Cheng, T. Yu, X. Jiang, and S. Liu, "Mixed lubrication analyses by a macro-micro approach and a full-scale mixed EHL model," *Journal of Tribology*, vol. 126, no. 1, pp. 81–91, 2004.

6

Effect of Size of Multiwalled Carbon Nanotubes Dispersed in Gear Oils for Improvement of Tribological Properties

Kodanda Rama Rao Chebattina,[1] V. Srinivas,[1] and N. Mohan Rao[2]

[1]*Department of Mechanical Engineering, GITAM University, Visakhapatnam, India*
[2]*Department of Mechanical Engineering, University College of Engineering, JNTU, Kakinada, India*

Correspondence should be addressed to V. Srinivas; vsvas1973@yahoo.com

Academic Editor: Patrick De Baets

The aim of the paper is to investigate the effect of size of multiwalled carbon nanotubes (MWCNTs) as additives for dispersion in gear oil to improve the tribological properties. Since long pristine MWCNTs tend to form clusters compromising dispersion stability, they are mildly processed in a ball mill to shorten the length and stabilized with a surfactant before dispersing in lubricant. Investigations are made to assess the effect of ball milling on the size and structure of MWCNTs using electron microscopy and Raman spectroscopy. The long and shortened MWCNTs are dispersed in EP 140 gear oil in 0.5% weight. The stability of the dispersed multiwalled carbon nanotubes is evaluated using light scattering techniques. The antiwear, antifriction, and extreme pressure properties of test oils are evaluated on a four-ball wear tester. It is found that ball milling of MWCNTs has a strong effect on the stability and tribological properties of the lubricant. From Raman spectroscopy, it is found that ball milling time of up to 10 hours did not produce any defects on the surface of MWCNTs. The stability of the lubricant and the antiwear, antifriction, and extreme pressure properties have improved significantly with dispersion shortened MWCNTs. Ball milling for longer periods produces defects on the surface of MWCNTs reducing their advantage as oil additives.

1. Introduction

Gear oils used in industries and automotive engines are often subjected to heavy loads, due to which they experience high temperatures and pressures causing higher friction and surface damage leading to failure of the system. To prevent failure, conventional engine and gear oils are dispersed with extreme pressure (EP) and antiwear (AW) additives that react chemically with the metal surfaces, forming easily sheared layers and thereby preventing severe wear and seizure. Allotropes of carbon such as graphite, fullerenes, carbon nanotubes, and graphene have attracted the interest of the researchers due to their special properties. The hybridization of the atomic orbital of carbons in carbon nanotubes and fullerenes is of type *sp*2, similar to that of graphite, making a perfect hexagonal array of atoms. The *sp*2 C–C bond of CNTs is considered one of the strongest in solid materials; thereby CNTs are expected to yield exceptionally good mechanical properties. Lubricants dispersed with allotropes of carbon are being extensively studied for their lower friction coefficients, thereby improving antiwear properties. MWCNTs possess large surface area compared to many inorganic nanomaterials and can be very easily surface-modified. Several novel studies are made on the effect of dispersion of multiwalled carbon nanotubes on the wear and friction characteristics of lubricants. The length of the nanotube synthesized by existing methods is known to be thousands of times larger than their width and thus limits their functionality for many applications. Owing to their large length to diameter ratios, MWCNTs despite use of surfactant tend to form agglomerates faster, thereby leading to them settling in the liquid medium. This aspect is a main challenge in obtaining stable dispersion in liquid medium. The most common and frugal method to reduce the size of MWCNTs is ball milling which shortens the length of MWCNTs and obtaining open ends. However, ball milling for extended period produces defects and damages the graphite structure rendering it useless for dispersion in lubricants. Studies are made on the effect of ball milling on the structure and defect generation. Pierard et. al. [1] studied the effect of ball milling on the structure of single-walled carbon nanotubes. Raman spectroscopy was employed to study the defects produced in various hours of ball milling time. It is found that there exists an optimum time to keep the tubular structure intact without defects. In case of single-walled carbon nanotubes, ball milling times of over 50 hours completely destroy the structure producing amorphous carbon. Dresselhaus et. al.[2], Cancado et. al. [3], and Paton et. al. [4] suggested methods to detect defects and evaluate purity of carbon nanotubes and graphene. It was proposed that the intensity of G-band and the D/G ratio can be highly useful to determine both the purity and the defect density of carbon nanotubes and graphene. Chen et. al. [5] first studied the effect of dispersed of ball-milled and stearic acid modified multiwalled carbon nanotubes (MWCNTs) on the stability and thereby improvement in the lubricating properties of liquid paraffin base oil. The friction and wear tests were conducted on a pin-on-plate wear-testing machine. It is found that ball-milled MWCNTs could form stable suspensions and improve the antiwear and antifriction properties. Engine oils dispersed with nanomaterials are investigated for tribological property enhancement by several researchers. A detailed review of the prior art is listed in Table 1.

As given in Table 1, most of the studies were done with dispersion of larger amounts of MWCNTS in base lubricants or paraffinic oils without additives. Only few studies were made on commercial grade formulated lubricating oils. Further in the studies, due to dispersion of MWCNTs in higher concentrations, the viscosity of the lubricant is found to be significantly enhanced. This enhancement in viscosity is the major reason for the improved tribological properties. Furthermore higher concentrations of any nanomaterials may lead to faster agglomeration rates and the stability of the suspension will get compromised. Since MWCNTs have long length ranging from 1 to 25 microns and diameter in nanometers, there is a strong ability of entanglement of individual nanotubes leading to formation of clusters. These clusters tend to become hard and make the MWCNTs lose their special properties. Moreover, the stability of nanofluid has not been assessed quantitatively in the studies made so far. Further the defects produced in MWCNTs due to ball milling and its consequential effects on tribological properties also need to be assessed. The present study is aimed at investigating improvements in antiwear, antifriction, and extreme pressure properties of formulated EP 140 grade gear engine oil dispersed with surface-modified and ball-milled MWCNTs. Ball milling is performed in inert gas medium and at a low intensity to prevent damage to the structure of MWCNTs. A simpler surface modification technique is used to stabilize the MWCNTs in oil medium and the stability of the suspensions is investigated over a period of 2 months. The effect of decrease in length of MWCNTs due to ball milling on the tribological properties has been studied. The repeatability of results over a period of 60 days is investigated to observe consistent performance and the average values are reported. The effect of additives in the oil along with surface-modified MWCNTs has resulted in use of lesser amount of MWCNTs in the lubricant dispersion compared to studies reported in the literature which is one of the novel features of the study. The paper also compares chemical and physical routes for dispersion of MWCNTs in lubricant by comparing performance of surface-modified pristine MWCNTs and surface-modified ball-milled MWCNTs, respectively.

2. Experimental

2.1. Materials. In the present study, multiwalled carbon nanotubes produced by CVD method have been procured from M/s Cheap Tubes Inc., USA. The size of MWCNTs is 20-40 nm in diameter, 25 microns in length, and 95% of purity. All other chemicals purchased are of GR grade. The surfactant is AR grade procured from M/s Sigma Aldrich India Pvt limited. GL4 (EP 140 grade) gear oil is selected as base lubricant.

2.2. Ball Milling of Multiwalled Carbon Nanotubes. As the length of the carbon nanotubes is thousands of times larger than their width, ball milling of MWCNTs is a common procedure to generate short and open-ended nanotubes. Ball milling apparatus consists of tungsten carbide lined bowls containing tungsten carbide lined ball. Arrangement is provided to perform ball milling under argon atmosphere to prevent oxidation of material during ball milling. The speed of rotation of bowls is set at 400 RPM and the ratio of balls to MWCNTs is taken as 10:1 to ensure minimal damage to the tubular structure of the MWCNTs. Higher speeds and ratios will increase the impact and thereby attrition of MWCNTs. Ball milling was performed for 4 and 20 hours at 400 RPM to avoid damage to the structure. After ball milling, the MWCNTs are heated in air at 600°C to remove amorphous carbon generated during ball milling process. These ball-milled MWCNTs are characterized using HRSEM and transmission electron microscopy to determine the average length of the MWCNTs.

2.3. Electron Microscopy. Figure 1(a) shows HRSEM image of pristine long length entangled MWCNTs. Figure 1(b) shows MWCNTs ball-milled for 5 hours with a small change in length of MWCNTs as compared to Figure 1(a). Figures 2(a) and 2(b) show images of 10- and 20-hour ball-milled MWCNTs.

It can be seen that the average length of 10-hour ball-milled MWCNTs is under 4 microns. From Figure 2(b) (TEM image) it can be observed that the length of 20-hour ball-milled CNTs has come down to around 150 nm size.

2.4. Raman Spectroscopy. Raman spectroscopy is employed to assess the formation of defects during ball millings and shown in Figure 3 for pristine MWCNTS, 5-, 10-, and 20-hour ball-milled MWCNTs. The sp2 structure of MWCNTs causes first order peaks D and G bands that are approximately located at 1350 cm^{-1} and 1580 cm^{-1}, respectively. Defect free MWCNTs due to intact hexagonal graphite structure make the G-band sharper. Defects in MWCNTs make the G-band peaks wider and shorter. On the other hand, D band peak represents lattice defects and finite crystal size. During defect

Table 1: Summary of research carried out by authors.

Authors	Base fluid, type of nano materials and concentration	Experimental apparatus	Major findings
Bhaumik et al. [6]	Mineral oil dispersed with graphite and MWCNTs in concentration 0.1 to 0.6 Wt%.	Pin on disk tribometer and Four-ball tester	It is found that MWCNTs outperformed graphite with wear reduced by 70-75% and load bearing capacity increasing by 20%.
Chen et. al. [5]	Paraffin oil dispersed with stearic acid modified MWCNTs in 0.45% Wt %	Ring on plate	Friction coefficient decreased by 10%, wear reduced by 30-40%. It was proposed that friction-reduction ability of nano-lubricant depends on both nano materials and obtaining stable dispersion of nano-particle in lubricant.
Cornelio et. al. [7]	Oil and water with SWCNTs & MWCNTs 0.01 to 0.05 Wt %	Twin-disk machine for measurements in rolling-sliding contact	The friction coefficients and wear losses measured were lower for either oil or water dispersed with nanotubes with friction coefficients reported as low as 0.063. It was proposed that decrease of friction and high wear resistance are due to the formation of an amorphous carbon film transferred from the CNT's on the surface.
Ghaednia et. al. [8]	Mineral base oil dispersed with CuO (9 nm) in 0.5 to 2 wt%	disk-on-disk friction and wear test	Friction coefficient decreased by 14 and 23% for the CuO nanoparticle concentrations of 1.0 and 2.0% wt, respectively. It was suggested that the reduction in the real area of contact due to dispersion of nanomaterials in lubricant is possible mechanism for reduction of friction
Hernandez Battez et. al. [9]	poly-alpha-olefin(PAO6) dispersed with CuO, ZnO and ZrO2 in 0.5 to 2% wt	Four-ball tester	The results indicate that the extreme pressure behavior of lubricant with nano particles is strongly dependent on the size and hardness of the nanoparticles. Particles with hardness less than surfaces in contact exhibited good EP behavior.
Hu et. al. [10]	Water dispersed with sodium dodecyl sulfate modified oxidized MWCNTs in 0.1% Wt (for application in metal working fluids)	Four-ball tester	It is found that oxidation of MWCNTs produced defects on the surface leading to formation of better suspension. There is a good improvement in tribological properties with surfactant modified MWCNTs compared to pristine MWCNTs indicating strong influence of stability of suspension.
Joly-Pottuz et. al.[11]	poly-alpha-olefin dispersed with carbon nano onions and graphite powders in 0.1% wt	pin-on-flat tribometer	Carbon nano-onions show better tribological properties than graphite powder. It is also found that tribofilm formed by carbon onions converts wear particles into ultrafine lubricious iron oxides, thus preventing further abrasive wear process.
Khalil et. al. [12]	Mobil gear 627 and paraffinic mineral oils in 0.1 to 2 wt%	Four-ball tribotester	A 50% reduction in friction and an increase in weld load by up to 100% are observed with dispersion of MWCNTs in lubricant. The Tribological performance is attributed to deposition of MWCNTs nanoparticles on the worn surface resulting in decreasing the shearing stress, thus improving the tribological properties.
Lee et. al. [13]	Raw mineral oil (Sun Oil, Japan) dispersed with fullerenes in 0.01 to 0.05% wt.	disk-on-disk tribotester	Fullerenes have reduced friction by 30% by reducing the metal surface contacts. Further, it is found that volume fraction is a key factor to control the friction and wear
Lee et. al. [14]	Commercial gear oil dispersed with nano graphite (55 nm) in 0.1 to 0.5% wt. Alkyl aryl sulfonate is used as dispersant	disk-on-disk tribotester	The results indicate dispersion of nano graphite in lubricant boosted the lubrication characteristics. The possible reason for improvement of properties is suggested to be due to nanoparticles acting as ball bearing spacers between the friction surfaces reducing the contact between the plates.

Table 1: Continued.

Authors	Base fluid, type of nano materials and concentration	Experimental apparatus	Major findings
Pena-Paras et. al. [15]	4 types of Metal working fluids dispersed with TiO_2, Al_2O_3, CuO and MWCNTs in 0.01 to 0.1% wt	A four-ball T-02U tribotester	All nano materials improved the tribological properties with MWCNTs giving best results. Tribo-sintering of nano materials on the rubbing surfaces during machining is the reason proposed for the performance improvement.
Puzyr et. al. [16]	Commercial oil dispersed with surface modified nano diamonds in 0.01% wt	block-on-ring test setup	Nano diamond additives in oils improved the anti- wear properties and decreased the oil temperature compared to base oils. The anti-wear mechanism of ND additives was attributed to the formation of a hard and porous layer between the contact surfaces.
Sarma et. al. [17]	SM grade engine oil dispersed with Cu and TiO2 nano particles in 0.025 to 0.1% wt	Pin on disk tester and engine test rig	It was found that copper nano particle dispersed lubricant gave best results on engine tests. 3 to 5% increase in thermal efficiency of the engine is observed.
Srinivas et. al. [18]	EP 140 Transmission oil dispersed with WS_2 and MoS_2 nano materials in 0.5% wt.	Four-ball tester	Lubricant dispersed with WS_2 nanoparticles gave higher weld load and load wear index (LWI) than that of lubricant dispersed with MoS_2 nanoparticles. The reason for better performance of WS_2 nano particles is attributed to their lower hardness resulting in better deposition on the rubbing surfaces under load.
Srinivas et.al. [19]	CI 4 Engine oil dispersed with long MWCNTs	Four-ball tester	The surface modification of multi walled carbon tubes plays prominent role in the improving stability and thereby anti-wear and anti-friction properties of engine oils.
Viesca et. al. [20]	Poly alpha olefin (PAO6) dispersed with carbon coated nano particles in 0.5 to 2% wt	Block on-ring tribometer and four-ball tester	carbon-coated copper nanoparticles decreases wear and increases the load-carrying capacity of polyalphaolefin

Table 2: Raman Spectra characteristics of MWCNTs.

MWCNTs	Intensity of D band, ID	Intensity of G band, IG	Intensity of G' band, IG'	Ratio of ID and IG
Pristine	1536	2100	1594	0.731
5 hour ball milled	1261	1708	1082	0.738
10 hour ball milled	1410	1855	974	0.76
20 hour ball milled	1413	1266	774	1.11

formation, due to breaking of the 2D translational symmetry the D band peak will increase and become wider. Another peak G' band which can be seen at 2700 cm^{-1} Raman Shift represents amorphous defects in MWCNTs.

Table 2, provides the intensities of D, G, and G' bands. In case of pristine MWCNTs, 5- and 10-hour ball-milled MWCNTs, there is no significant difference in the intensities of D, G, and G' bands with ratio of intensities of D and G bands remaining marginally the same. In case of 20-hour ball-milled MWCNTs, a significant decrease in intensity of G and G' bands is observed with increase of intensity of D band indicating mild destruction of graphite structure and formation of amorphous defects. In all cases, the peaks of all bands are sharp indicating either no defects or mild defect (in case of 20-hour ball-milled MWCNTs).

2.5. Surface Modification of MWCNTs. Pristine MWCNTs tend to agglomerate and form large particles clusters in liquid medium. Moreover, after the ball milling process, the MWCNTs tend to be compressed by the balls forming larger aggregates of entangled MWCNTs. To disentangle them and make them stable in the lubricant medium, it is required to modify the surface of MWCNTs with a surfactant to create stearic repulsions between individual nanotubes. To stabilize the nanoparticles in the liquid medium, a surfactant SPAN 80 (Sorbitan monooleate) is used to modify surface of MWCNTs during the preparation of oil. SPAN 80 is a nonionic surfactant with a hydrophilic-lipophilic balance of 4.6 which is ideally suitable for oils. It adsorbs on the surface of MWCNTs reducing their surface energy, thereby preventing agglomeration

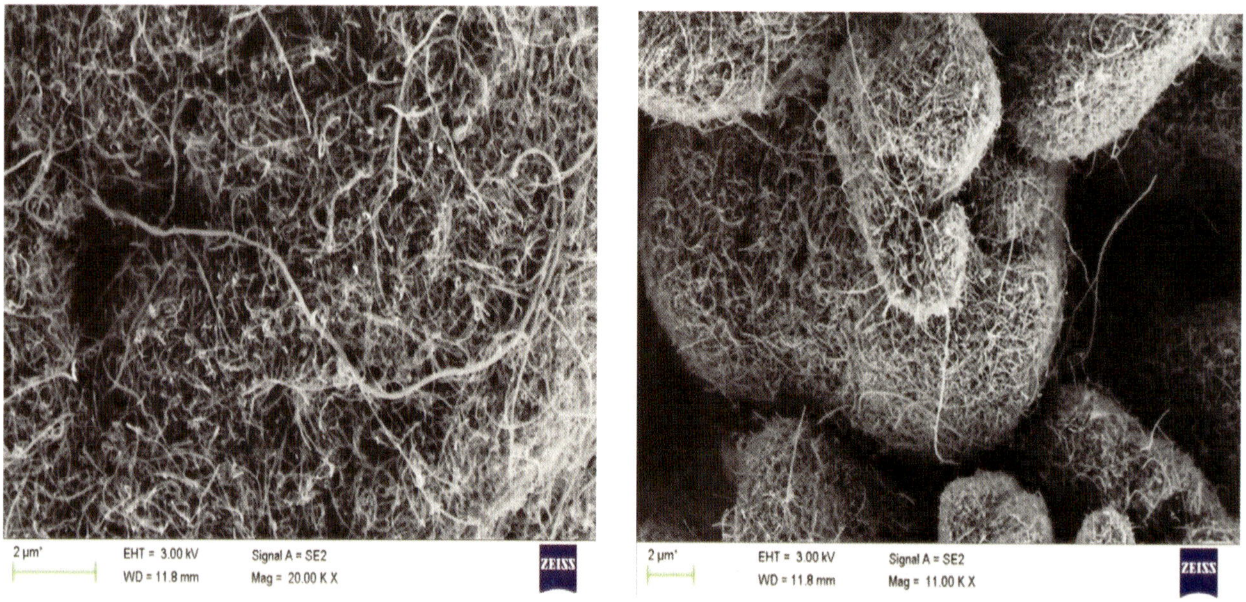

(a) (b)

Figure 1: HRSEM images of pristine and 5-hour ball-milled MWCNTS. (a) Pristine MWCNTS, (b) 5-hour ball-milled MWCNTS.

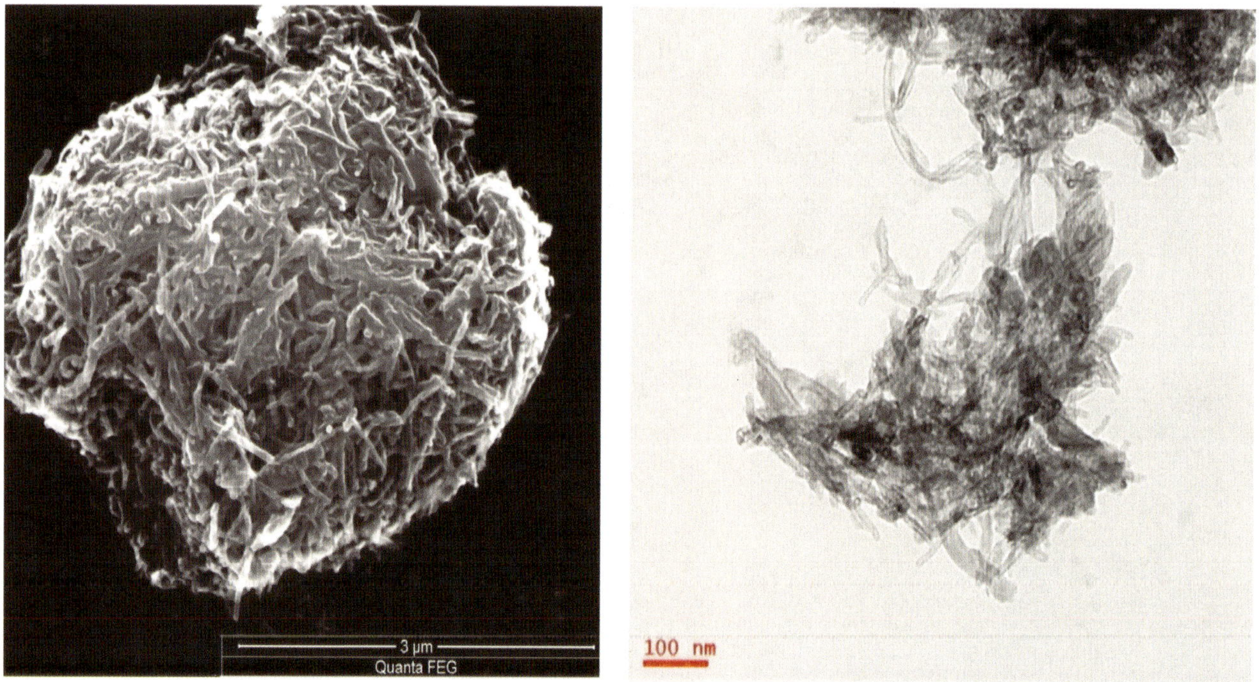

(a) (b)

Figure 2: HRSEM images of 10-hour and 20-hour ball-milled MWCNTs. (a) 10-Hour ball-milled MWCNTs, (b) 20-hour ball-milled MWCNTs.

and settling of nanoparticles. To prepare surface-modified MWCNTs, SPAN 80 and multiwalled carbon nanotubes (both pristine and ball-milled) are taken in the ratio of 2:1 and ultrasonicated in a solvent for 30 minutes, which creates a mechanochemical reaction. This reaction coats the surfactant on to the surface of the MWCNTs.

2.6. FTIR Spectroscopy. The surface-modified MWCNTs are characterized for functional groups on the surface using

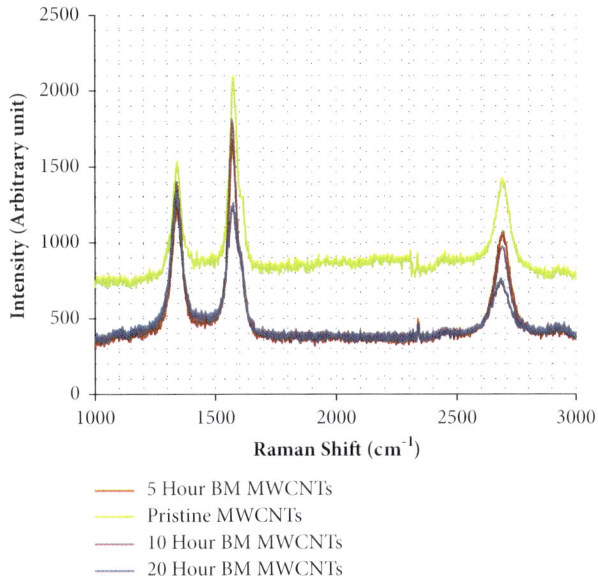

FIGURE 3: Raman spectroscopy of pristine and ball-milled MWCNTs.

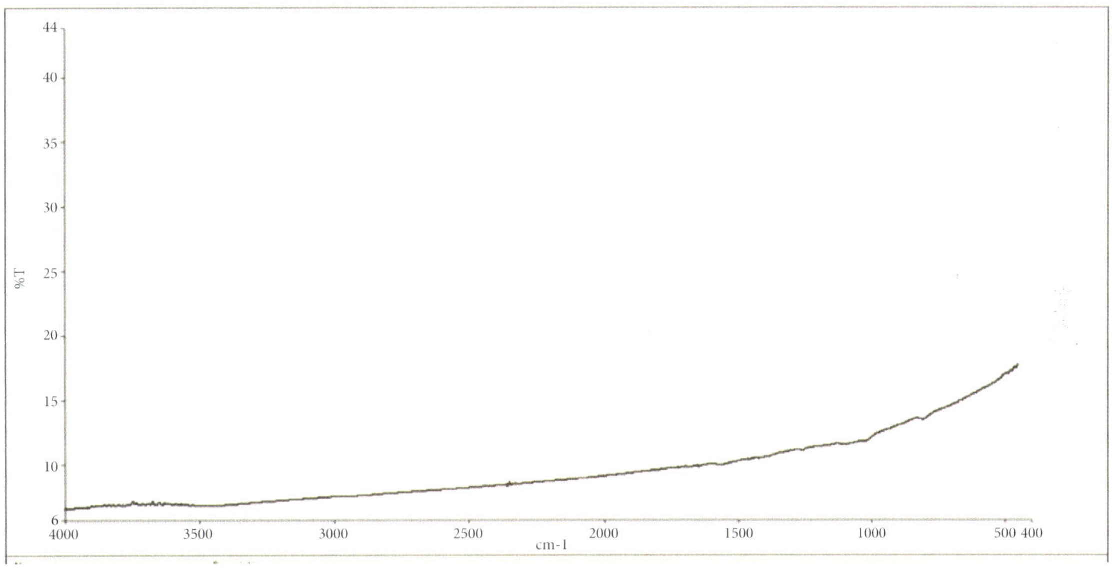

FIGURE 4: FTIR spectroscopy of pristine MWCNTs.

Fourier transforms infrared spectroscopy as shown in Figures 4 and 5. Figure 4 shows pristine MWCNTs with no characteristic peak detected. Figure 4 shows the FTIR image of surface-modified MWCNTs with characteristics peaks between 1463 and 1486 cm^{-1} wavelength indicating lipophilic groups attached to the surface.

2.7. Preparation and Evaluation of Stability of Lubricants with MWCNTs. The surface-modified long and ball-milled MWCNTs in 0.5 weight percent are dispersed in lubricating oil by processing it in a probe ultrasonicator (**Hielscher UP400S**) for about 30 minutes. The stability of the nanofluid for a period of 60 days is monitored by light scattering techniques. Ball-milled MWCNTs when dispersed in lubricating oils could form better stable suspension compared to long MWCNTs and gave better tribological properties.

The stability of the lubricants dispersed with MWCNTs is evaluated using light scattering techniques by means of zeta sizer (Horiba SZ 100). The zeta potential of the samples, an indicator of dispersion stability of MWCNTs in the lubricating oil medium, has been analyzed over a period of 60 days. A zeta potential value of $\pm$ 40 indicates good stability. As the viscosity of the gear oil is high, the oil samples are diluted using toluene before charging it into the zeta sizer to improve the transmittance of the oil so that accurate values

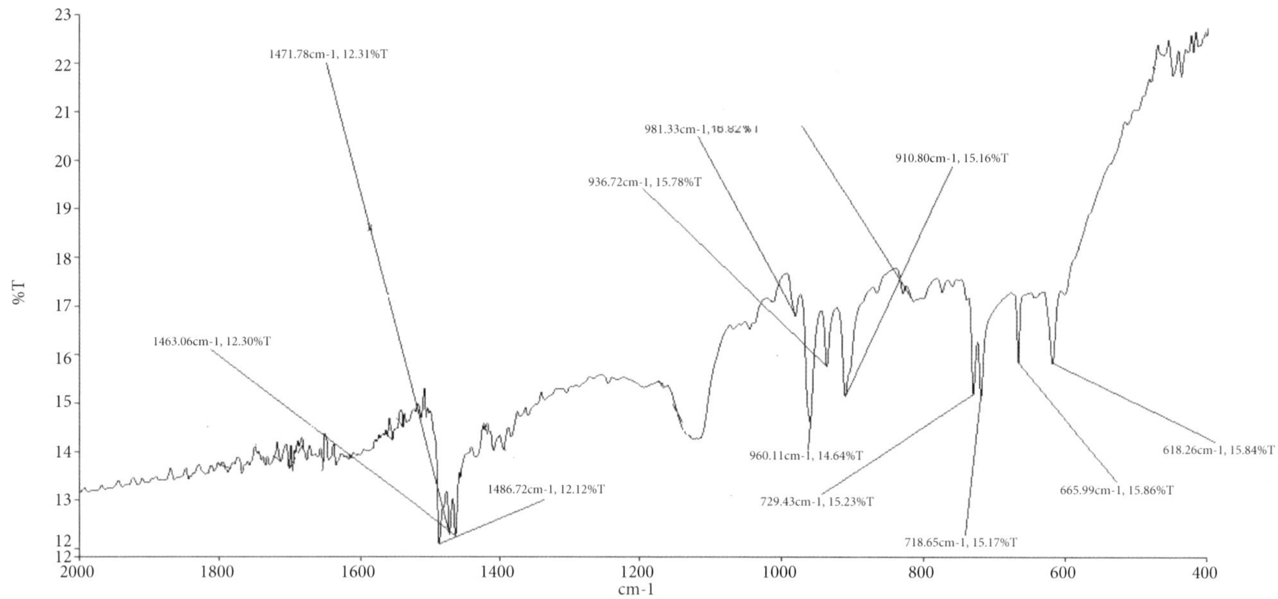

FIGURE 5: FTIR spectroscopy of surface-modified MWCNTs.

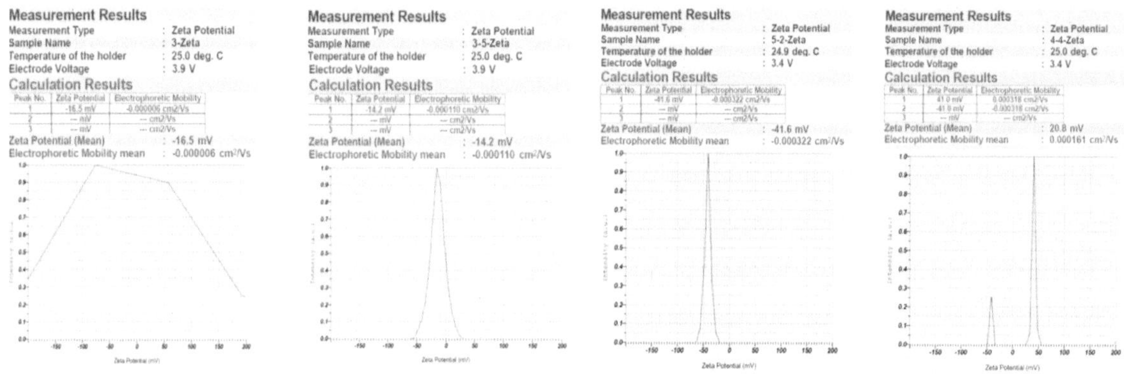

(a) (b) (c) (d)

FIGURE 6: Zeta potential variation of nanofluids during 60 days. (a) Lubricants with long MWCNTs immediately after preparation, (b) lubricant with long MWCNTs after 60 days, (c) lubricant with 5-hour ball-milled MWCNTs immediately after preparation, (d) lubricant with 5-hour ball-milled MWCNTs after 60 days.

can be obtained. Higher values of zeta potential values are found when the oil samples are dispersed with ball-milled MWCNTs. The variation of zeta potential immediately after preparation and 60 days after preparation is shown in Figures 6 and 7.

Figures 6(a) and 6(b) show the variation of zeta potential of lubricant dispersed with long MWCNTs and 5-hour ball-milled MWCNTs, respectively. As can be seen the zeta potential with long MWCNTs is much less indicating low stability. With five-hour ball-milled MWCNTs there is an improvement in the stability compared to stability of long MWCNTs. Figures 7(a) and 7(b) show the variation of zeta potential of lubricant dispersed with 10-hour and 20-hour ball-milled MWCNTs. As can be seen there is a significant improvement in the stability with 20-hour MWCNTs showing best stability. This can be attributed to lower agglomeration rates due to short length of MWCNTs.

2.8. Physicochemical Properties of Test Oils. The gear oils are manufactured by blending base stocks and additive components to meet the requirements of standards. Basic physicochemical properties should be in compliance with international standards for statutory purposes. The main physicochemical properties to be assessed for a gear oil are viscosity, viscosity index, pour point, flash point, total acid number, and copper strip corrosion resistance. All the properties for test oils are evaluated in 5 replicable experiments and the average values are reported.

As can be observed from Table 3, the effect of ball milling and surface modification has practically no effect on the

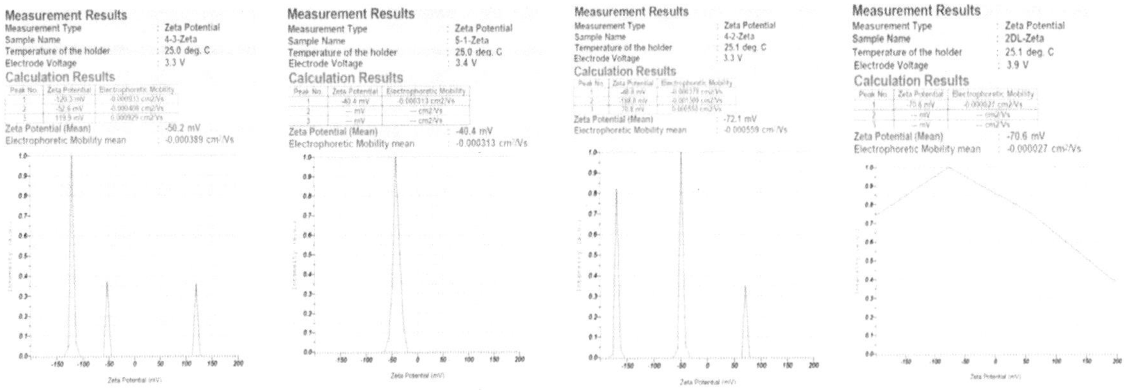

(a) (b) (c) (d)

FIGURE 7: Zeta potential variation of nanofluids. (a) Lubricants with 10-hour ball-milled MWCNTs immediately after preparation, (b) lubricant with 10-hour ball-milled MWCNTs after 60 days, (c) lubricant with 20-hour ball-milled MWCNTs immediately after preparation, (d) lubricant with 20-hour ball-milled MWCNTs after 60 days.

physicochemical properties of the test oils. The viscosity and viscosity index are unchanged with dispersion of nanomaterials. Surface modification has no effect on pour point, flash point, and total acid number of the test lubricants.

2.9. Tests for Tribological Properties. The test oils are tested for improvement in tribological properties on a four-ball tester. The weight percentage of surface-modified and ball-milled MWCNTs is maintained as 0.5 wt %. The standard code of the tests, a rotating steel ball, is pressed against three steel balls firmly held together under a load and immersed in lubricant. The test parameters of load, duration, temperature, and rotational speed are set in accordance with standard test procedure.

In wear test done as per ASTM D 4172, the average scar diameter formed on the bottom of three balls shows the ability of the lubricant to prevent wear. A larger diameter indicates poor antiwear behavior. The test is carried out for one hour at a load of 40 kgf and speed of 1200 RPM with temperature of oil maintained at 75°C.

The friction test is carried out to find the friction coefficient offered by the lubricant as per ASTM D 5183 code. Initially the balls are subjected to "wear in" for one hour at a load of 40 kgf and speed of 600 RPM with temperature of oil maintained at 75°C. After "wear in", the used lubricating oil is discarded and balls are cleaned. Fresh lubricant sample is taken in the ball cup with the same worn test balls in place. The test is again started under the above conditions with load varying from an initial load of 10 kgf and increasing by 10 kgf at the end of each successive 10 min interval until there is a sharp rise in the Frictional Torque which indicates incipient seizure. This is called seizure load which is an important factor in determining the effectiveness of the lubricant.

ASTM D 2783 specifies the extreme pressure properties of lubricant in terms of weld load which is the ultimate load at which the lubricant evaporates due to high pressure and temperature resulting in all the four balls welded to each other. The standard also specifies another parameter called "load wear index" which indicates the behavior of lubricant in resisting the aforementioned weld conditions. The test is used to determine the load carrying properties of a lubricant at high test loads usually encountered in gears. In this test on four-ball tester a series of 10 tests of 10-second duration are carried with varying load under the following conditions:

temperature of oil: room temperature,
speed of rotation: 1760 RPM,
duration: 10 s,
load applied: 32 kgf to weld load.

A total of 10 readings are considered in the test and the corrected load is calculated for all ten readings. The load wear index is a single parameter that shows the overall EP behavior in a range between well below seizure and welding is calculated from the corrected load.

$$\text{Corrected load} = \frac{\mathrm{LD_H}}{\mathrm{X}} \tag{1}$$

where L is the applied load, kgf, X is the average scar diameter on the worn balls in mm and Hertz scar diameter, and $D_H = 8.73 \times 10^{-2}(\mathrm{L})^{1/3}$ in mm. The load wear index is calculated from the expression LWI = (A/10) (kgf), where A is the sum of the corrected loads determined for the ten applied loads immediately preceding the weld load.

3. Results and Analysis

All the tests are conducted in ten repeatable trails over a period of 60 days to ascertain the influence of nanoparticles in terms of repeatable and reproducible results, and the average values are reported. The results of wear test conducted as per ASTM D4172 are as given below for different test oils. From Table 4, it can be observed that the average wear scar of lubricant with MWCNTs is much less than that of base lubricant. Further, with dispersion of long MWCNTs, the range of wear scar over 60 days of testing is higher compared to other oils due to poor lubricant suspension.

Table 5 shows the results of friction test in terms of average friction coefficient and seizure load. With dispersion

Table 3: Physicochemical properties of test oils.

Test oil	Viscosity at 40°C (cSt)	Viscosity at 100°C (cSt)	Viscosity index	Flash Point °C	Pour Point °C	Total Acid Number (mg KOH/g)	Copper strip Corrosion test
Base oil (EP140 gear oil)	408.4	27.9	94	194	-3	0.25	2a
Base oil +0.5% pristine MWCNTs	409.1	27.75	93	192	-3	0.26	2a
Base oil + 0.5% 10 hour ball milled MWCNTs	410.2	28.6	95	194	-3	0.25	2a
Base oil + 0.5% 20 hour ball milled MWCNTs	411.5	28.4	95	194	-3	0.26	2a

Table 4: Results of wear test conducted as per ASTM D 4172 at 40 kgf load.

Test Oils	Average wear scar in microns
Base lubricant (EP 140 transmission oil)	352.353
Base lubricant + long MWCNTs	347.071
Base lubricant + 5 hour Ball milled MWCNTs	335.085
Base lubricant + 10 hour Ball milled MWCNTs	317.954
Base lubricant + 20 hour Ball milled MWCNTs	323.547

Table 5: Results of friction test conducted as per ASTM D 5183.

Test oil	seizure load, kgf	Average coefficient of friction
Base lubricant (EP 140 transmission oil)	220	0.0901
Base lubricant + long MWCNTs	230	0.0839
Base lubricant +5 hour Ball milled MWCNTs	240	0.0784
Base lubricant + 10 hour Ball milled MWCNTs	260	0.0746
Base lubricant + 20 hour Ball milled MWCNTs	240	0.0783

of ball-milled MWCNTs, there is a good improvement in both seizure load and the friction coefficient.

The 10 sets of results of repeatable friction characteristics are plotted in graphs in Figure 8. In case of lubricant dispersed with long MWCNTs, although the performance improved during the initial days, there is steady decrease in the performance characterized by increase in friction coefficient due to poor stability over a period of time. Shortened MWCNTs due to their better stability gave a consistence performance over a period of 60 days with lubricant dispersed 10-hour ball-milled MWCNTs giving best performance. The variation of friction torque with time is plotted in Figure 9. From the graph it can be seen that the effect of nanomaterials is more significant at higher loads. In boundary lubrication regime there is significant contact between surfaces separated by a thin lubricant film. Increase of normal load would sweep the lubricant out of the contact region, reducing the lubricant film thickness between surfaces and increasing the chance of contact between surfaces in motion. Due to short length, the ball-milled MWNTs dispersed in lubricant could effortlessly glide and roll between the two contact surfaces in motion like spacers. This increases the pressure limits of the lubricant, thereby significantly reducing friction coefficient and improving the seizure load. Lubricant dispersed with ball-milled MWCNTs has shown consistent performance on the torque-time plot compared to base lubricant.

The extreme pressure properties, namely, last nonseizure load, weld load, and load wear index of test oils under consideration, are summarized in Table 6.

With dispersion of MWCNTs, there is an improvement in last nonseizure load, weld load, and load wear index. A graph showing the variation of wear scar diameter with load is shown in Figure 10. The region between points 0 to 1 is normally designated as antiwear region in which all the lubricants exhibited similar behavior. Points 1 and 1′ indicated the last nonseizure load (LNSL) up to which the wear scar formed is uniform.

The region above last nonseizure load is called extreme pressure zone in which the efficacy of additives in the gear oil comes into play. In this region, the pressures and temperature are very high and the additives are supposed to withstand these extremities. MWCNTs due to their good mechanical properties could form a barrier between the surfaces withstanding the extremities to improve weld load.

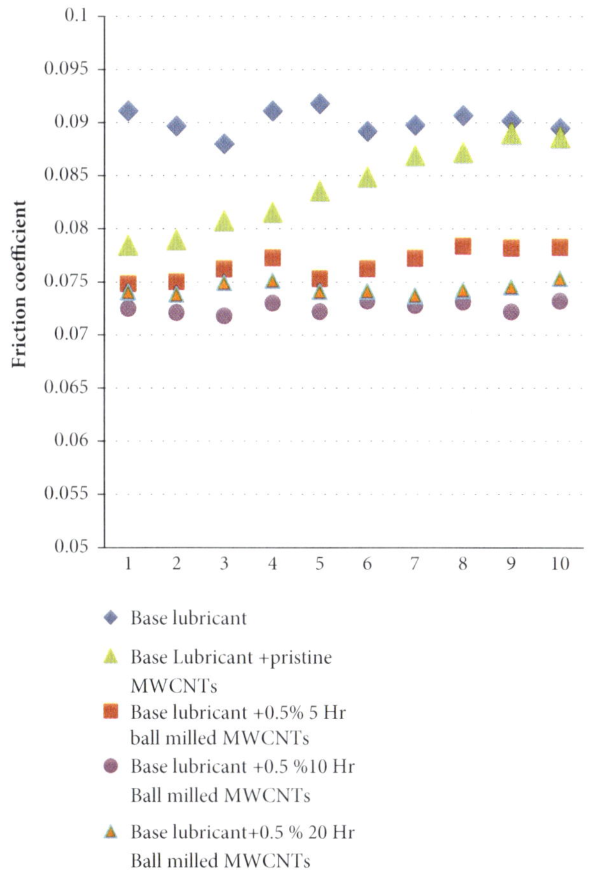

Figure 8: Variation of friction coefficient in 10 repeatable tests carried out during 2 months.

TABLE 6: Results of EP test conducted as per ASTM D 2783.

Test oil	Last non seizure load, kgf	weld load, kgf	load wear index
Base lubricant (EP 140 transmission oil)	120	250	64.82
Base lubricant + long MWCNTs	120	315	69.54
Base lubricant +5 hour Ball milled MWCNTs	140	315	71.23
Base lubricant + 10 hour Ball milled MWCNTs	140	400	75.65
Base lubricant +20 hour Ball milled MWCNTs	120	250	72.81

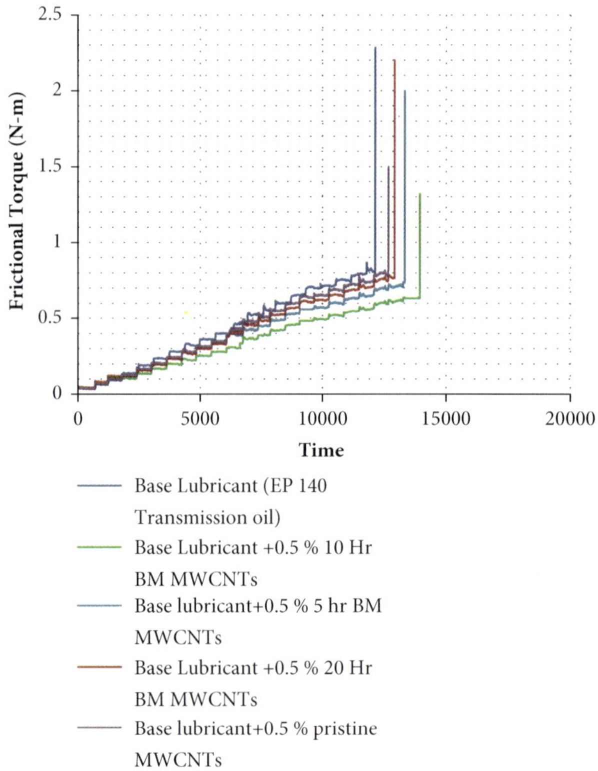

FIGURE 9: Variation of friction torque with time during friction test.

There is a reduction in the wear scar with dispersion of pristine and ball-milled MWCNTs. The effect of dispersion of shortened MWCNTs is found to be quite visible in this region. Lubricant dispersed with 10-hour ball-milled MWCNTs improved the LNSL and could reduce the wear scars even in the extreme pressure regions. Lubricant dispersed with ball-milled MWCNTs due to decrease of the friction coefficient offered lower wear scars on the test balls. Further after last nonseizure load, it can be observed that the wear scar diameter with dispersion of nanomaterials is significantly low, resulting in improvement in load wear index as well as weld load. This is due to better dispersion owing to short length. Here the ball milling timing of MWCNTs plays an important role in defining the optimum length with minimum damage to the tube structure. The optimum length for attaining good stability and tribological properties can be assessed as 1 to 5 microns, which can be attained with 10 hours of ball milling at 400 RPM speed. 20-hour ball-milled MWCNTs could improve the load wear index by performing well in the antiwear region but, due to defect formation during ball milling, exhibited lower performance in EP region. This can be attributed to the formation of defects on the surface (Figure 3 and Table 2) leading to loss of special properties of MWCNTs.

4. Conclusions

(1) The ball milling of multiwalled carbon tubes prior to dispersion in lubricant plays an important role in improvement of stability, antiwear, antifriction, and extreme pressure properties of gear oil.

(2) Ball milling could shorten the MWCNTs making them stable in the lubricant for a period of more than 60 days.

(3) From Raman spectroscopy it can be observed that ball milling time of up to 10 hours did not produce any defects on the surface of MWCNTs but 20-hour ball milling produced mild defects on the surface.

(4) Long MWCNTs, although surface-modified, when dispersed in lubricants exhibited poor stability and could

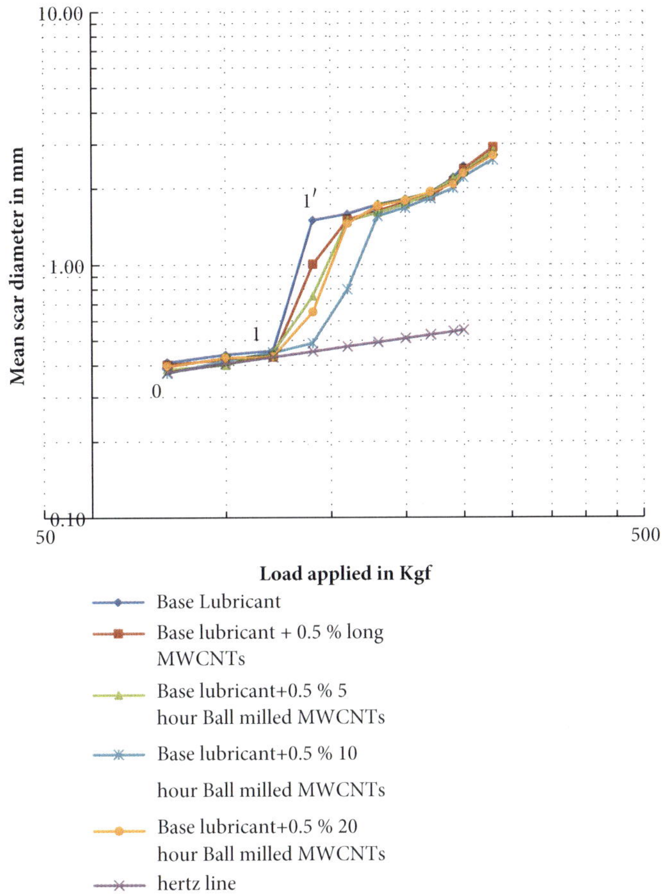

FIGURE 10: Variation of wear scar diameter with applied load during EP test.

marginally improve the antifriction and extreme pressure properties.

(5) The physicochemical properties remain unaltered with dispersion of surface-modified and ball-milled MWCNTs.

(6) There is a good improvement in the wear scar diameters and friction coefficients with dispersion of shortened MWCNTs as short-length MWCNTs could slide between mating surfaces reducing contact.

(7) The load wear index and weld load of lubricants dispersed with shortened MWCNTs have improved significantly as the performance of short MWCNTs is good in both antiwear and extreme pressure region.

(8) At an optimum length of MWCNTs in the range of 1 to 3 microns, the lubricant could give best results while too short MWCNTs due to formation of defects despite forming good suspension could not give best performance.

Nomenclature

D_H: Hertz scar diameter
EP: Extreme pressure
AW: Antiwear
L: Load applied
LNSL: Last nonseizure load
LWI: Load wear index
MWCNTs: Multiwalled carbon nanotubes
SL: Seizure load
X: Average scar diameter.

Conflicts of Interest

The authors declare that they have no conflicts of interest.

Acknowledgments

The authors gratefully acknowledge the support received from Hindustan Petroleum Corporation Ltd., India, for conducting the tests. The authors acknowledge the assistance from IIT Madras, IIT Kharagpur, and Osmania University, Hyderabad, in characterization. The authors sincerely thank the management of GITAM deemed University, India, for the support extended.

References

[1] N. Pierard, A. Fonseca, J.-F. Colomer et al., “Ball milling effect on the structure of single-wall carbon nanotubes,” *Carbon*, vol. 42, no. 8-9, pp. 1691–1697, 2004.

[2] M. S. Dresselhaus, A. Jorio, A. G. Souza Filho, and R. Saito, "Defect characterization in graphene and carbon nanotubes using Raman spectroscopy," *Philosophical Transactions of the Royal Society A*, vol. 368, pp. 5355–5377, 2010.

[3] L. G. Cançado, A. Jorio, E. H. M. Ferreira et al., "Quantifying defects in graphene via Raman spectroscopy at different excitation energies," *Nano Letters*, vol. 11, no. 8, pp. 3190–3196, 2011.

[4] K. R. Paton, E. Varrla, C. Backes et al., "Scalable production of large quantities of defect-free few-layer graphene by shear exfoliation in liquids," *Nature Materials*, vol. 13, no. 6, pp. 624–630, 2014.

[5] C. S. Chen, X. H. Chen, H. U. Jing et al., "Effect of multi-walled carbon nanotubes on tribological properties of lubricant," *Transactions of Nonferrous Metals Society of China*, vol. 15, no. 2, pp. 300–305, 2005.

[6] S. Bhaumik, S. Prabhu, and K. J. Singh, "Analysis of tribological behavior of carbon nanotube based industrial mineral gear oil 250 cSt viscosity," *Advances in Tribology*, vol. 2014, Article ID 341365, 8 pages, 2014.

[7] J. A. C. Cornelio, P. A. Cuervo, L. M. Hoyos-Palacio, J. Lara-Romero, and A. Toro, "Tribological properties of carbon nanotubes as lubricant additive in oil and water for a wheel-rail system," *Journal of Materials Research and Technology*, vol. 5, no. 1, pp. 68–76, 2016.

[8] H. Ghaednia, R. L. Jackson, and J. M. Khodadadi, "Experimental analysis of stable CuO nanoparticle enhanced lubricants," *Journal of Experimental Nanoscience*, vol. 10, no. 1, pp. 1–18, 2013.

[9] A. Hernández Battez, R. González, D. Felgueroso et al., "Wear prevention behaviour of nanoparticle suspension under extreme pressure conditions," *Wear*, vol. 263, no. 7-12, pp. 1568–1574, 2007.

[10] Y. Hu, Y. Peng, and H. Wang, "Tribological behaviors of surfactant-functionalized carbon nanotubes as lubricant additive in water," *Tribology Letters*, vol. 25, no. 3, pp. 247–253, 2006.

[11] L. J. Pottuz, B. Vacher, N. Ohmae, J. M. Martin, and T. Epicier, "Anti-wear and friction reducing mechanisms of carbon nano-onions as lubricant additives," *Tribology Letters*, vol. 30, no. 1, pp. 69–80, 2008.

[12] W. Khalil, A. Mohamed, M. Bayoumi, and T. A. Osman, "Tribological properties of dispersed carbon nanotubes in lubricant," *Fullerenes, Nanotubes and Carbon Nanostructures*, vol. 24, no. 7, pp. 479–485, 2016.

[13] J. Lee, S. Cho, Y. Hwang, C. Lee, and S. H. Kim, "Enhancement of lubrication properties of nano-oil by controlling the amount of fullerene nanoparticle additives," *Tribology Letters*, vol. 28, no. 2, pp. 203–208, 2007.

[14] C. Lee, Y. Hwang, Y. Choi, J. Lee, C. Choi, and J. Oh, "A study on the tribological characteristics of graphite nano lubricants," *International Journal Of Precision Engineering and Manufacturing*, vol. 10, no. 1, pp. 85–90, 2009.

[15] L. Peña-Parás, J. Taha-Tijerina, A. García et al., "Antiwear and extreme pressure properties of nanofluids for industrial applications," *Tribology Transactions*, vol. 57, no. 6, pp. 1072–1076, 2014.

[16] A. P. Puzyr, A. E. Burov, G. E. Selyutin, V. A. Voroshilov, and V. S. Bondar, "Modified nanodiamonds as antiwear additives to commercial oils," *Tribology Transactions*, vol. 55, no. 1, pp. 149–154, 2012.

[17] P. K. Sarma, V. Srinivas, V. D. Rao, and A. K. Kumar, "Experimental study and analysis of lubricants dispersed with nano Cu and TiO_2 in a four-stroke two wheeler," *Nanoscale Research Letters*, vol. 6, pp. 233–243, 2011.

[18] V. Srinivas, R. N. Thakur, A. K. Jain, and M. Saratchandra Babu, "Tribological studies of transmission oil dispersed with molybdenum disulfide and tungsten disulfide nanoparticles," *Journal of Tribology*, vol. 139, no. 4, pp. 041301–041306, 2017.

[19] V. Srinivas, Rao. Kodanda Rama, and C. N. Mohan Rao, "Lubricating and physico-chemical properties of CI-4 plus engine oil dispersed with surface modified multi-walled carbon nanotubes," *Tribology-Materials, Surfaces & Interfaces*, vol. 12, no. 2, pp. 107–114, 2018.

[20] J. L. Viesca, A. Hernández Battez, R. González, R. Chou, and J. J. Cabello, "Antiwear properties of carbon-coated copper nanoparticles used as an additive to a polyalphaolefin," *Tribology International*, vol. 44, no. 7-8, pp. 829–833, 2011.

7

Tribological Behavior of Laser Textured Hot Stamping Dies

Andre Shihomatsu,[1] Sergio Tonini Button,[1] and Iris Bento da Silva[2]

[1] *Department of Manufacturing and Materials Engineering, School of Mechanical Engineering, University of Campinas, 13083-970 Campinas, SP, Brazil*
[2] *Department of Mechanical Engineering, School of Engineering of Sao Carlos, University of Sao Paulo, 13566-590 Sao Carlos, SP, Brazil*

Correspondence should be addressed to Sergio Tonini Button; sergio1@fem.unicamp.br

Academic Editor: Patrick De Baets

Hot stamping of high strength steels has been continuously developed in the automotive industry to improve mechanical properties and surface quality of stamped components. One of the main challenges faced by researchers and technicians is to improve stamping dies lifetime by reducing the wear caused by high pressures and temperatures present during the process. This paper analyzes the laser texturing of hot stamping dies and discusses how different surfaces textures influence the lubrication and wear mechanisms. To this purpose, experimental tests and numerical simulation were carried out to define the die region to be texturized and to characterize the textured surface topography before and after hot stamping tests with a 3D surface profilometer and scanning electron microscopy. Results showed that laser texturing influences the lubrication at the interface die-hot sheet and improves die lifetime. In this work, the best texture presented dimples with the highest diameter, depth, and spacing, with the surface topography and dimples morphology practically preserved after the hot stamping tests.

1. Introduction

In recent years, researches on tribology of metal forming processes for the automotive industry are focused on the improvement of products surface quality and process performance. In hot stamping the improvement of the die lifetime is a relevant factor for obtaining better products, lower costs, and high productivity, so it is essential to understand how the severe conditions of pressure and temperature influence the lubrication at the interface dies-sheet and, consequently, the wear mechanisms which can cause die lifetime reduction.

Hot stamping of hardenable steels consists in heating the blank above the austenitizing temperature and then transferring it to the press where it is stamped and finally cooled and tempered between the dies, improving productivity and the mechanical properties of the stamped part like ultimate strength and impact toughness [1].

Hot stamping is a relatively fast process if compared to other metal forming processes, so the manufacturing costs would be significantly increased if production has to be interrupted to premature and frequent changes of worn dies. To resist the severe working conditions of hot stamping, dies must present high hot hardness to minimize wear at high temperatures, low sensitivity to cracks caused by thermal shocks, low sensitivity to local tempering and softening, high mechanical strength to avoid mechanically induced cracks, low chemical affinity with the sheet alloy being stamped, and high oxidation resistance at high temperatures to avoid cracking by corrosion [2].

Therefore, many researches have been presented to improve dies lifetime by developing new materials for dies and blanks and coatings for the steel sheets [3–5].

The wear mechanisms at the tools-workpiece interface are very complicated and therefore have been studied for years by many researchers for several hot metal forming processes. Among the factors which may influence the way die surfaces can be damaged, the most studied are the process conditions (temperature, speed, and contact pressure), die surface modification (coating, texturing, and heat treatment), and workpiece surface improvement by coating and heat treatment [6, 7].

Recently, several studies have been presented to evaluate the influence of coating steel sheets on the performance of hot stamping dies [3, 4, 8–13].

In the work of Boher et al. [10] a high strength boron steel coated with Al-Si was hot stamped with dies made of three steels with different contents of carbon, chromium, vanadium, molybdenum, manganese, and silicon. The first die made of X50CrMoV5 steel is heat-treated and nitride-coated to present a fine and homogeneous martensite microstructure. The second material, nonnitrided 60CrMoTiV16 steel, presents good resilience and good wear resistance and a martensitic microstructure with fine and spherical precipitates and polyhedral large precipitates rich in titanium, tungsten, and silicon, which probably cause a significant fluctuation of hardness.

Finally, the third material called SG3, also nonnitrided, is fully martensitic with some fine and globular carbides rich in iron, tungsten, and silicon. From the results obtained with a deep-drawing process simulator (DDPS) they concluded that the most important wear mechanism was the material transfer caused by the adhesion of particles of the sheet coating during the hot sliding. The surface damage presented a rapid kinetics influenced by the hardness and particles shape and distribution of each die steel.

The modification of the surface texture by many manufacturing processes is being extensively studied [14–19] and the laser texturing (LT) is one of the most potentials to improve lubrication by lubricant retention and reduce wear and die damage to get products with high surface quality [20–26]. The topography of sliding surfaces can induce extra hydrodynamic pressure and increase the lubricant film stiffness, and the dimples or grooves can act as reservoirs for lubricants and trap debris to reduce abrasive wear [27].

Ibatan et al. [15] show the recent development on surface texturing to improve the tribological performance of bearing sliders. They comment that laser texturing is one of the most advanced surface texturing techniques to produce microdimple patterns by short processing times. Otherwise, LT involves focused pulsed laser, which may generate microdimple patterns surrounded by a solidified melt rim. Besides, material melting and vaporization create heat-affected zone areas on the solidified melt rim changing the local microstructures and mechanical properties demanding appropriate pulse energy and pulse frequency.

Ramesh et al. [16] also studied the influence of different textures on the lubrication testing different combinations of width, depth, and density of microgrooves produced by casting, tested in a high pressure tribometer (HPT) under mixed and hydrodynamic lubrication established by varying the normal load and the sliding speed. They concluded that, in some combinations, textured surfaces presented worst results than the nontextured tools. The best combination to reduce friction and wear was obtained with the groove depth equal to the film thickness, the grooves density around 20–30%, and the groove width between 100 and 200 μm.

Geiger et al. [20] evaluated the influence of laser texturing on the lifetime of cold forging dies to minimize the wear caused by poor lubrication. They also studied the influence of two coatings on the textured surface: TiN and TiN + MoS_2. They concluded that textures with TiN + MoS were the best to increase die lifetime because the dimples present in the textures retain lubricant, which is expelled during the process reducing the friction and minimizing abrasion.

Andersson et al. [21] analyzed the influence of laser texturing with some lubricant usually applied in the metal forming industry. The textured surfaces presented two densities of dimples and two ratios for the depth/diameter of the dimples. Three lubricants with different viscosities were applied. They concluded that laser texturing decreased significantly the friction and die wear, being the best combination of texture and lubrication that one with the lowest density of dimples, the largest ratio depth to diameter, and the most viscous lubricant.

In this work, we study the tribological effects of laser texturing the surface of hot stamping dies to evaluate the influence of the textures on the lubrication and wear by varying the diameter, the depth, and the density of the dimples and by evaluating the textures before and after the stamping tests by microhardness measurements, 3D surface topography, and scanning electron microscopy (SEM).

2. Materials and Methods

2.1. Preliminary Considerations. To define the conditions of laser texturing and lubrication, this work was based on some conclusions presented in the literature, which led to the hypotheses listed below considering the parameters shown in Figure 1.

As observed by Rapoport et al. [24], Ding et al. [27], and other researchers, the shape and the dimensions of the dimples are altered by the metallurgical transformations which occur during and after laser texturing and involve uncontrolled thermal expansion and contraction caused by repeated melting and solidification cycles which induce residual stresses. Crests (or bulges) may be formed, which, according to Ding et al. [27], are harmful to the formation of the tribofilm and can favor surfaces with galling.

Therefore, some hypotheses may be presented on the influence of the laser texturing parameters on the wear and lubrication:

(a) The increase in the number of dimples increases the retention of lubricant, so the density of dimples (γ) must be as high as possible to retain more lubricant.

(b) The distance between two adjacent dimples (δ) must be as small as possible to guarantee forming a contin uous film of lubricant with thickness (t).

(c) Larger dimple diameters (d) with small distance between dimples (δ) can build up a bulge with a crest; therefore the crest height (H) must be limited to avoid metal-to-metal contact.

(d) To guarantee some lubricant to be expelled to the interface during sliding

  (i) the diameter of the crest (D) must be limited to avoid the excessive reduction of δ which could cause metal-to-metal contact;

  (ii) the dimple depth (h) must be less than the dimple diameter (d);

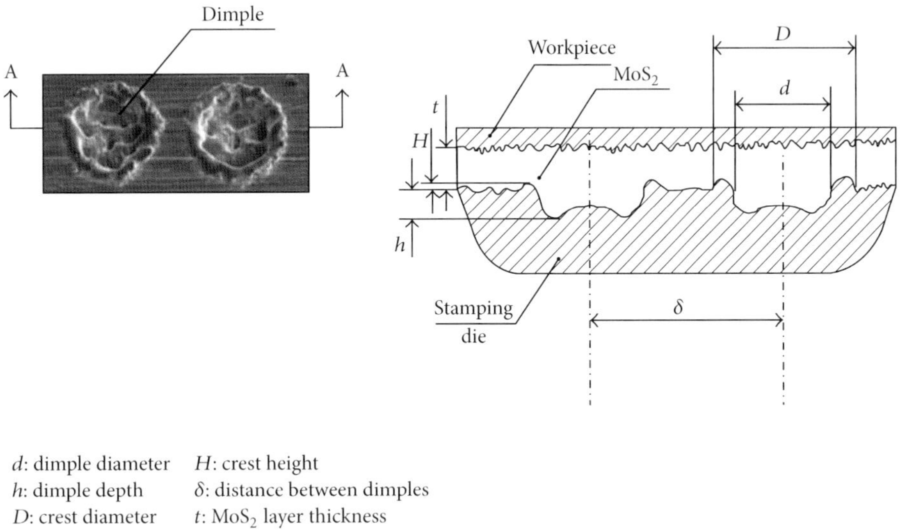

FIGURE 1: Parameters for the laser surface texturing of hot stamping dies.

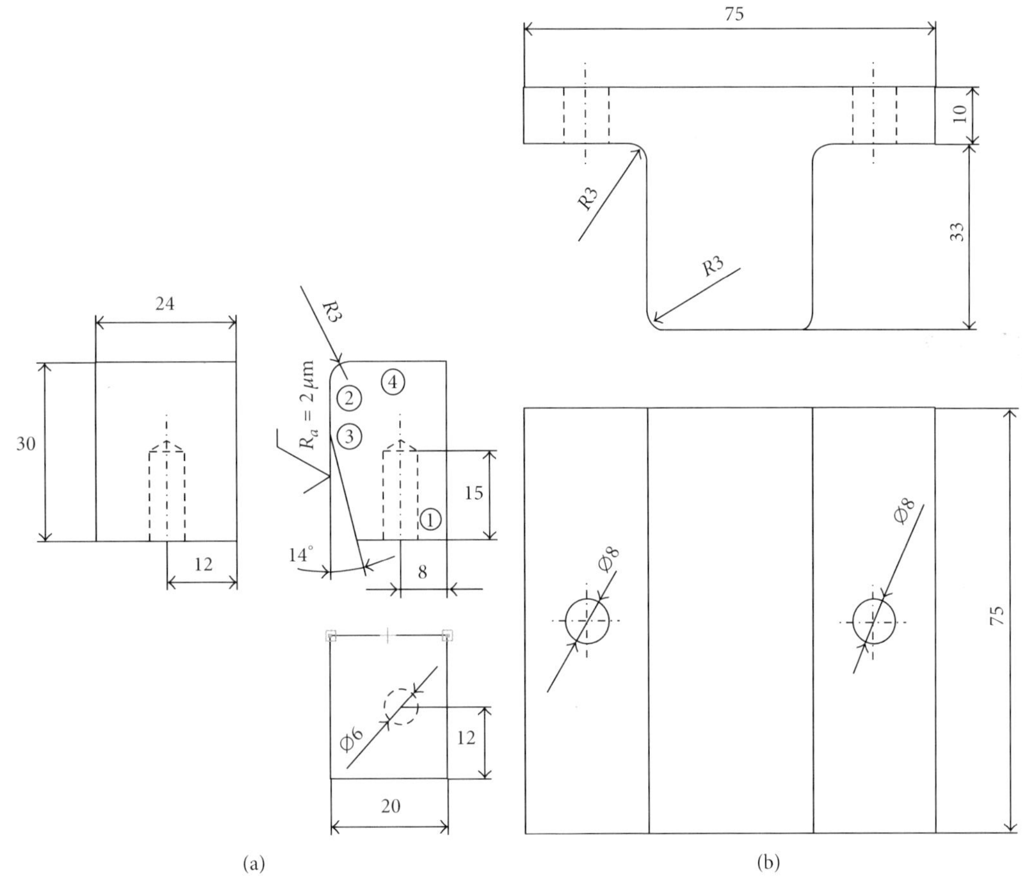

FIGURE 2: Dimensions (in mm) of the dies (a) and punch (b) used in the hot stamping tests.

(iii) the dimple depth (h) added to the height of the crest (H) must be less than the diameter of the crest (D).

2.2. Numerical Simulation. To define the region of the dies to be laser textured, hot stamping of a steel part "U" shaped was simulated using the software Forge 2008 based on the Finite Element Method (FEM) in which the wear of the dies is modeled by modified Archard's equations [28].

Figure 2 shows the tooling (punch and dies) designed for the hot stamping tests and Figure 3 shows the punch and dies with the sheet blank positioned and discretized for the

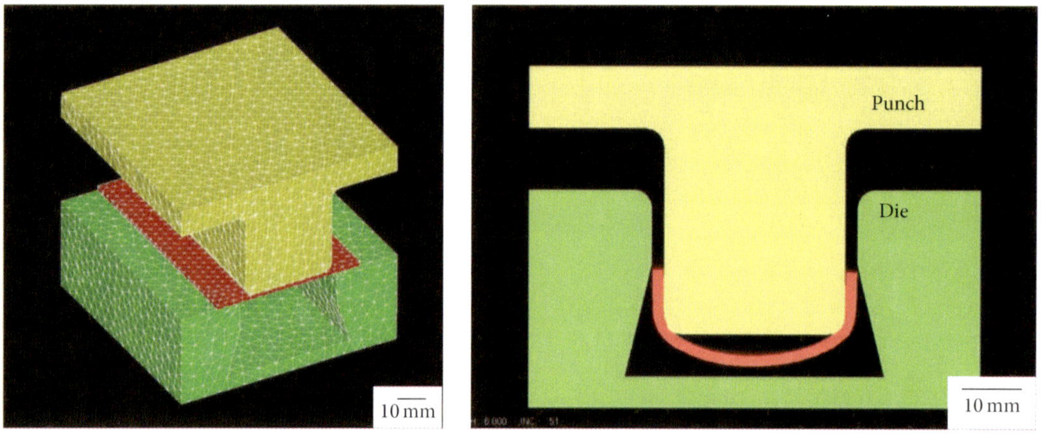

(a) (b)

FIGURE 3: Numerical simulation—tooling and blank: before hot stamping (a) and after hot stamping (b).

TABLE 1: Inputs for the numerical simulation.

	Material	Initial temperature	Type and number of finite elements	Mean size of finite elements
Blank	Steel DIN 27MnCrB5—1.8 mm thick	900°C	Tetrahedral 14,623	2.46 mm
Tooling	Steel AISI H13	25°C	Triangular 2,635	7.85 mm
Process parameters	Press speed 10 mm/s	Solid lubricant MoS_2	Heat transfer coefficients and materials properties [29]	

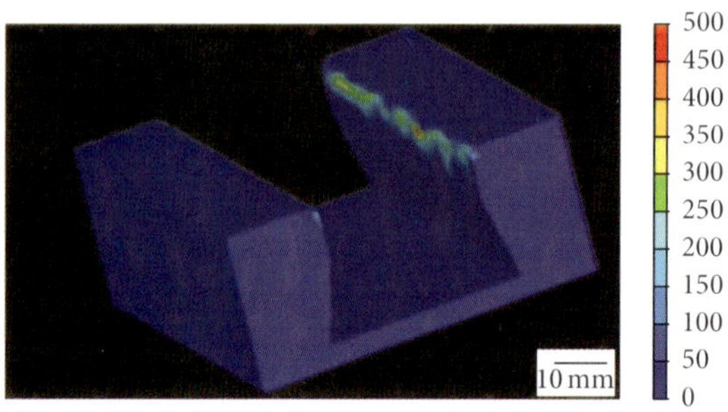

FIGURE 4: Die wear in MPa·mm—numerical results.

numerical test. Table 1 shows the conditions applied in the simulations.

As shown in Figure 4, the regions of the dies that presented the largest damage, in MPa·mm, were the die radius and part of the vertical surface below. Boher et al. [10] obtained similar results confirming that those regions are interesting to analyze the influence of laser texturing.

2.3. Laser Texturing. The parameters to laser texturing the regions shown in Figure 4 were adopted considering the results obtained by Ramesh et al. [16] who concluded that the dimple depth (h) should be near the film thickness in the interface die-blank. In this work, h is assumed to be 30 μm for two textures and 150 μm for the other two to evaluate opposite conditions to supplying solid lubricant for the interface die-workpiece.

Three distances between adjacent dimples (177, 266, 420, and 531 μm) and three dimple diameters (100, 150, and 300 μm) were chosen to evaluate the influence of dimples density equal to 10 or 25%, as suggested by Hu et al. [25] who concluded that a dimple density of 15 to 25% is recommended to increase the amount of solid lubricant coating in the interface dies-workpiece and, consequently, to increase tool life.

All the surfaces were textured with a solid-state lamp-pumped Nd:YAG laser, operating at 1,064 nm wavelength with a power of 100 W. The dimples dimensions were defined and controlled by the processing software available in the laser equipment.

Table 2 shows the conditions used to laser-texture each of the five textures analyzed in this work presented in Figure 5. All the dies were milled, quenched, and tempered to 52HRC, ground, and finally laser textured. A sixth die only ground (R_a = 2.0 μm) was also tested and analyzed.

2.4. Hot Stamping Tests. The punch and the dies were assembled in a hydraulic press with nominal capacity of 300 kN and work speed of 10 mm/s. Before each stamping test, the dies were cleaned and lubricated with solid MoS_2.

The blank of DIN 27MnCrB5 steel (1.8 mm thick, 65 mm wide, and 82 mm long) was heated in an electric furnace at 900°C for 10 minutes and then transferred and hot stamped

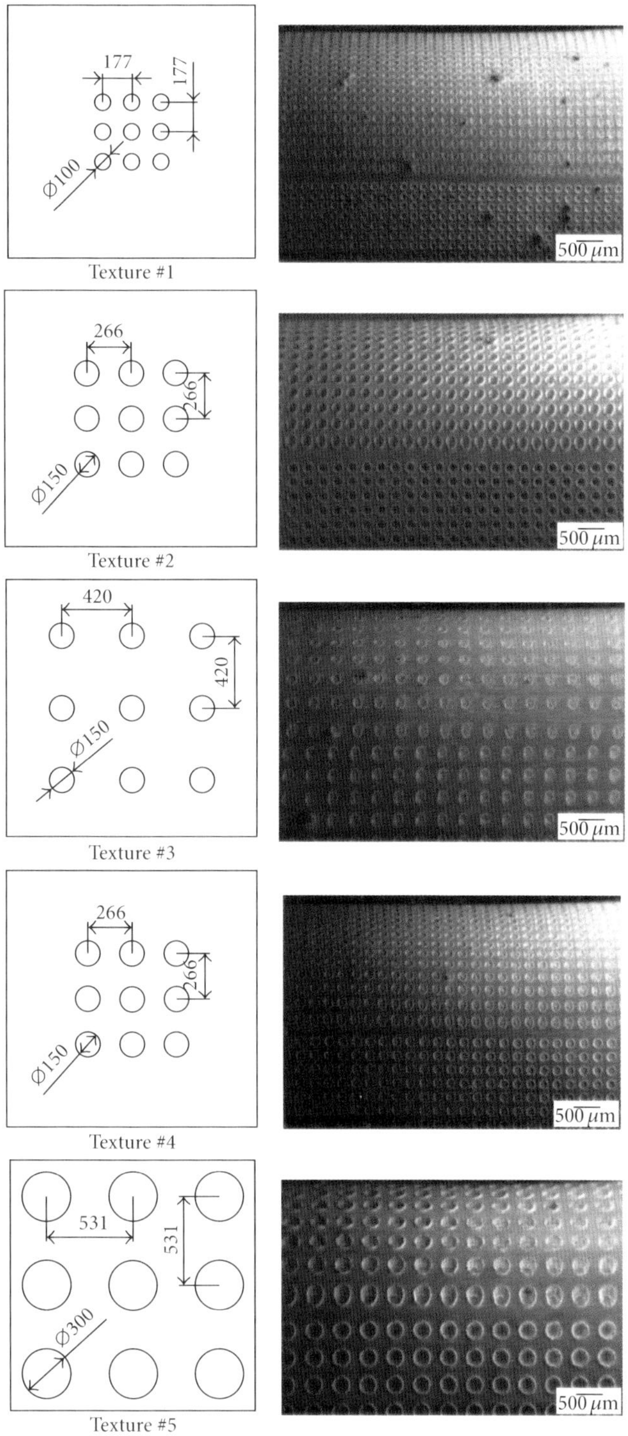

FIGURE 5: Sketches and SEM images at the die radius of the five textures before stamping tests.

in the press to form a "U" shaped part. Each texture was tested 100 times.

2.5. Characterization of the Laser Textured Dies. The textured surfaces of the dies were analyzed before and after the hot stamping tests to evaluate the surface topography, the microhardness, and the surface integrity. After hot stamping all the surfaces were cleaned with ethanol in ultrasound equipment before the analyses.

The surface topography was analyzed with the 3D profilometer WYKO NT100 from Veeco Instruments. The microhardness Vickers was measured three times in four positions (① to ④ in Figure 2(a)) for each texture, with the microdurometer Buehler 2100 with a load of 10 N.

TABLE 2: Dimensions of the texture parameters shown in Figure 1.

Texture #	Dimple depth (h) [μm]	Distance (δ) [μm]	Dimple diameter (d) [μm]	Density $\pi\cdot(d/2\delta)^2$ [%]
1	30	177	100	25
2	150	266	150	25
3	30	420	150	10
4	30	266	150	25
5	150	531	300	25

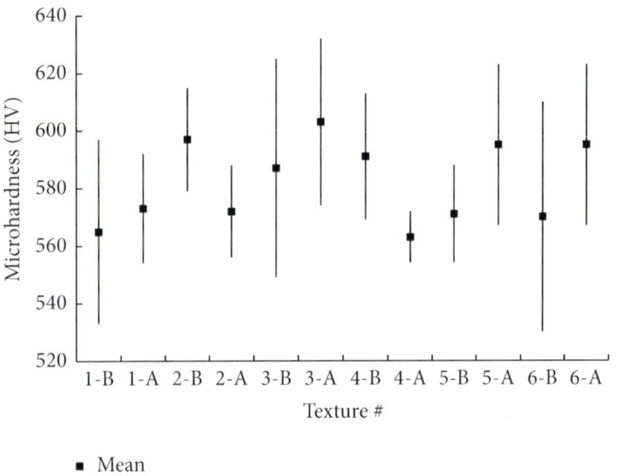

FIGURE 6: Microhardness Vickers measured in point (2) near the die radius for all textures before (B) and after (A) hot stamping tests: maximum, mean, and minimum values.

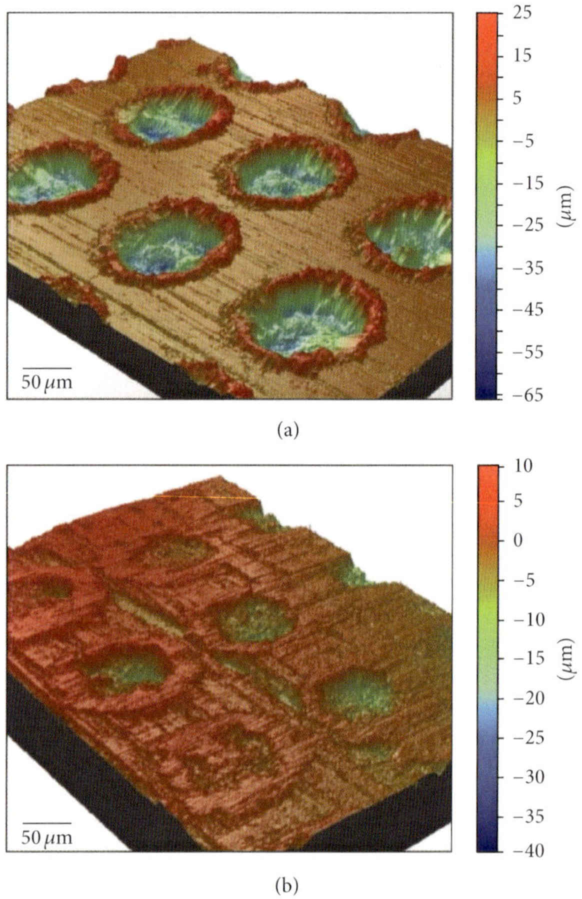

FIGURE 7: Surface topography—texture #1, die radius: (a) before and (b) after the stamping tests.

The scanning electron microscope Zeiss EVO MA15 was used to evaluate the surface integrity of the textured dies, respectively, before and after the stamping tests. The surfaces were also photographed with a digital camera Sony DSC-W350.

3. Results and Discussion

3.1. Microhardness Vickers. Table 3 presents the results of microhardness Vickers for all textures measured before (B) and after (A) the hot stamping tests and Figure 6 shows the results for position ② in Figure 2(a), near the die radius.

Comparing statistically the means and standard deviations for each point of the six textures before and after the stamping tests it may be concluded that there is not a significant difference, assuming a significance level α equal to 0.05, and the same was observed when comparing the hardness at each point among the six textures, before or after the stamping tests. Therefore, the laser texturing and the stamping tests did not change the die hardness and, consequently, did not contribute to the wear of the textured surfaces.

3.2. Surface Topography. To evaluate how laser texturing affected the dimples geometry and dimensions, the diameters of six dimples were measured twice for each texture, totalizing twelve measurements. The same six dimples were taken to measure three times the distance between adjacent dimples and dimples depth. Table 4 presents the results of the measurements from the profilometer analysis before the stamping tests.

The mean and dispersion of each dimension near the nominal values, except for the dimple depth of texture #5 and the dimple diameter of textures #3 and #5, confirm that the conditions applied in the laser process allowed a controlled texturing in most of the conditions.

The dispersion of dimple diameter of texture #3 kept the minimum dimple density near the nominal 10% justifying the worst lubrication condition, as discussed in the analysis of the surface topography, and confirming the conclusions of Rapoport et al. [24].

Although texture #5 presented a mean measured dimple depth smaller than the nominal and a mean dimple diameter larger than the nominal, the minimum dimple density was not reduced and the maximum density was increased to 43% that is near the optimum density (40 to 50%) found by Rapoport et al. [24] and 43% found by Hu et al. [25]. Therefore, the highest dimple density of texture #5 favored the trapping of lubricant and its availability to lubricate the interface, as will be discussed in the analysis of Figure 11(b).

Figures 7–12 show the results of the profilometer in the region of the die radius for the six surfaces before and after the stamping tests.

TABLE 3: Microhardness Vickers of all textures before and after the hot stamping tests.

Texture #	Position	Before hot stamping tests		After hot stamping tests	
		Mean [HV]	Standard deviation	Mean [HV]	Standard deviation
1	1	565	20	538	25
	2	565	32	573	19
	3	597	19	592	30
	4	574	23	534	32
2	1	486	40	590	32
	2	597	18	572	16
	3	534	35	576	4
	4	525	23	568	23
3	1	571	33	501	30
	2	587	38	603	29
	3	581	17	581	19
	4	580	29	616	21
4	1	607	19	563	12
	2	591	22	563	9
	3	584	51	565	29
	4	530	20	569	32
5	1	514	13	568	35
	2	571	17	595	28
	3	564	34	568	19
	4	583	32	585	18
As grinded	1	548	26	592	28
	2	570	40	595	28
	3	564	25	568	19
	4	568	27	585	35

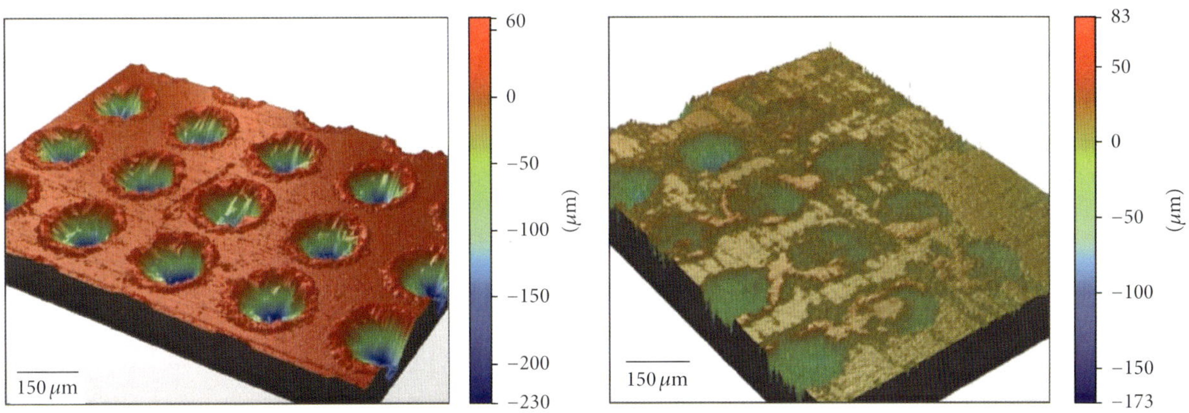

(a) (b)

FIGURE 8: Surface topography—texture #2, die radius: (a) before and (b) after the stamping tests.

Bulges are evident in textures #1, #2, and #4 before the stamping tests, with the highest crests being observed in texture #2 (Figure 8(a)), which presents the highest dimple depth, both nominal and measured (about 150 μm). Because texture #2 presents smaller dimple diameter and spacing between adjacent dimples than texture #5 (with the same nominal depth) and consequently should contain less lubricant and be more susceptive to galling, it could present a significant change of the surface as is actually observed in Figure 8(b).

TABLE 4: Nominal and measured dimensions of the texture parameters shown in Figure 1 before stamping tests.

Texture #	Dimple depth (h) [μm]		Distance (δ) [μm]		Dimple diameter (d) [μm]		Density $\pi \cdot (d/2\delta)^2$ [%]	
	Nominal	Measured (mean/standard deviation)	Nominal	Measured (mean/standard deviation)	Nominal	Measured (mean/standard deviation)	Nominal	Measured (minimum/maximum)
1	30	25 ± 5	177	175 ± 5	100	106 ± 10	25	22/37
2	150	151 ± 10	266	265 ± 4	150	171 ± 8	25	29/37
3	30	27 ± 1	420	424 ± 6	150	206 ± 45	10	11/28
4	30	23 ± 7	266	269 ± 6	150	155 ± 10	25	25/31
5	150	93 ± 3	531	543 ± 17	300	363 ± 28	25	28/43

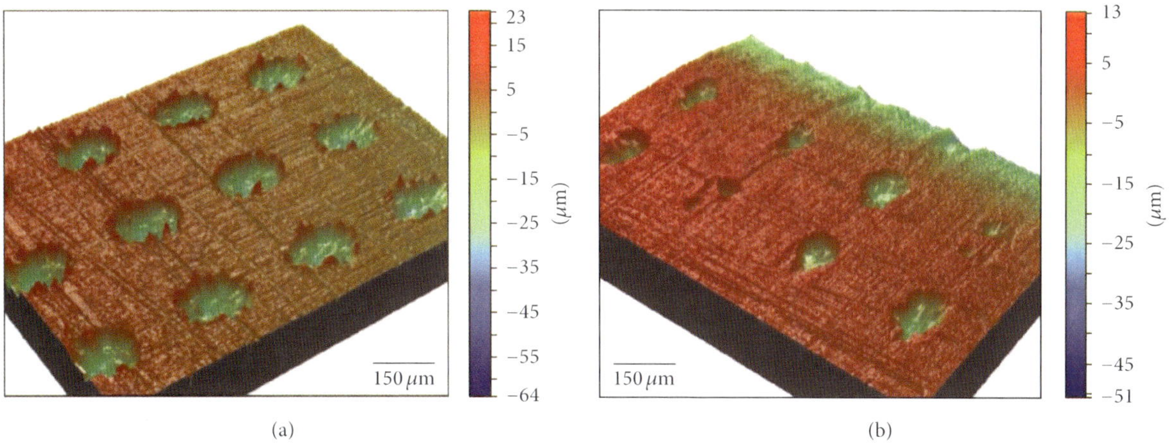

FIGURE 9: Surface topography—texture #3, die radius: (a) before and (b) after the stamping tests.

FIGURE 10: Surface topography—texture #4, die radius: (a) before and (b) after the stamping tests.

Textures #3 and #5 presented few small crests, while textures #1 and #4 presented bulges with similar height and distribution. Textures #1, #3, and #4 presented similar dimple depths before the stamping tests.

After the hot stamping tests, texture #1 presented a significant change of topography, and dimple depth was significantly reduced (Figure 7(b)). The distance between dimples was also reduced by the approximation of bulges probably deformed by the metal-to-metal contact between the crests and the workpiece surface caused by the insufficient lubricant carried out from the small and shallow dimples.

Texture #2 also experienced a significant change: the topography was completely modified, the dimple depth presented the highest reduction among the five textures, and the bulges were completely deformed flattening the surface (Figure 8(b)).

Before the stamping tests, texture #3 (Figure 9(a)) had few bulges, the same dimple depth and dimple diameter of texture #4, but had a spacing 60% larger than texture #4 (Figure 10(a)). After the stamping tests, as observed in Figure 9(b), texture #3 had most of the dimples deformed by wear during the stamping tests, and some practically disappeared, maybe because little lubricant was available in the shallow dimples and insufficient to lubricate the large surface among dimples with the lower density.

Otherwise, texture #4 (Figure 10(b)) experienced a significant reduction in dimple depth, and the bulges were still present and deformed but did not decrease the spacing between the dimples.

Texture #5, beyond presenting few bulges before stamping tests (Figure 11(a)), also presented high dimple diameter and spacing, which, respectively, prevented galling and increased the amount of lubricant trapped in the dimples, resulting after the hot stamping tests in a texture without crests and with preserved dimples since neither the depth nor the spacing was significantly reduced (Figure 11(b)).

The ground surface with average roughness of 2.0 μm (Figure 12(a)) was significantly affected by the stamping tests (Figure 12(b)), the surface roughness was reduced, and parallel scratches were formed perpendicular to the original ones, probably by plowing caused by direct metal-to-metal contact, confirming that smooth surfaces are not able to trap lubricant and avoid wear, as observed by other

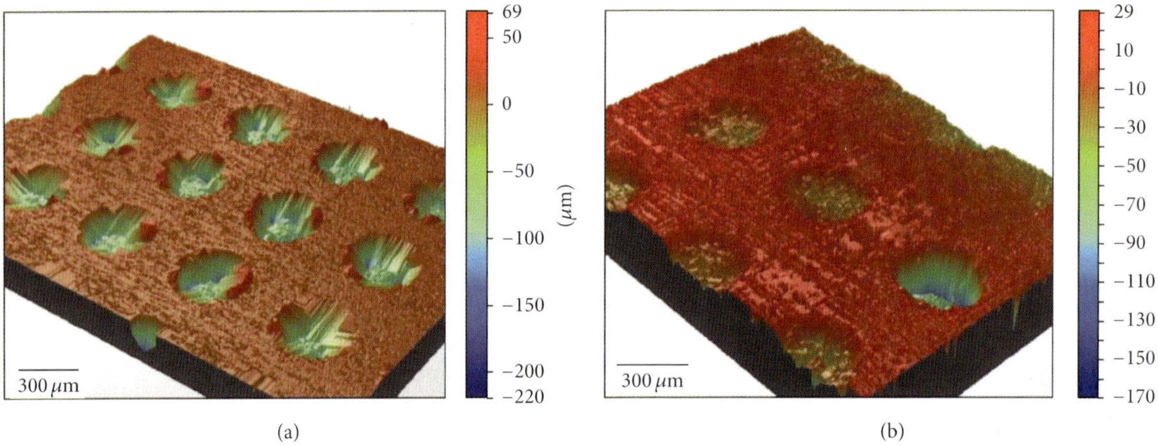

FIGURE 11: Surface topography—texture #5, die radius: (a) before and (b) after the stamping tests.

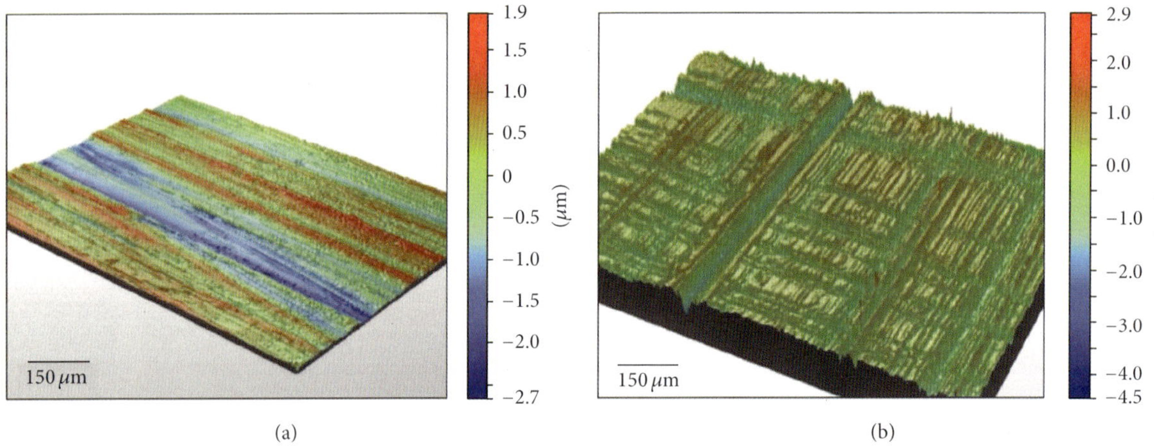

FIGURE 12: Surface topography—ground surface, die radius: (a) before and (b) after the stamping tests.

authors who compared textured and untextured surfaces [16, 27].

3.3. Scanning Electron Microscopy. Figures 13–18 show the SEM surfaces, in the region of the die radius, for the five laser textured dies and the ground surface before and after the stamping tests, with two magnifications: 200 and 700x.

SEM images are coherent to the results of the surface topography. Textures #1, #2, and #3 (Figures 13(b), 14(b), and 15(b)) were completely modified by hot stamping tests, with bulges being almost completely deformed and flattened, probably by the wear by adhesion and galling caused by metal-to-metal contact and abrasion generated by oxides and other hard particles, as observed by Ghiotti et al. [8] and Wieland and Merklein [9] and by other researchers who also studied the tribology in hot stamping.

Most dimples of all textures are filled with particles of the solid lubricant or oxides from the workpiece at elevated test temperature, as can be observed in Figure 19, which shows particles in a dimple of texture #5.

Otherwise, textures #4 and #5 (Figures 16(b) and 17(b)) presented the most preserved dimples after hot stamping tests, as observed and commented on before in the surface topography analysis, with the bulges being still present in texture #4.

Texture #5 practically presents the same dimples morphology after the stamping tests, except for the initial few bulges that were slightly flattened.

Finally, the images of the ground surface (Figure 18(b)) are also very similar to the surface topography (Figure 12(b)) with scratches perpendicular to the original ones, probably caused by abrasive wear, and apparently with a smoother surface than before the hot stamping tests.

4. Conclusions

The influence of dimples dimensions (diameter, depth, and spacing) on the performance of laser textured hot stamping

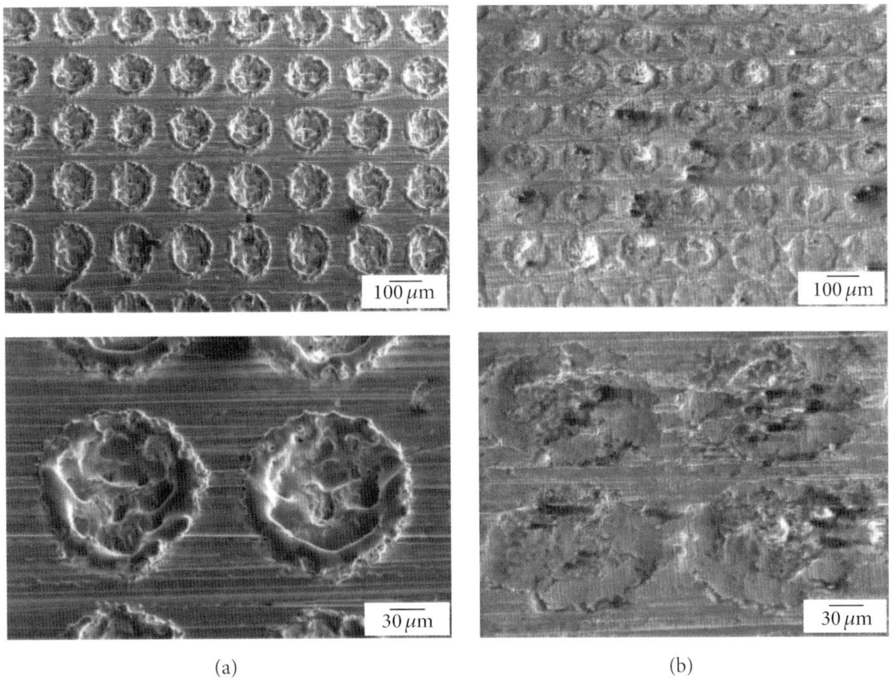

FIGURE 13: SEM images of die radius—texture #1: (a) before and (b) after hot stamping tests.

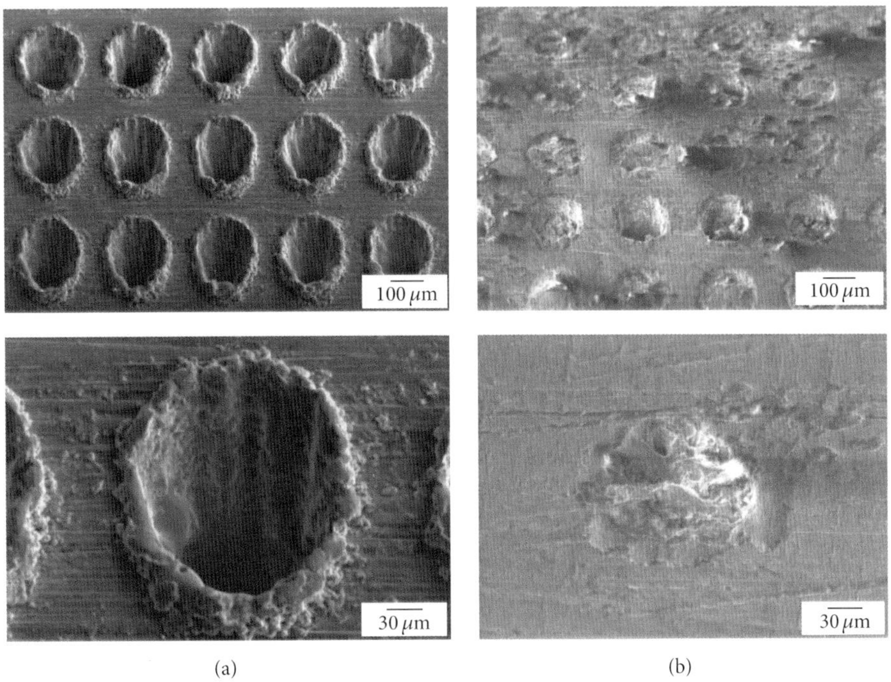

FIGURE 14: SEM images of die radius—texture #2: (a) before and (b) after hot stamping tests.

dies has been studied, and the main conclusions are as follows:

(i) Laser texturing is a suitable process to improve the tribological performance of hot stamping dies, since all the textures evaluated in this work were better than the untextured surface to prevent metal-to-metal contact.

(ii) The processing conditions used to laser texturing gave good results in terms of dimple dimensions and spacing for most of the tested textures.

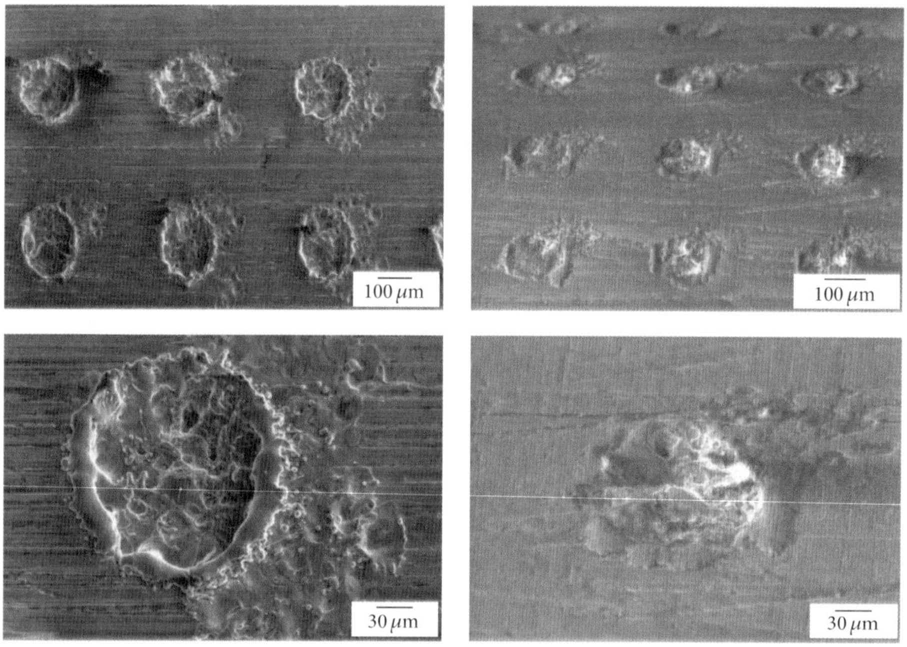

(a) (b)

FIGURE 15: SEM images of die radius—texture #3: (a) before and (b) after hot stamping tests.

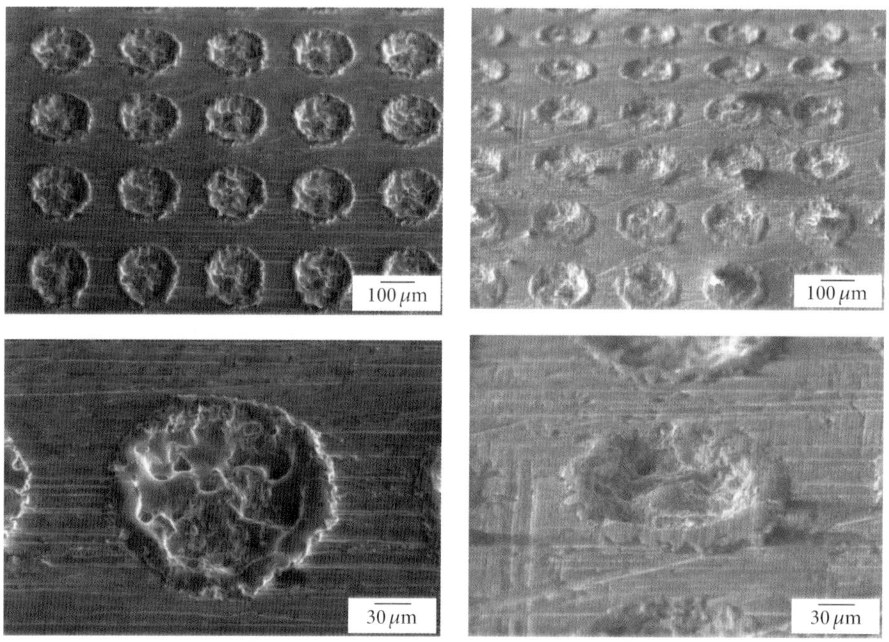

(a) (b)

FIGURE 16: SEM images of die radius—texture #4: (a) before and (b) after hot stamping tests.

(iii) Increasing the number of dimples increases the retention of lubricant, so the density of dimples must be as high as possible to retain more lubricant. However, the distance between two adjacent dimples cannot be too small to be interfered by the bulges built up during laser texturing and deformed in the hot stamping tests.

(iv) The dimple diameter must be the highest as possible with a high spacing to avoid metal-to-metal contact due to the high bulges.

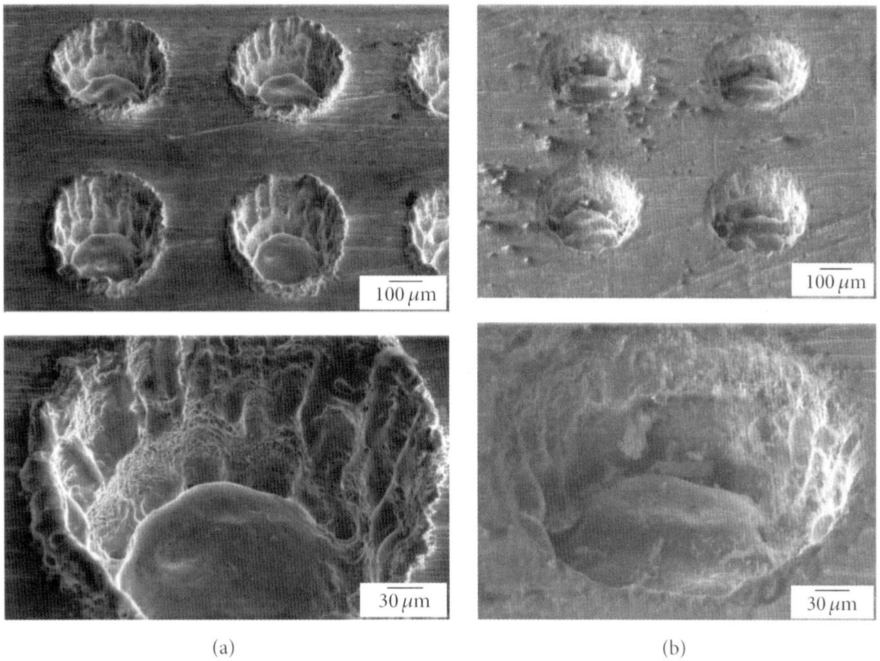

Figure 17: SEM images of die radius—texture #5: (a) before and (b) after hot stamping tests.

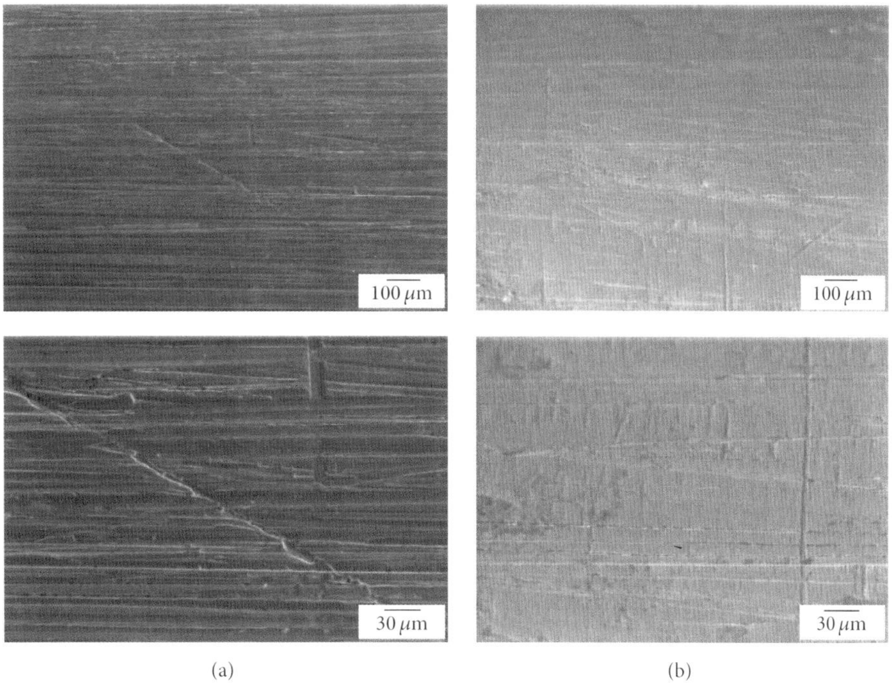

Figure 18: SEM images of die radius—ground surface: (a) before and (b) after hot stamping tests.

(v) Textures with the smallest dimples diameter, depth, and spacing presented the worst performance with the topography completely modified by adhesive and abrasive wear.

(vi) The best performance was obtained by the texture with the highest dimples density, near to 40%, as concluded by other researchers, with the highest dimple diameter and spacing, and with a depth of one-quarter of the dimple diameter. The surface topography and dimples morphology were practically preserved after the stamping tests, except for the bulges that were slightly flattened.

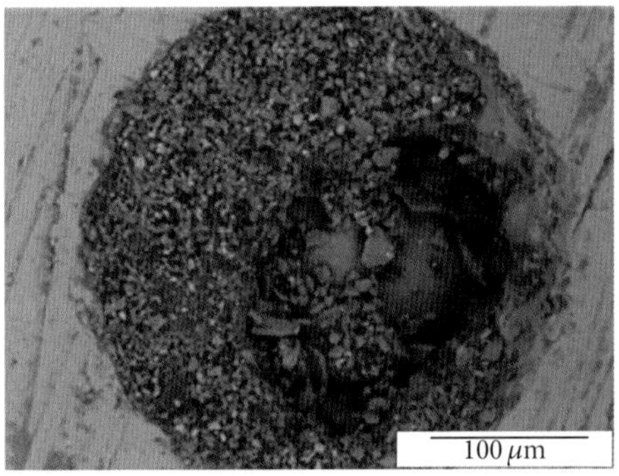

FIGURE 19: SEM image of particles inside a dimple on the die radius of texture #5 after hot stamping tests.

Competing Interests

The authors declare that they have no competing interests.

Acknowledgments

The authors wish to thank FAPESP (Process 2011/12927-6) for the financial support to this work.

References

[1] H. Karbasian and A. E. Tekkaya, "A review on hot stamping," *Journal of Materials Processing Technology*, vol. 210, no. 15, pp. 2103–2118, 2010.

[2] A. Schubert, R. Neugebauer, D. Sylla, M. Avila, and M. Hackert, "Manufacturing of surface microstructures for improved tribological efficiency of powertrain components and forming tools," *CIRP Journal of Manufacturing Science and Technology*, vol. 4, no. 2, pp. 200–207, 2011.

[3] A. Ghiotti, S. Bruschi, and F. Borsetto, "Tribological characteristics of high strength steel sheets under hot stamping conditions," *Journal of Materials Processing Technology*, vol. 211, no. 11, pp. 1694–1700, 2011.

[4] M. R. Lovell and Z. Deng, "Characterization of interfacial friction in coated sheet steels: influence of stamping process parameters and wear mechanisms," *Tribology International*, vol. 35, no. 2, pp. 85–95, 2002.

[5] M. Merklein, J. M. Allwood, B.-A. Behrens et al., "Bulk forming of sheet metal," *CIRP Annals—Manufacturing Technology*, vol. 61, no. 2, pp. 725–745, 2012.

[6] X. Tian, Y. Zhang, and J. Li, "Investigation on tribological behavior of advanced high strength steels: influence of hot stamping process parameters," *Tribology Letters*, vol. 45, no. 3, pp. 489–495, 2012.

[7] F. Findik, "Latest progress on tribological properties of industrial materials," *Materials and Design*, vol. 57, pp. 218–244, 2014.

[8] A. Ghiotti, F. Sgarabotto, and S. Bruschi, "A novel approach to wear testing in hot stamping of high strength boron steel sheets," *Wear*, vol. 302, no. 1-2, pp. 1319–1326, 2013.

[9] M. Wieland and M. Merklein, "Wear behavior of uncoated and coated tools under complex loading conditions," *Tribology in Industry*, vol. 34, no. 1, pp. 11–17, 2012.

[10] C. Boher, S. Le Roux, L. Penazzi, and C. Dessain, "Experimental investigation of the tribological behavior and wear mechanisms of tool steel grades in hot stamping of a high-strength boron steel," *Wear*, vol. 294-295, pp. 286–295, 2012.

[11] S. Hernandez, J. Hardell, H. Winkelmann, M. R. Ripoll, and B. Prakash, "Influence of temperature on abrasive wear of boron steel and hot forming tool steels," *Wear*, vol. 338-339, pp. 27–35, 2015.

[12] A. Ghiotti, S. Bruschi, F. Sgarabotto, and P. F. Bariani, "Tribological performances of Zn-based coating in direct hot stamping," *Tribology International*, vol. 78, pp. 142–151, 2014.

[13] L. Pelcastre, J. Hardell, and B. Prakash, "Galling mechanisms during interaction of tool steel and Al-Si coated ultra-high strength steel at elevated temperature," *Tribology International*, vol. 67, pp. 263–271, 2013.

[14] D. J. Wentink, D. Matthews, N. M. Appelman, and E. M. Toose, "A generic model for surface texture development, wear and roughness transfer in skin pass rolling," *Wear*, vol. 328-329, pp. 167–176, 2015.

[15] T. Ibatan, M. S. Uddin, and M. A. K. Chowdhury, "Recent development on surface texturing in enhancing tribological performance of bearing sliders," *Surface and Coatings Technology*, vol. 272, pp. 102–120, 2015.

[16] A. Ramesh, W. Akram, S. P. Mishra, A. H. Cannon, A. A. Polycarpou, and W. P. King, "Friction characteristics of microtextured surfaces under mixed and hydrodynamic lubrication," *Tribology International*, vol. 57, pp. 170–176, 2013.

[17] J. G. Parreira, C. A. Gallo, and H. L. Costa, "New advances on maskless electrochemical texturing (MECT) for tribological purposes," *Surface and Coatings Technology*, vol. 212, pp. 1–13, 2012.

[18] M. L. Magri, A. E. Diniz, and S. T. Button, "Influence of surface topography on the wear of hot forging dies," *International Journal of Advanced Manufacturing Technology*, vol. 65, no. 1–4, pp. 459–471, 2013.

[19] D. Jucius, A. Guobiene, and V. Grigaliūnasa, "Surface texturing of polytetrafluoroethylene by hot embossing," *Applied Surface Science*, vol. 256, no. 7, pp. 2164–2169, 2010.

[20] M. Geiger, U. Popp, and U. Engel, "Excimer laser micro texturing of cold forging tool surfaces—influence on tool life," *CIRP Annals—Manufacturing Technology*, vol. 51, no. 1, pp. 231–234, 2002.

[21] P. Andersson, J. Koskinen, S. Varjus et al., "Microlubrication effect by laser-textured steel surfaces," *Wear*, vol. 262, no. 3-4, pp. 369–379, 2007.

[22] T. Hu, L. Hu, and Q. Ding, "Effective solution for the tribological problems of Ti-6Al-4V: combination of laser surface texturing and solid lubricant film," *Surface and Coatings Technology*, vol. 206, no. 24, pp. 5060–5066, 2012.

[23] J. Li, D. Xiong, H. Wu, Y. Zhang, and Y. Qin, "Tribological properties of laser surface texturing and molybdenizing duplex-treated stainless steel at elevated temperatures," *Surface and Coatings Technology*, vol. 228, supplement 1, pp. S219–S223, 2013.

[24] L. Rapoport, A. Moshkovich, V. Perfilyev et al., "Friction and wear of MoS_2 films on laser textured steel surfaces," *Surface and Coatings Technology*, vol. 202, no. 14, pp. 3332–3340, 2008.

[25] T. Hu, Y. Zhang, and L. Hu, "Tribological investigation of MoS_2 coatings deposited on the laser textured surface," *Wear*, vol. 278-279, pp. 77–82, 2012.

[26] A. Rosenkranz, L. Reinert, C. Gachot, and F. Mücklich, "Alignment and wear debris effects between laser-patterned steel surfaces under dry sliding conditions," *Wear*, vol. 318, no. 1-2, pp. 49–61, 2014.

[27] Q. Ding, L. Wang, and L. Hu, "Tribology optimization by laser surface texturing: from bulk materials to surface coatings," in *Laser Surface Engineering*, chapter 16, pp. 405–422, Woodhead Publishing, 2015.

[28] J. F. Archard, "Contact and rubbing of flat surfaces," *Journal of Applied Physics*, vol. 24, no. 8, pp. 981–988, 1953.

[29] M. H. F. Batalha and S. T. Button, "Hot formability of DIN 27MnCrB5 steel sheets under controlled thinning," *Journal of Materials Processing Technology*, vol. 228, pp. 98–105, 2016.

8

True Stability of Lubricants Determined using the Ball-on-Disk Test

Angela Maria Tortora and Deepak Halenahally Veeregowda

Application Development Laboratory, Ducom Instruments Europe B.V., 9713 GX Groningen, Netherlands

Correspondence should be addressed to Deepak Halenahally Veeregowda; deepak.v@ducom.com

Academic Editor: Huseyin Çimenoğlu

True stability of lubricants can be determined when there is minimum change in the contact area and also the intervention of wear debris in the contact zone. Here, we have used the ball-on-disk instrument with the migrating point contact, that is, relative motion between the ball and disk condition to fix the contact area and minimize the wear debris at the contact zone. The jump in the friction coefficient indicates the film failure, which appeared earlier for the motor oil 5W30 compared to 5W40. Such profile was not recorded in absence of relative motion. Therefore, 5W40 was considered to have a better lubricant stability than 5W30. Applying the same test condition to the natural lubricants shows that glycerol has better lubricant stability than glycerol-water mixture. Superior true lubricant stability by glycerol and 5W40 can be related to its high viscosity. However, they were less wear resistant compared to low viscosity lubricants like 5W30 and glycerol-water. We suspect the role of microscopic wear debris at the contact zone for this behavior although it should have been avoided in the migrating point contact condition. Overall, ball-on-disk instrument with a migrating point contact condition is an effective technique to determine the stability of lubricants.

1. Introduction

An effective lubrication can be achieved with a lubricating film that is stable during the hydrodynamic, mixed, and boundary conditions. Four ball [1], ball-on-disk [2], and block-on-ring instruments [3] are some of the well-established industrial standard techniques used in tribological characterization of the lubricants. However, its application in studying the stability of the lubricants has been limited due to the fixed motion between the contacts [4]. In these techniques, true contact area during the test is increasing due to wear and it reduces the true contact pressure. Such a drop in the true contact pressure can facilitate higher load carrying capability of a lubricant. Moreover, the wear debris accumulated at the contact zone can increase or decrease the loading carrying capability of a lubricant [5]. Overall, the true stability of a lubricant is misrepresented due to the decrease in contact pressure and accumulation of wear debris. Therefore, it is important to fix the contact area, avoid intervention of wear debris at the contact zone, and determine the true stability of the lubricants.

The aim of this research is to use a modified ball-on-disk technique to investigate the friction, wear, and true stability of the lubricating films. Two types of contact conditions in the ball-on-disk test, (i) migrating point, that is, relative motion between the ball and disk, and (ii) fixed point and fixed point-variable speed contact, that is, no relative motion between the ball and disk, are used in this study. Synthetic lubricants like the motor oil 5W40 and 5W30 and natural lubricants like the glycerol and glycerol-water mixture were tested using the ball-on-disk. The friction coefficient of the lubricants was measured using the ball-on-disk test, the disk wear was qualitatively and quantitatively determined using a noncontact profilometer. The friction coefficient profile from the ball-on-disk test was used to evaluate the stability of the lubricants.

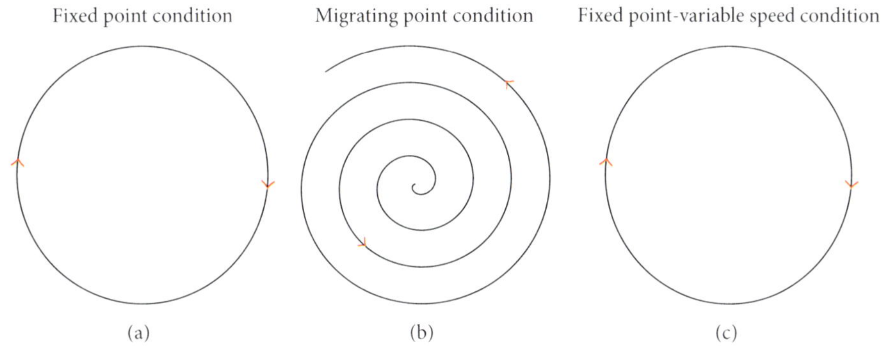

FIGURE 1: Different contact conditions in the ball-on-disk test. (a) fixed point condition and circular geometry, (b) migrating point condition and spiral geometry, and (c) fixed point-variable speed condition and circular geometry.

2. Materials and Methods

2.1. Lubricants. Synthetic lubricants were multigrade motor oils of viscosity grade 5W30 and 5W40 (from the same manufacturer) and the natural lubricants were glycerol (99.9% pure, Sigma Aldrich, Germany) and glycerol-water mixture (glycerol diluted in water to have a final concentration of 50% w/w).

2.2. Friction Coefficient of Lubricants. Friction coefficient (COF) of the lubricants was determined using the ball-on-disk instrument (TR 20, Ducom Instruments Pvt. Ltd., India). Silicon Nitride (Si_3N_4) ceramic balls of 6 mm in diameter and Inconel 617 (IN617) disks of 60 mm in diameter were used in this study. Ball was pressed on the rotating disk at a normal force of 6.5 N. Ball position could be either fixed or reciprocating. The reciprocating motion was achieved through a stepper motor, pulley, and a linear guide arrangement. Servo motor was used to vary the speed of the disk between 1000 rpm and 1 rpm. Disk was enclosed within a chamber that can hold a lubricant of 2 mL in volume. Friction force at the point contact between the ball and disk was measured using a button type load cell. The friction coefficient of a lubricant was determined by dividing the friction force with the respective normal force.

Friction coefficient of the lubricants was determined at different contact conditions. All these contact conditions were software controlled using the labview based Winducom 2010 software (Ducom Instruments Pvt. Ltd., India). The duration of the test was fixed at 60 s and the total distance travelled was fixed at 86 m.

2.3. Fixed Point Contact Condition. The ball position was fixed over the disk at a wear track diameter of 28 mm and the disk was rotating at a fixed speed of 1000 rpm. The geometric shape of this wear track was circular as illustrated in Figure 1(a). In this condition there were multiple passes over a single wear track.

2.4. Migrating Point Contact Condition. The ball was linearly moved away from the center of the disk at a sliding speed of 0.5 mm/s while the disk was rotating at a speed of 1000 rpm. The geometric shape of this wear track was spiral as illustrated in Figure 1(b). In this condition there were no multiple passes over a single wear track.

2.5. Fixed Point-Variable Speed Contact Condition. The ball position was fixed over a disk at a wear track diameter of 28 mm and the disk was rotating at a variable speed. The disk speed decreased exponentially from 1000 rpm to 70 rpm within the testing duration of 60 s. The geometric shape of this wear track was circular as illustrated in Figure 1(c). In this condition there were multiple passes over a single wear track and it was at different sliding speeds.

2.6. Surface Topography and Wear Area. Three-dimensional images of the wear track on the disk surface were captured using a noncontact profilometer (Proscan 2000, Scantron Industrials Ltd., UK). A white light chromatic sensor with a maximum measuring range of 0.3 mm was used in the profilometer. The scan rate and the scan area were fixed at 300 points per second and 2 mm × 2 mm, respectively. The disks after the friction test were cleaned with 70% ethanol and later the wear track was imaged. The depth and width of the wear track on the disk were calculated by using the Proscan software (Scantron Industrials Ltd., UK). The geometry of the wear track was parabolic and its area was calculated using

$$\text{Wear area} = \frac{2}{3}\left(\text{depth} \times \text{width}\right). \quad (1)$$

3. Results

3.1. Friction Coefficient of Synthetic Lubricants. In the fixed point contact condition (Figure 2(a)) the run-in friction coefficient for motor oil 5W30 and 5W40 was 0.381 and 0.260. After the run-in phase, the friction coefficient of the motor oil 5W40 stabilized earlier than the motor oil 5W30. The stable friction coefficient of 5W40 was 0.022 and it was two times lower than motor oil 5W30.

In the migrating point contact condition (Figure 2(b)) there was no run-in phase for the motor oils. For 5W30, at 23rd s of testing the friction coefficient suddenly jumped from 0.073 to 0.383. Similar jump in the friction was recorded for

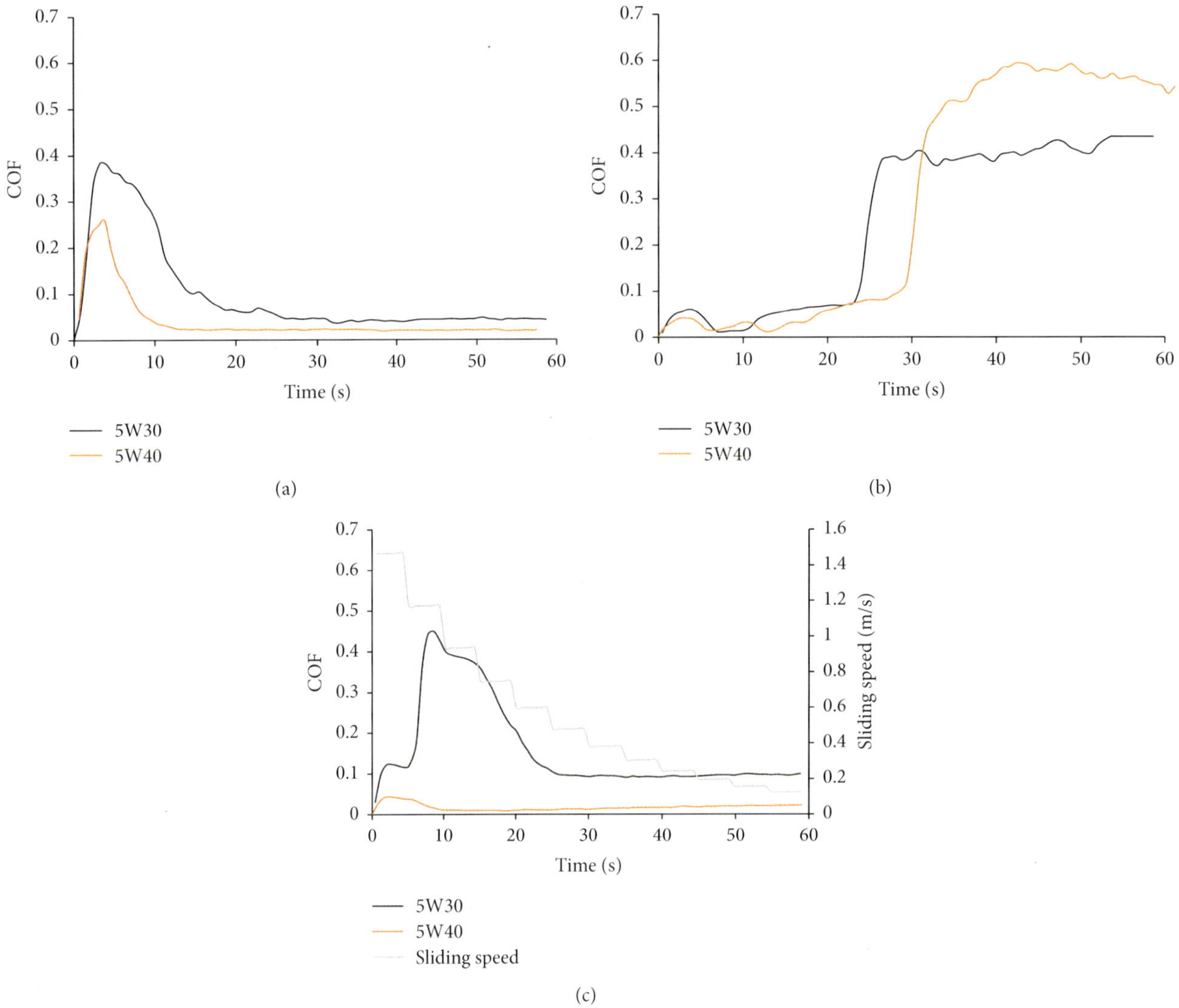

Figure 2: Friction coefficient (COF) of motor oil 5W30 and 5W40. (a) fixed point condition and circular geometry, (b) migrating point condition and spiral geometry, and (c) fixed point- variable speed condition and circular geometry.

5W40 at 30th s and the friction after failure was higher than 5W30.

In the fixed point-variable speed contact condition (Figure 2(c)) the friction coefficient for the motor oil 5W30 decreased from 0.442 to 0.097 when the sliding speed decreased from 1.30 m/s to 0.47 m/s. The friction coefficient for the motor oil 5W40 was always lower than 5W30. Its friction coefficient decreased from 0.098 to 0.025 when the sliding speed decreased from 1.30 m/s to 1.17 m/s.

3.2. Friction Coefficient of Natural Lubricants. In the fixed point contact condition (Figure 3(a)) the run-in friction coefficient for glycerol-water mixture was 0.383. And, there was no run-in phase when the glycerol was used as a lubricant. The stable friction coefficient for glycerol-water mixture and glycerol was 0.133 and 0.033.

In the migrating point contact condition (Figure 3(b)) there was no run-in phase for the glycerol and glycerol-water mixture. For glycerol-water mixture, at 11th s of testing the friction coefficient jumped from 0.019 to 0.540. Similar jump in the friction coefficient for glycerol was recorded at the 32nd s of testing.

In the fixed point-variable speed contact condition (Figure 3(c)) the friction coefficient of the glycerol-water mixture decreased from 0.529 to 0.180 when the sliding speed decreased from 1.30 m/s to 0.42 m/s. The friction coefficient of glycerol was stable at 0.023 when the sliding speed decreased from 1.30 m/s to 0.33 m/s.

3.3. Surface Topography of Wear Tracks and Wear Area. Wear track on the disk after the fixed point contact testing (Figures 4(a) and 4(d)) was smooth. It was only the disk lubricated by the motor oil 5W30 that had a huge pile-up along the sides of the wear track. The worn area on the disk lubricated by the motor oil 5W40 and 5W30 was 1250 μm^2 and 1560 μm^2, respectively.

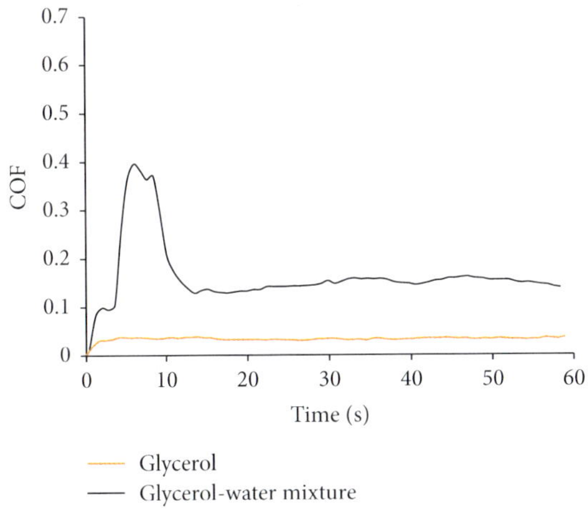

(a)

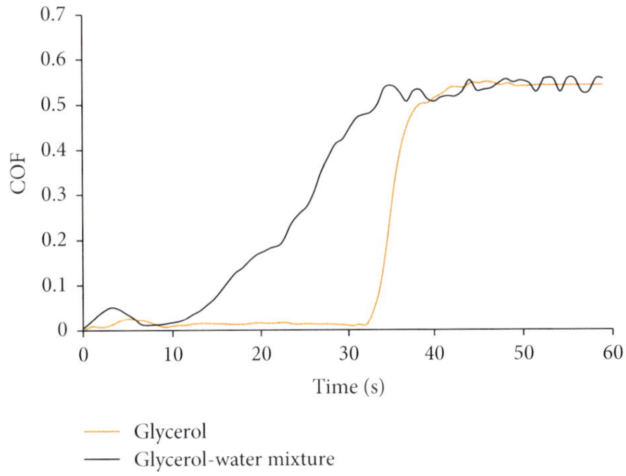

(b)

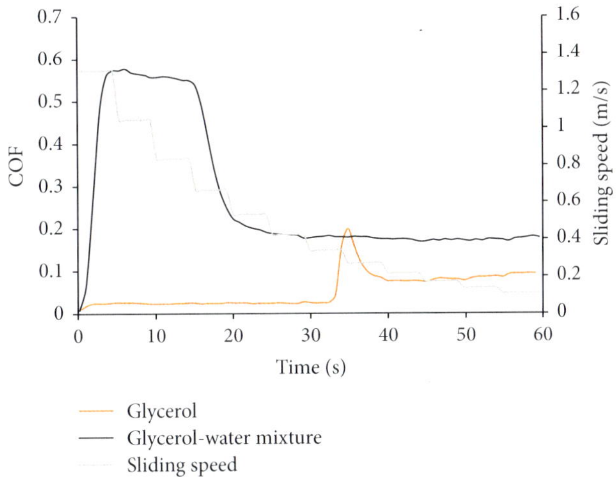

(c)

FIGURE 3: Friction coefficient of glycerol and glycerol-water mixture. (a) fixed point condition and circular geometry, (b) migrating point condition and spiral geometry, and (c) fixed point-variable speed condition and circular geometry.

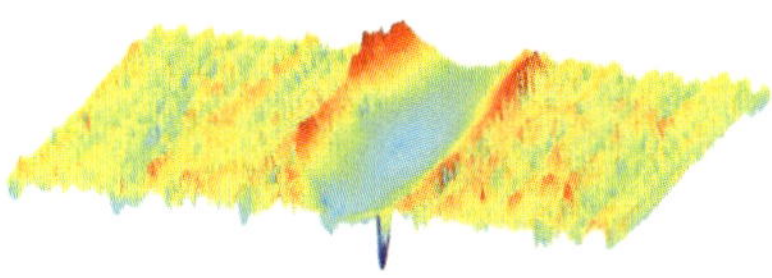

(a) Fixed point; 1564 μm^2 (wear area)

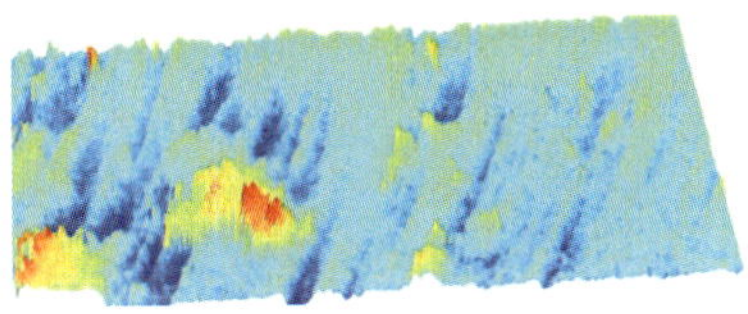

(b) Migrating point; 413 μm^2

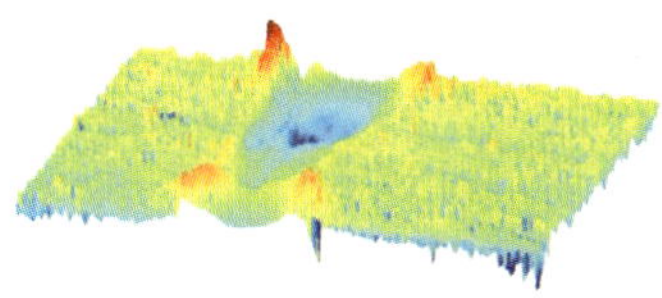

(c) Fixed point-variable speed; 1585 μm^2

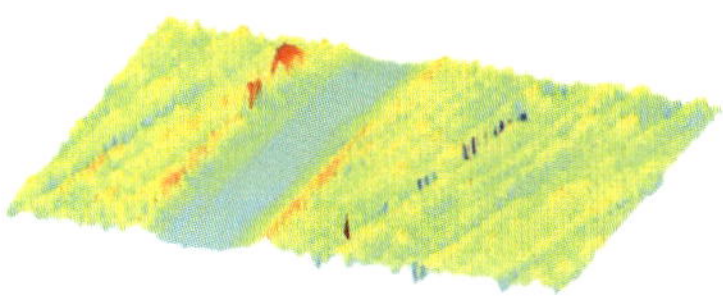

(d) Fixed point; 1250 μm^2 (wear area)

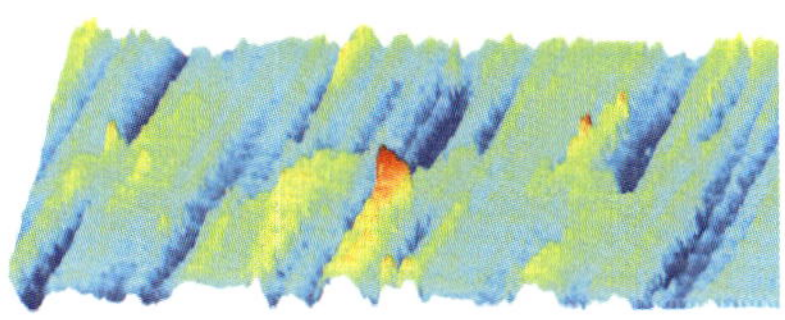

(e) Migrating point; 1866 μm^2

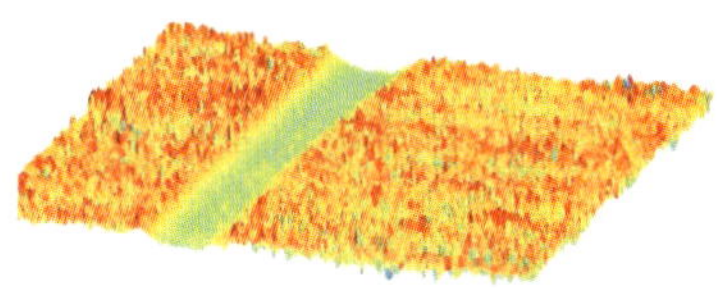

(f) Fixed point-variable speed; 360 μm^2

FIGURE 4: Surface topography images (2 mm × 2 mm) of the wear tracks on the disk lubricated by the motor oil 5W30 or 5W40. (a, d) fixed point condition and circular geometry, (b, e) migrating point condition and spiral geometry, and (c, f) fixed point-variable speed condition and circular geometry.

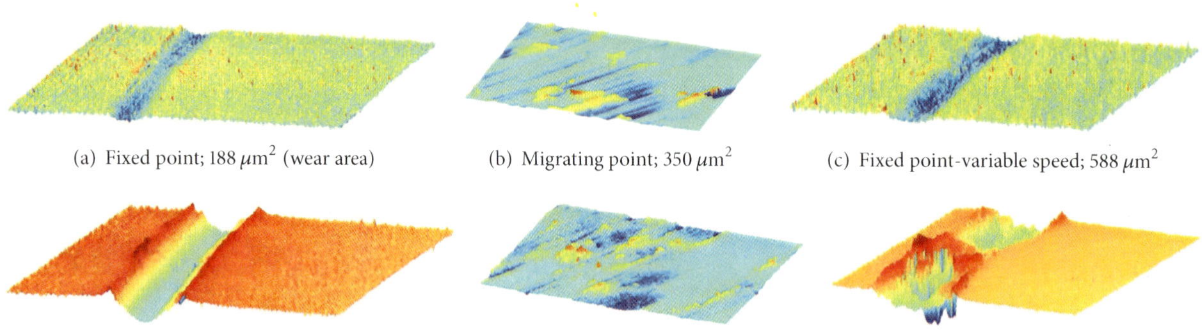

(a) Fixed point; 188 μm^2 (wear area)

(b) Migrating point; 350 μm^2

(c) Fixed point-variable speed; 588 μm^2

(d) Fixed point; 1664 μm^2 (wear area)

(e) Migrating point; 354 μm^2

(f) Fixed point-variable speed; 2188 μm^2

Figure 5: Surface topography of the wear tracks on the disk lubricated by glycerol or glycerol-water mixture. (a, d) fixed point condition and circular geometry, (b, e) migrating point condition and spiral geometry, and (c, f) fixed point-variable speed condition and circular geometry.

There were several discontinuous scratches on the disk after the migrating point contact testing (Figures 4(b) and 4(e)). The area of a scratch on the disk lubricated by the motor oil 5W40 was 4.5 times higher compared to motor oil 5W30.

After the fixed point contact-variable speed testing (Figures 4(c) and 4(f)), the disk lubricated by 5W30 had wider grooves (Figure 4(c)). The wear on the disk lubricated by 5W40 was smooth and its area was 4.4 times lower compared to the motor oil 5W30.

Wear track on the disk lubricated by either glycerol or glycerol-water mixture was smooth after the fixed point contact testing (Figures 5(a) and 5(d)). The worn area due to the glycerol-water mixture was 9 times higher compared to glycerol.

There were several discontinuous, shorter, and deeper scratches after the migrating point contact testing (Figures 5(b) and 5(e)). The area of a scratch on the disk lubricated either by the glycerol or glycerol-water mixture was the same.

After the fixed point-variable speed contact testing the wear track was smooth when lubricated by glycerol whereas it had deeper grooves when lubricated by glycerol-water mixture (Figures 5(c) and 5(f)). The worn area on the disk lubricated by the glycerol-water mixture was 3.8 times higher compared to the glycerol.

4. Discussion

Stability of lubricants is influenced by the wear at the sliding interface. Increase in wear due to plastic deformation will increase the contact area and, theoretically, it decreases the true contact pressure. In such conditions the lubricants tend to show more stability and therefore it is difficult to determine the true stability of a lubricant. Such a situation is expected in the fixed point contact condition where an increase in the friction force is compensated by the decrease in contact pressure. As a result the stable friction profile is observed after the run-in phase. However, the same set of lubricants showed an unstable friction profile when the contact area was fixed (or true contact pressure was fixed) during the test in the migrating point contact condition. The unstable friction profiles include a sudden jump and increase in the friction coefficient. This jump is due to the maximum asperities in contact after the failure of all the lubrication regimes such as hydrodynamic, mixed, and boundary, at the ball and disk interface. Similar profiles are also recorded during the scuffing test where the stability of the lubricants is determined by deliberately increasing the contact pressure until the lubricating films breakdown [6]. In our study, the lubricating films from 5W40 and glycerol failed at a later stage, hence, higher true stability in 5W40 and glycerol compared to 5W30 and glycerol-water mixture. Therefore, we suggest that the true stability of the lubricants can be better evaluated using a migrating point contact condition compared to the fixed point contact condition in the ball-on-disk instrument.

Lubrication efficiency will largely depend on the lubricant viscosity [7], structure, and composition of surface active additives [8]. High viscous lubricants can form a thicker fluid film that separates the contacting bodies; hence, it decreases the friction and the wear in the system [7]. Such a lubrication condition can be related to the low friction and wear behavior of 5W40 and glycerol, in general. However, the mechanism of lubrication after the film failure is complex and it cannot be explained based on the lubricant viscosity. In migrating point contact condition, after the failure of 5W40 lubricant, its friction was higher compared with 5W30. Furthermore, this high friction was translated into more wear despite its higher true stability (Figures 2(b), 4(b), and 4(e)). We hypothesize that the mechanism can be related to the microscopic tribofilms formed after the failure of 5W40 which can be more abrasive compared to 5W30. Note that migrating point contact only minimizes the accumulation of wear debris; however it does not completely block the wear debris accumulating in the contact.

In contrast, the friction profile of glycerol after its failure overlaps with the friction of the glycerol-water mixture. Also, the wear profiles are similar (Figures 3(b), 5(b), and 5(e)). Therefore, we assume that the microscopic tribofilms formed after the failure are similar in nature. However, further investigation on the composition and structure of the tribofilm formed after the failure is necessary to confirm our hypothesis.

Glycerol can be an alternative to the mineral or synthetic base oil used in the lubricants [9]. In our study, we show that glycerol as a lubricant is more wear resistant and has higher true lubricant stability compared to motor oil 5W40 with synthetic base oil. However the high viscosity in glycerol is an issue, as more energy is needed to shear glycerol, increasing the temperature of the system. To overcome this disadvantage, water is added to the glycerol to decrease its viscosity. Shi et al. [9] showed that the viscosity of glycerol-water mixture (50% w/w) was 100 times lower and the wear after glycerol-water lubrication was 12 times higher than pure glycerol. The same concentration of glycerol-water mixture is used in our study and the results confirm that there is more wear and also high friction. Severity of the wear after glycerol-water lubrication is represented by a huge pile-up on the edges of the wear tack (Figure 5(d)). Spalling that represents the fatigue failure was also observed on the wear tracks of the disk lubricated by the glycerol-water mixture during the fixed point-variable speed contact condition (Figure 5(f)). Therefore, the extent of failure by glycerol-water mixture lubrication is catastrophic compared with pure glycerol lubrication.

5. Conclusions

From this tribological investigation of the synthetic and natural lubricants at different point contact conditions, it can be concluded that

(i) true lubricant stability is best studied using the migrating point contact condition in ball-on-disk testing, where there is minimum change in the true contact area and intervention of wear debris;

(ii) low friction coefficient is related to low wear; however there is no relationship between lubricant stability and wear;

(iii) 5W40 has better lubricant stability compared to 5W30; however the tribofilm formed after the failure of 5W40 increases the friction and wear;

(iv) glycerol shows the lowest friction and its film stability is better than the synthetic oils like 5W40 and 5W30;

(v) addition of water to the glycerol increases the friction and wear; also, it reduces the true film stability of pure glycerol.

Conflict of Interests

The authors declare that there is no conflict of interests regarding the publication of this paper.

References

[1] ASTM, "Standard test method for measurement of extreme-pressure properties of lubricating fluids (Four-Ball method)," ASTM D2783-03, 2009.

[2] ASTM, "Standard test method for wear testing with a pin-on-disk apparatus," ASTM G99-05, ASTM International, West Conshohocken, Pa, USA, 2010.

[3] ASTM, "Standard test method for measurement of extreme-pressure properties of lubricating fluids (Timken method)," ASTM D2782-02, ASTM International, West Conshohocken, Pa, USA, 2014.

[4] O. Bridgeman, "The present status of the extreme pressure lubrication problem," SAE Technical Paper 340011, 1934.

[5] J. Enthoven and H. A. Spikes, "Infrared and visual study of the mechanisms of scuffing," *Tribology Transactions*, vol. 39, no. 2, pp. 441–447, 1996.

[6] ASTM, "Standard test method for evaluating the scuffing load capacity of oils," ASTM D5182-97, ASTM International, West Conshohocken, Pa, USA, 2014.

[7] M. Castro and J. Seabra, "Scuffing and lubricant film breakdown in FZG gears part II. New PV scuffing criteria, lubricant and temperature dependent," *Wear*, vol. 215, no. 1-2, pp. 114–122, 1998.

[8] H. Y. Chu, W. C. Hsu, and J. F. Lin, "The anti-scuffing performance of diamond nano-particles as an oil additive," *Wear*, vol. 268, no. 7-8, pp. 960–967, 2010.

[9] Y. Shi, I. Minami, M. Grahn, M. Björling, and R. Larsson, "Boundary and elastohydrodynamic lubrication studies of glycerol aqueous solutions as green lubricants," *Tribology International*, vol. 69, pp. 39–45, 2014.

9

Performance of a Ferrofluid based Rough Parallel Plate Slider Bearing: A Comparison of Three Magnetic Fluid Flow Models

Jimit R. Patel and G. M. Deheri

Department of Mathematics, Sardar Patel University, Vallabh Vidyanagar, Anand, Gujarat 388 120, India

Correspondence should be addressed to Jimit R. Patel; patel.jimitphdmarch2013@gmail.com

Academic Editor: Michael M. Khonsari

Efforts have been made to present a comparison of all the three magnetic fluid flow models (Neuringer-Rosensweig model, Shliomis model, and Jenkins model) so far as the performance of a magnetic fluid based parallel plate rough slider bearing is concerned. The stochastic model of Christensen and Tonder is adopted for the evaluation of effect of transverse surface roughness. The stochastically averaged Reynolds-type equation is solved with suitable boundary conditions to obtain the pressure distribution resulting in the calculation of load carrying capacity. The graphical results establish that for a bearing's long life period the Shliomis model may be employed for higher loads. However, for lower to moderate loads, the Neuringer-Rosensweig model may be deployed.

1. Introduction

The analysis of hydrodynamic lubrication of a nonporous slider bearing is a classical one. The exact solution of Reynolds equation for slider bearings with different simple film geometries is discussed in a number of research papers and books [1–4]. The squeeze film slider bearing is used in clutch plates, automobile transmissions, and domestic appliances. Due to this, many investigators [4–8] dealt with the problem of squeeze film slider bearing.

During the last few years a noticeable amount of interest in tribology research has been devoted to the study of effect of surface roughness on the hydrodynamic lubrication because most of the bearing surfaces become rough after having some run-in and wear.

Bearing surface could be roughened through the manufacturing process, the wear, and the impulsive damage. Surface roughness evaluation is very essential for many fundamental problems such as contact deformation, friction, heat and electric current conduction, tightness of contact joints, and positional accuracy. To study the effect of surface roughness, Christensen and Tonder [9–11] utilized a stochastic concept and developed an averaging film model to lubricated surfaces with transverse roughness. On the basis of Christensen and Tonder's stochastic model of roughness, Ting [12] presented a simple mathematical analog for determination of the squeeze film behavior between two parallel annular disks. Prakash and Tiwari [13] analyzed the effect of surface roughness on the response of a squeeze film between two circular plates when one plate had a porous facing. Guha [14] investigated the concept of a stochastic process which was applied to the problem of surface roughness in the dynamic characteristics of hydrodynamic journal bearings of finite width. Turaga et al. [15] studied a Reynolds-type equation for finite hydrodynamic bearings (plane slider and cylindrical journal) with rough surfaces (longitudinal, transverse, and isotropic roughness). Gururajan and Prakash [16] examined the effect of surface roughness in hydrodynamic narrow porous journal hearings operating under steady conditions. It was shown that the results were significantly different than those for the case of an infinitely long journal bearing. Gadelmawla et al. [17] proposed the mathematical formulation of surface roughness parameter. Sinha and Adamu [18] discussed the thermal and roughness effects on the performance characteristics of an infinite tilted pad slider bearing. It was observed that, for a nonparallel slider bearing, the load carrying capacity due to the combined effect was less than the load capacity due to the roughness effect for both models. Adamu and Sinha [19] analyzed the thermal and roughness effects on different characteristics of an infinitely long tilted pad slider bearing considering heat conduction through both the pad and the slider.

In fact, ferrofluid can solve many difficult sealing, lubricating, detection, heat transfer, and damping problems. Ferrofluid based devices have found applications in high-vacuum equipment, laser systems, computers, inertia dampers, loudspeakers, material separation, domain detection, and many other areas. Among the many applications in engineering are those taking advantage of the possibility of collecting and holding firmly small quantities of such fluids in a region with highly focused magnetic fields.

Neuringer and Rosensweig [20] developed a simple flow model to describe the steady flow of magnetic fluids in the presence of slowly changing external magnetic fields. A good number of papers are available in the literature for the study of different types of bearing using Neuringer and Rosensweig flow model, for example, Tipei [21] in short bearing, Agrawal [22], Shah and Bhat [23], and Deheri and Patel [4] in slider bearing, journal bearing by Nada and Osman [24] and Patel et al. [25], and circular plates by Shah and Bhat [26] and Deheri and Abhangi [27]. Later on, the flow model of Neuringer and Rosensweig was modified by Jenkins [28] with Maugin's modification. It was found that Neuringer-Rosensweig model modified pressure while Jenkins flow model modified both the pressure and the velocity of the ferrofluid. The steady-state performance of bearings with Jenkins model based magnetic fluids was discussed by Agrawal [22], Ram and Verma [29], and Shah and Bhat [30]. Ahmad and Singh [31] analyzed the theoretical model of a magnetic fluid lubrication of porous-pivoted slider bearing with slip velocity. Patel and Deheri [32, 33] dealt with the effect of slip velocity and surface roughness on the performance of Jenkins model based magnetic squeeze film in curved rough circular plates. It was manifested that for any type of improvement in the performance characteristics the slip parameter was required to be reduced even if variance (−ve) occurred and suitable magnetic strength was in force. Also, all the above studies established that the load carrying capacity of the bearing system increased with increasing magnetization. Patel and Deheri [34] investigated the combined effect of slip velocity and surface roughness on the performance of Jenkins model based magnetic squeeze film in curved rough annular plates. It was observed that the effect of transverse surface roughness remained adverse in general; Jenkins model based ferrofluid lubrication provided some measures in mitigating the adverse effect and this became more manifest when the slip parameter was reduced and negatively skewed roughness occurred.

In the last few years, the fact that the magnetic fluid when used as a lubricant turns in an enhanced performance has drawn significant attentions all over the world. On the other hand Shliomis [35] proposed a ferrofluid flow model, in which the effects of rotation of magnetic particles, their magnetic moments, and the volume concentration were included. Kumar et al. [36] analyzed the effect of a ferrofluid squeeze film for spherical and conical bearings with a constant external magnetic field applied in the direction transverse to that of fluid flow. Singh and Gupta [37] theoretically discussed the effect of ferrofluid on the dynamic characteristics of curved slider bearings using Shliomis model. On the ground of the ferrohydrodynamic model proposed by Shliomis [35], Lin [38] discussed the influence of fluid inertia forces on the ferrofluid squeeze film between a sphere and a plate in the presence of external magnetic fields. It was observed from the above studies that the volume concentration and the intensity of magnetic field provided an increase in the load carrying capacity and the time of approach. All these researchers have found the steady-state characteristics of the bearings lubricated with magnetic fluids, resorting to the model proposed by Shliomis [35]. Patel and Deheri [32, 33] discussed the effect of different porous structures on the performance of a Shliomis model based magnetic squeeze film in rotating rough porous curved circular plates. It was shown that the adverse effect of transverse roughness could be compensated by the positive effect of magnetization in the case of negatively skewed roughness, suitably choosing the rotation ratio and the curvature parameters.

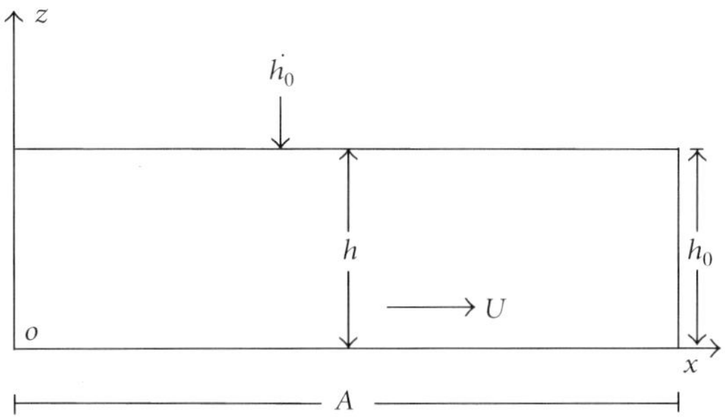

FIGURE 1: Configuration of the bearing system.

Till today there is no investigation regarding the comparison of the three well-known magnetic fluid flow models, so far as bearing performance is concerned. So it was deemed appropriate to launch an investigation into the performance of a ferrofluid based rough parallel plate slider bearing, considering all the three ferrofluid flow models. This adds a new dimension as the roughness also is factored into the discussion.

In fact, the aim of the current study is to analyze the effect of a ferrofluid based rough parallel plate slider bearing considering the comparison of three magnetic fluid flow models, namely, Neuringer-Rosensweig model, Shliomis model, and Jenkins model.

2. Analysis

The physical configuration of the parallel plate slider bearing is presented in Figure 1. The slider moves with a uniform velocity U in the x direction and squeeze velocity $\dot{h}_0$ in the z direction, h_0 being the minimum film thickness. The length is A and B is the breadth of the bearing in the y direction.

The bearing surfaces are considered to be transversely rough. In view of the discussions of Christensen and Tonder [9–11], the thickness h of the lubricant film is taken as

$$h = \overline{h} + h_s, \tag{1}$$

where $\overline{h}$ denotes the mean film thickness and h_s is the deviation from the mean film thickness characterizing the random roughness of the bearing surfaces. h_s is assumed to be stochastic in nature and governed by the probability density function

$$f(h_s) = \begin{cases} \dfrac{35}{32c}\left(1 - \dfrac{h_s^2}{c^2}\right)^3, & -c \le h_s \le c \\ 0, & \text{elsewhere,} \end{cases} \quad (2)$$

where c is the maximum deviation from the mean film thickness. The mean α, the standard deviation σ, and the parameter ε, which is the measure of symmetry of the random variable h_s, are defined by the relations

$$\alpha = E(h_s),$$
$$\sigma^2 = E\left[(h_s - \alpha)^2\right], \quad (3)$$
$$\varepsilon = E\left[(h_s - \alpha)^3\right],$$

where E denotes the expected value given by

$$E(R) = \int_{-c}^{c} Rf(h_s)\,dh_s. \quad (4)$$

The details can be seen from Christensen and Tonder [9–11].

The lubricant film is assumed to be isoviscous and incompressible and the flow is laminar. A magnetic fluid is used as the lubricant in present study. The magnetic field is oblique to the lower plate as considered in Agrawal [22]. Prajapati [39] discussed the effect of various forms of the magnitude of the magnetic field on the squeeze film performance. Following Prajapati [39] and Bhat [40] the magnitude of the magnetic field is considered as

$$M^2 = Kx(A - x), \quad (5)$$

where K is suitably chosen to suit the dimensions of both sides and strength of the magnetic field, the details of which can be found in Bhat and Deheri [7] and Patel and Deheri [41].

In 1964, Neuringer and Rosensweig developed a simple model to describe the steady flow of magnetic fluids in the presence of slowly changing external magnetic fields. The model consists of the following equations:

$$\rho(\overline{q}\nabla)\overline{q} = -\nabla p + \eta\nabla^2\overline{q} + \mu_0(\overline{M}\nabla)\overline{H}, \quad (6)$$

$$\nabla\overline{q} = 0, \quad (7)$$

$$\nabla \times \overline{H} = 0, \quad (8)$$

$$\overline{M} = \overline{\mu}\overline{H}, \quad (9)$$

$$\nabla\left(\overline{H} + \overline{M}\right) = 0, \quad (10)$$

where ρ is the fluid density, $\overline{q}$ represents the fluid velocity in the film region, $\overline{H}$ denotes the external magnetic field, $\overline{\mu}$ is the magnetic susceptibility of the magnetic field, p represents the film pressure, η denotes the fluid viscosity, and μ_0 is the permeability of the free space. The details of all these parameters have been treated in Bhat [40] and Prajapati [39].

Using (8)-(9), (6) becomes

$$\rho(\overline{q}\nabla)\overline{q} = -\nabla\left(p - \frac{\mu_0\overline{\mu}}{2}M^2\right) + \eta\nabla^2\overline{q}. \quad (11)$$

Therefore, the modified Reynolds equation for parallel plate slider bearing for Neuringer and Rosensweig model is obtained as

$$\frac{d}{dx}\left[h^3\frac{d}{dx}\left(p - \frac{\mu_0\overline{\mu}}{2}M^2\right)\right] = 12\eta\dot{h}_0. \quad (12)$$

Shliomis [35] pointed out that magnetic particles of a magnetic fluid could relax in two ways when the applied magnetic field changed. One was by the rotation of magnetic particles in the fluid and the other one by rotation of the magnetic moment within the particles. The particle rotation is given by Brownian relaxation time parameters τ_B while the intrinsic rotational process is described by the relaxation time parameters τ_S. Assuming steady flow and neglecting inertial and second derivatives of $\overline{S}$, the equations governing the flow become

$$-\nabla p + \eta\nabla^2\overline{q} + \mu_0\left(\overline{M}\nabla\right)\overline{H} + \frac{1}{2\tau_S}\nabla \times \left(\overline{S} - I\overline{\Omega}\right) = 0, \quad (13)$$

$$\overline{S} = I\overline{\Omega} + \mu_0\tau_S\left(\overline{M} \times \overline{H}\right), \quad (14)$$

$$\overline{M} = M_0\frac{\overline{H}}{H} + \frac{\tau_B}{I}\left(\overline{S} \times \overline{M}\right), \quad (15)$$

where $\overline{S}$ is the internal angular momentum and I is the sum of moments of inertia of the particles per unit volume:

$$\overline{\Omega} = \frac{1}{2}\nabla \times \overline{q} \quad (16)$$

together with

$$\nabla\overline{q} = 0,$$
$$\nabla \times \overline{H} = 0, \quad (17)$$
$$\nabla\left(\overline{H} + \overline{M}\right) = 0.$$

As discussed in [40], $\overline{q}$ is the fluid viscosity in the film region, $\overline{H}$ is external magnetic field, $\overline{\mu}$ is magnetic susceptibility of the magnetic field, p is the film pressure, η is the fluid viscosity, and μ_0 is the permeability of the free space.

By making use of (14) in (13) and (15), one finds that

$$-\nabla p + \eta\nabla^2\overline{q} + \mu_0\left(\overline{M}\nabla\right)\overline{H} + \frac{1}{2}\mu_0\nabla \times \left(\overline{M} \times \overline{H}\right) = 0, \quad (18)$$

$$\overline{M} = M_0\frac{\overline{H}}{H} + \tau_B\left(\overline{\Omega} \times \overline{M}\right). \quad (19)$$

Neglecting $\tau_B\ \tau_S$ terms, substitution of $\overline{M}$ in the above equation leads to

$$-\nabla p + \left(\eta + \frac{\mu_0}{4}\tau_B\overline{M}\,\overline{H}\right)\nabla^2\overline{q} + \mu_0\left(\overline{M}\nabla\right)\overline{H} + \frac{1}{2} \cdot \mu_0\tau_B\left[\nabla\left(\overline{\Omega}\,\overline{H}\right) \times \overline{M} + \left(\overline{\Omega}\,\overline{H}\right)\nabla \times \overline{M} - \nabla\left(\overline{M}\,\overline{H}\right) \times \overline{\Omega}\right] = 0. \quad (20)$$

From (19), it is easily observed that an initial approximation to $\overline{M}$ is

$$\overline{M} = M_0 \frac{\overline{H}}{H}. \tag{21}$$

Substituting the value of $\overline{M}$ on the right side of (19), a second approximation to $\overline{M}$ is found to be

$$\overline{M} = M_0 \frac{\overline{H}}{H} + \frac{M_0}{H} \tau_B \left(\overline{\Omega} \times \overline{H}\right). \tag{22}$$

Again, substituting this value of $\overline{M}$ on the right side of (19), a third approximation to $\overline{M}$ is availed as

$$\overline{M}\,\overline{H} = M_0 H + \frac{M_0}{H} \tau_B^2 \left\{\left(\overline{\Omega}\,\overline{H}\right)^2 - \Omega^2 H^2\right\}. \tag{23}$$

Making use of this model, Bhat [40] and Patel and Deheri [32, 33], the modified Reynolds-type equation for Shliomis model for a one-dimensional flow as in a slider bearing with the slider moving with a uniform velocity U in the x direction and for impermeable slider bearing the governing equation is

$$\begin{aligned} \frac{d}{dx}\left(h^3 \frac{dp}{dx}\right) &= 12\eta_a \dot{h}_0 + 6\eta_a U \frac{dh}{dx} \\ &\quad - \frac{3N\tau_B^3}{16\eta_a} U^2 \frac{d}{dx}\left(h \frac{dp}{dx}\right) \\ &\quad - \frac{3N\tau_B^3}{320\eta_a^3} \frac{d}{dx}\left(h^5 \left(\frac{dp}{dx}\right)^3\right). \end{aligned} \tag{24}$$

The details are explained in Bhat [40].

In [28], Jenkins proposed a model to describe the flow of a ferrofluid. In view of Maugin's modification, equations of the model for steady flow are [28, 29]

$$\begin{aligned} \rho\left(\overline{q} \cdot \nabla\right)\overline{q} &= -\nabla p + \eta \nabla^2 \overline{q} + \mu_0 \left(\overline{M} \cdot \nabla\right)\overline{H} + \frac{\rho J^2}{2} \nabla \\ &\quad \times \left[\frac{\overline{M}}{M} \times \left\{\left(\nabla \times \overline{q}\right) \times \overline{M}\right\}\right] \end{aligned} \tag{25}$$

together with (7)–(10), where J is a material constant. From (6) and (25) it is observed that Jenkins model is a generalization of Neuringer-Rosensweig model with an additional term

$$\begin{aligned} &\frac{\rho J^2}{2} \nabla \times \left[\frac{\overline{M}}{M} \times \left\{\left(\nabla \times \overline{q}\right) \times \overline{M}\right\}\right] \\ &\quad = \frac{\rho J^2 \overline{\mu}}{2} \nabla \times \left[\frac{\overline{H}}{H} \times \left\{\left(\nabla \times \overline{q}\right) \times \overline{H}\right\}\right] \end{aligned} \tag{26}$$

which modifies the velocity of the fluid.

In view of Bhat [40] and Patel and Deheri [32, 33], the modified Reynolds equation for Jenkins model of a one-dimensional flow, in Cartesian coordinates, turns out to be

$$\begin{aligned} &\frac{d}{dx}\left(\frac{h^3}{\left(1 - \rho J^2 \overline{\mu} H / 2\eta\right)} \frac{d}{dx}\left(p - \frac{\mu_0 \overline{\mu}}{2} M^2\right)\right) \\ &\quad = 6\eta U \frac{dh}{dx} + 12\eta \dot{h}_0. \end{aligned} \tag{27}$$

Under the usual assumptions of hydrodynamic lubrication [8, 39, 40] and the stochastic modelling of Christensen and Tonder [9–11] the modified Reynolds' equation governing the pressure distribution takes the form for Neuringer-Rosensweig model, Shliomis model, and Jenkins model, respectively, as

$$\frac{d}{dx}\left[g(h) \frac{d}{dx}\left(p - \frac{\mu_0 \overline{\mu}}{2} M^2\right)\right] = 12\eta \dot{h}_0, \tag{28}$$

$$\begin{aligned} &\frac{d}{dx}\left(g(h) \frac{dp}{dx}\right) \\ &\quad = 12\eta_a \dot{h}_0 + 6\eta_a U \frac{d}{dx}\left(g(h)^{1/3}\right) \\ &\qquad - \frac{3N\tau_B^3}{16\eta_a} U^2 \frac{d}{dx}\left(g(h)^{1/3} \frac{dp}{dx}\right) \\ &\qquad - \frac{3N\tau_B^3}{320\eta_a^3} \frac{d}{dx}\left(g(h)^{5/3} \left(\frac{dp}{dx}\right)^3\right), \end{aligned} \tag{29}$$

$$\begin{aligned} &\frac{d}{dx}\left(\frac{g(h)}{\left(1 - \rho J^2 \overline{\mu} H / 2\eta\right)} \frac{d}{dx}\left(p - \frac{\mu_0 \overline{\mu}}{2} M^2\right)\right) \\ &\quad = 6\eta U \frac{d}{dx}\left(g(h)^{1/3}\right) + 12\eta \dot{h}_0, \end{aligned} \tag{30}$$

where

$$g(h) = h^3 + 3h^2\alpha + 3\left(\sigma^2 + \alpha^2\right)h + 3\sigma^2\alpha + \alpha^3 + \varepsilon. \tag{31}$$

The concerned boundary conditions are

$$p(0) = p(A) = 0. \tag{32}$$

Introducing the nondimensional quantities,

$$\overline{h} = \frac{h}{h_0} = 1,$$

$$X = \frac{x}{A},$$

$$P = \frac{h_0^2 p}{\eta U A},$$

$$\beta_1 = -\frac{U h_0}{2\dot{h}_0 A},$$

$$N = \mu_0 M_0 H_0,$$

$$\mu^* = \frac{K \mu_0 \overline{\mu} h_0^2 A}{\eta U},$$

$$\tau = \frac{N \tau_B}{4\eta},$$

$$\eta_a = \eta(1+\tau),$$
$$\delta = \frac{3N\tau_B^3U^2}{16\eta(1+\tau)h_0^2},$$
$$\overline{J}^2 = \frac{\rho J^2\overline{\mu}\sqrt{K}A}{2\eta},$$
$$\overline{\sigma} = \frac{\sigma}{h_0},$$
$$\overline{\alpha} = \frac{\alpha}{h_0},$$
$$\overline{\varepsilon} = \frac{\varepsilon}{h_0^3} \quad (33)$$

and for Shliomis model

$$P = -\frac{h_0^3 p}{\eta A^2 \dot{h}_0} \quad (34)$$

and using (25), (19)–(23) reduce to

$$\frac{d}{dX}\left[g(\overline{h})\frac{d}{dX}\left(P - \frac{\mu^*}{2}X(1-X)\right)\right] = -\frac{6}{\beta_1},$$
$$\frac{d}{dX}\left(g(\overline{h})\frac{dP}{dX}\right) = -12(1+\tau) - \delta\left(g(\overline{h})^{1/3}\frac{d^2P}{dX^2}\right) - \frac{\delta}{80(1+\tau)^2\beta_1^2}\frac{d}{dX}\left\{g(\overline{h})^{5/3}\left(\frac{dP}{dX}\right)^3\right\},$$
$$\frac{d}{dX}\left(\frac{g(\overline{h})}{\left(1-\overline{J}^2\sqrt{X(1-X)}\right)} \cdot \frac{d}{dX}\left(P - \frac{\mu^*}{2}X(1-X)\right)\right) = 6\frac{d}{dX}\left(g(\overline{h})^{1/3}\right) - \frac{6}{\beta_1}, \quad (35)$$

where

$$g(\overline{h}) = \overline{h}^3 + 3\overline{h}^2\alpha + 3\left(\overline{\sigma}^2 + \overline{\alpha}^2\right)\overline{h} + 3\overline{\sigma}^2\overline{\alpha} + \overline{\alpha}^3 + \overline{\varepsilon}. \quad (36)$$

Solving (35) under the boundary conditions (32), one can find the dimensionless pressure for Neuringer-Rosensweig model, Shliomis model, and Jenkins model, respectively:

$$P = \frac{\mu^*}{2}X(1-X) + \frac{3}{g(\overline{h})\beta_1}\left(X - X^2\right), \quad (37)$$

$$P = \frac{(1+\tau)}{g(\overline{h}) + \delta g(\overline{h})^{1/3}}\left[6\left(X-X^2\right) + \frac{27\delta g(\overline{h})^{5/3}}{80\beta_1^2\left\{g(\overline{h}) + \delta g(\overline{h})^{1/3}\right\}^3}\left\{(2X-1)^4 - 1\right\}\right], \quad (38)$$

$$P = \frac{\mu^*}{2}X(1-X) + \frac{X(1-X)}{g(\overline{h})\beta_1}\left[3 - 2\overline{J}^2\sqrt{X(1-X)}\right]. \quad (39)$$

The dimensionless load carrying capacity takes the form for all three cases, respectively, as

$$W = \frac{h_0^2 w}{\eta U A^2 B} = \frac{\mu^*}{12} + \frac{1}{2g(\overline{h})\beta_1}, \quad (40)$$

$$W = -\frac{h_0^3 w}{\eta A^3 \dot{h}_0 B} = \frac{(1+\tau)}{g(\overline{h}) + \delta g(\overline{h})^{1/3}}\left[1 - \frac{27\delta g(\overline{h})^{5/3}}{100\beta_1^2\left\{g(\overline{h}) + \delta g(\overline{h})^{1/3}\right\}^3}\right], \quad (41)$$

$$W = \frac{\mu^*}{12} + \frac{32 - 3\pi\overline{J}^2}{64g(\overline{h})\beta_1}. \quad (42)$$

3. Results and Discussions

It is noticed that the nondimensional pressure distribution is given by (37)–(39) while (40)–(42) determine the nondimensional load carrying capacity. It is well known that the magnetization increases the viscosity of the lubricant which results in increased pressure and, consequently, the load carrying capacity. As (40)–(42) are linear with respect to magnetization parameter, an increase in the magnetization would lead to increased load carrying capacity.

The effect of magnetization parameter on the load carrying capacity is displayed in Table 1 and Figures 2–6. It is seen that the magnetization parameter causes increased load carrying capacity.

The following short version has been taken for the three models: NR stands for Neuringer-Rosensweig model, Sh stands for Shliomis model, and the symbol for Jenkins model is Je.

The influence of standard deviation on load carrying capacity is presented in Figures 7–10. As such, the transverse surface roughness affects the bearing performance adversely.

TABLE 1: Variation of load carrying capacity with respect to μ^*/τ and $\overline{\sigma}$.

$\mu^*/\tau \rightarrow$	0.1	0.2	0.3	0.4	0.5	$\overline{\sigma} \downarrow$
NR	1.246478	1.254811	1.263145	1.271478	1.279811	
Sh	1.078984	1.177073	1.275163	1.373252	1.471342	0.01
Je	1.242833	1.251167	1.2595	1.267833	1.276167	
NR	1.245169	1.253502	1.261835	1.270169	1.278502	
Sh	1.078539	1.176587	1.274636	1.372685	1.470734	0.02
Je	1.241528	1.249861	1.258194	1.266528	1.274861	
NR	1.242993	1.251326	1.259659	1.267993	1.276326	
Sh	1.077797	1.175779	1.273761	1.371742	1.469724	0.03
Je	1.239358	1.247691	1.256025	1.264358	1.272691	
NR	1.239959	1.248292	1.256626	1.264959	1.273292	
Sh	1.076763	1.17465	1.272538	1.370425	1.468313	0.04
Je	1.236333	1.244667	1.253	1.261333	1.269667	
NR	1.23608	1.244414	1.252747	1.26108	1.269414	
Sh	1.075437	1.173204	1.270971	1.368738	1.466506	0.05
Je	1.232466	1.2408	1.249133	1.257466	1.2658	

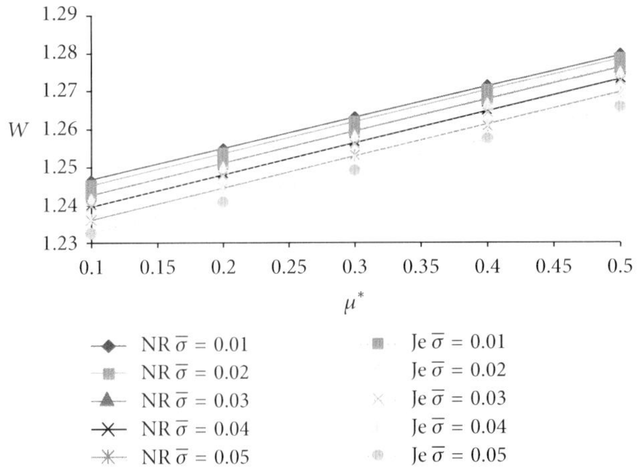

FIGURE 2: Variation of load carrying capacity with respect to μ^* and $\overline{\sigma}$.

Probably this may be due to the reason that the motion of the lubricant gets retarded by roughness. However, the standard deviation fails to reduce the load significantly.

In Figures 11-12, the effect of skewness on load carrying capacity is presented. The positively skewed roughness decreases the load carrying capacity while the nondimensional load increases due to the negatively skewed roughness. It is interesting to note that variance also follows the path of skewness so far as the trends of load carrying capacity are concerned. Thus, the negatively skewed roughness-variance (−ve) combination induces a positive effect.

In order to boost the bearing performance this paper suggests that the combined positive effect of negatively skewed roughness and variance (−ve) may be channelized for bearing design.

A comparison has been given below regarding the performance characteristics of a magnetic fluid based parallel

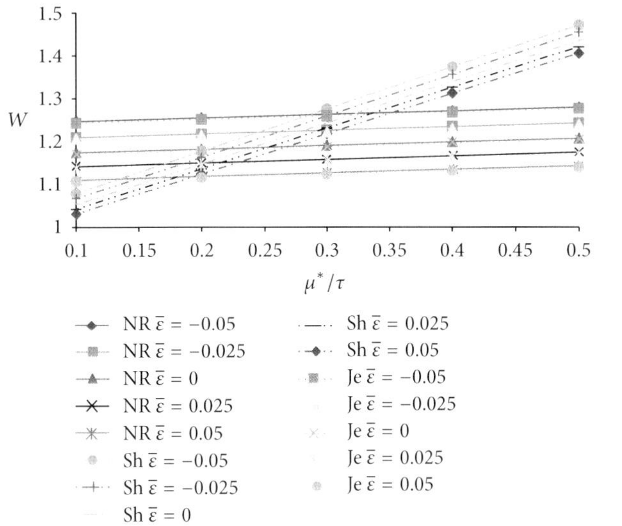

FIGURE 3: Variation of load carrying capacity with respect to μ^*/τ and $\overline{\varepsilon}$.

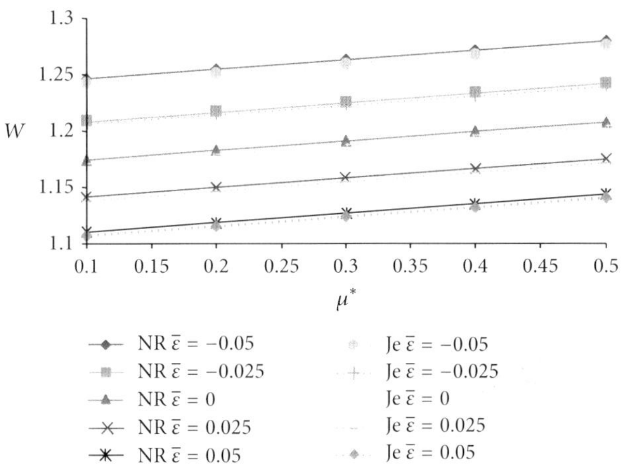

FIGURE 4: Variation of load carrying capacity with respect to μ^* and $\overline{\varepsilon}$.

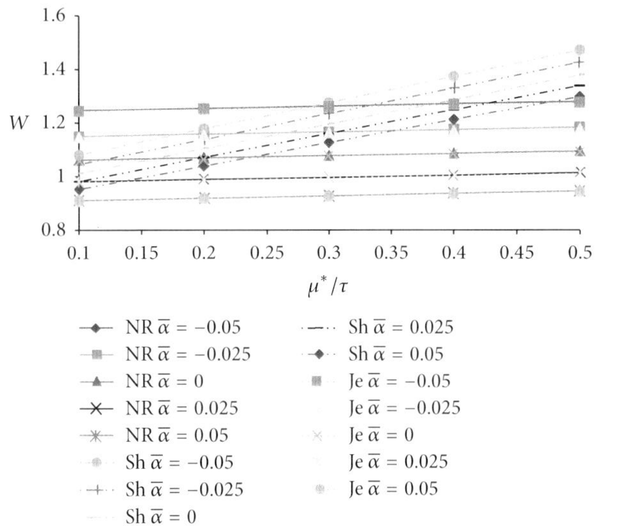

FIGURE 5: Variation of load carrying capacity with respect to μ^*/τ and $\overline{\alpha}$.

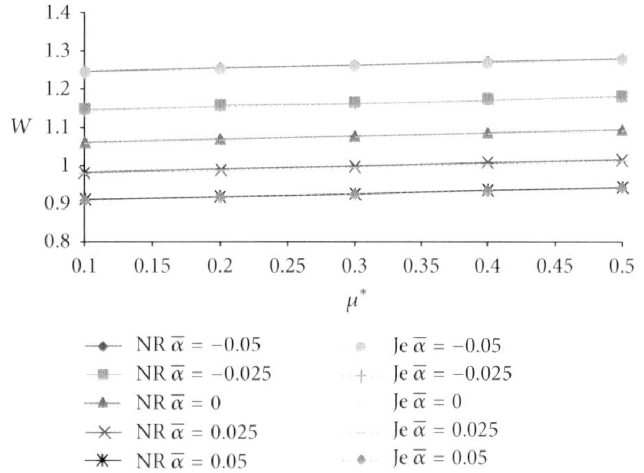

FIGURE 6: Variation of load carrying capacity with respect to μ^* and $\overline{\alpha}$.

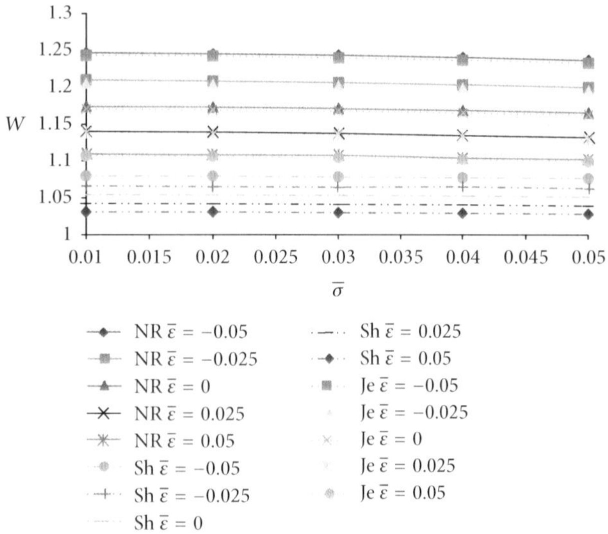

FIGURE 7: Variation of load carrying capacity with respect to $\overline{\sigma}$ and $\overline{\varepsilon}$.

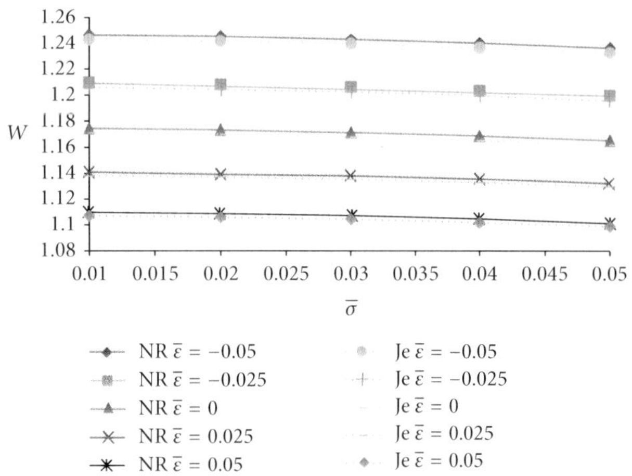

FIGURE 8: Variation of load carrying capacity with respect to $\overline{\sigma}$ and $\overline{\varepsilon}$.

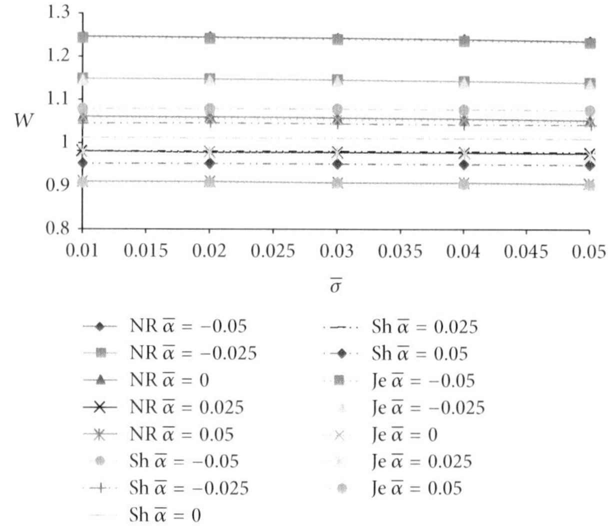

FIGURE 9: Variation of load carrying capacity with respect to $\overline{\sigma}$ and $\overline{\alpha}$.

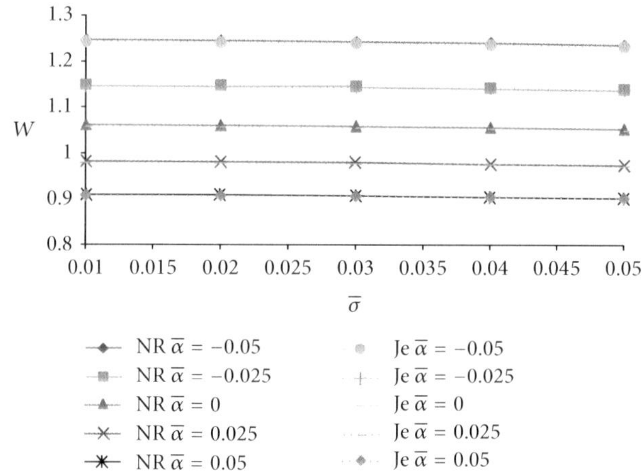

FIGURE 10: Variation of load carrying capacity with respect to $\overline{\sigma}$ and $\overline{\alpha}$.

plate slider bearing. The three models, namely, Neuringer-Rosensweig model, Shliomis model, and Jenkins model, of magnetic fluid flow are considered for the comparison. A comparison of the graphical representation yields the following:

(1) The magnetic fluid lubrication improves the bearing performance in case of all the three models because the magnetization induces an increase in the viscosity of the lubricant.

(2) However, the Shliomis model remains more effective as compared to the other two models so far as the effect of skewness is concerned. Further, the Neuringer-Rosensweig model and Jenkins model provide the same type of trends when skewness is taken into account.

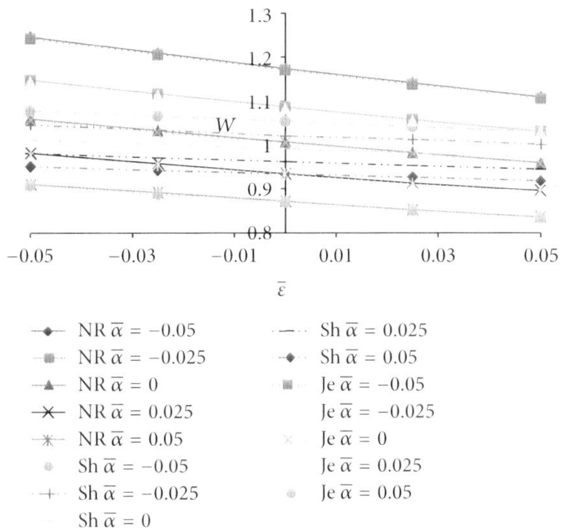

FIGURE 11: Variation of load carrying capacity with respect to $\bar{\varepsilon}$ and $\bar{\alpha}$.

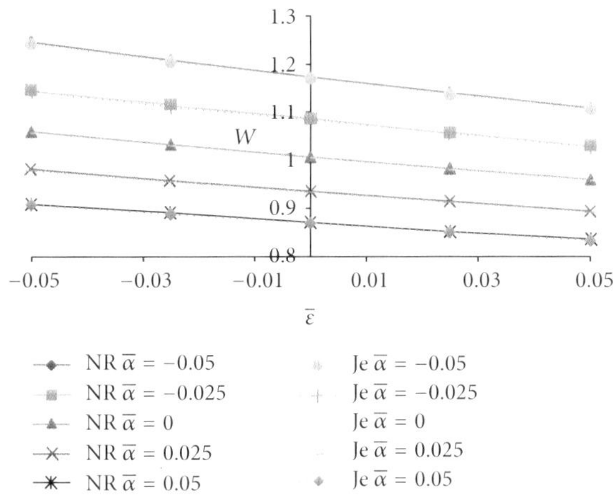

FIGURE 12: Variation of load carrying capacity with respect to $\bar{\varepsilon}$ and $\bar{\alpha}$.

(3) With respect to the effect of variance on the bearing performance once again the Shliomis model goes ahead of the other two models.

(4) Up to some extent the effect of standard deviation remains predominant in Shliomis model as compared to the other two models.

(5) In comparison with the other two models, the standard deviation causes severe load reduction in Jenkins model.

(6) The effect of negatively skewed roughness (which increases the load carrying capacity) remains more effective in Neuringer-Rosensweig model. The same is the case with respect to variance (−ve).

A close observation of the analysis presented here indicates that the load carrying capacity is enhanced by at least 7-8% as compared to the conventional fluid based rough slider bearing system.

4. Conclusion

This paper confirms that when the bearing has run for a longer time, the Neuringer-Rosensweig model may be preferred in order to counter the effect of roughness. However, the Shliomis model may be considered for higher loads, as compared to the other two models for this type of bearing system. The effects of Neuringer-Rosensweig model and Jenkins model are almost alike for nominal roughness. In addition, this type of bearing system supports a good amount of load even in the absence of flow, unlike the case of a traditional lubricant based bearing system. This investigation strongly indicates that the roughness aspect must be considered while designing the bearing system even if Shliomis model is in place.

Competing Interests

The authors declare that they have no competing interests.

References

[1] A. Cameron, *Principles of Lubrication*, Longmans, London, UK, 1966.

[2] S. K. Basu, S. N. Sengupta, and B. B. Ahuja, *Fundamentals of Tribology*, Prentice-Hall, New Delhi, India, 2005.

[3] B. C. Majumdar, *Introduction to Tribology of Bearings*, S. Chand and Company Limited, New Delhi, India, 2008.

[4] G. M. Deheri and J. R. Patel, "Magnetic fluid based squeeze film in a rough porous parallel plate slider bearing," *Annals of Faculty Engineering Hunedoara: International Journal of Engineering*, vol. 9, no. 3, pp. 443–448, 2011.

[5] J. Prakash and S. K. Vij, "Hydrodynamic lubrication of porous slider," *Journal of Mechanical Engineering Science*, vol. 15, no. 3, pp. 232–234, 1973.

[6] M. V. Bhat, "Hydrodynamic lubrication of porous composite slider bearings," *Japanese Journal of Applied Physics*, vol. 17, no. 3, p. 479, 1978.

[7] M. V. Bhat and G. M. Deheri, "Porous slider bearing with squeeze film formed by a magnetic fluid," *Pure and Applied Mathematika Sciences*, vol. 39, no. 1-2, pp. 39–43, 1995.

[8] G. M. Deheri, P. I. Andharia, and R. M. Patel, "Transversely rough slider bearings with squeeze film formed by a magnetic fluid," *International Journal of Applied Mechanics and Engineering*, vol. 10, no. 1, pp. 53–76, 2005.

[9] H. Christensen and K. C. Tonder, "Tribology of rough surfaces: stochastic models of hydrodynamic lubrication," Tech. Rep. 10/69-18, SINTEF, 1969.

[10] H. Christensen and K. C. Tonder, "Tribology of rough surfaces: parametric study and comparison of lubrication models, SINTEF," Report 22/69-18, 1969.

[11] H. Christensen and K. Tonder, "The hydrodynamic lubrication of rough bearing surfaces of finite width," *Journal of Lubrication Technology*, vol. 93, no. 3, pp. 324–329, 1971.

[12] L. L. Ting, "A Mathematical analog for determination of porous annular disk squeeze film behavior including the fluid inertia effect," *Transactions of the ASME. Journal of Basic Engineering*, vol. 94, no. 2, pp. 417–421, 1972.

[13] J. Prakash and K. Tiwari, "Roughness effects in porous circular squeeze-plates with arbitrary wall thickness," *Journal of Lubrication Technology*, vol. 105, no. 1, pp. 90–95, 1983.

[14] S. K. Guha, "Analysis of dynamic characteristics of hydrodynamic journal bearings with isotropic roughness effects," *Wear*, vol. 167, no. 2, pp. 173–179, 1993.

[15] R. Turaga, A. S. Sekhar, and B. C. Majumdar, "Stochastic FEM analysis of finite hydrodynamic bearings with rough surfaces," *Tribology Transactions*, vol. 40, no. 4, pp. 605–612, 1997.

[16] K. Gururajan and J. Prakash, "Effect of surface roughness in a narrow porous journal bearing," *Journal of Tribology*, vol. 122, no. 2, pp. 472–475, 2000.

[17] E. S. Gadelmawla, M. M. Koura, T. M. A. Maksoud, I. M. Elewa, and H. H. Soliman, "Roughness parameters," *Journal of Materials Processing Technology*, vol. 123, no. 1, pp. 133–145, 2002.

[18] P. Sinha and G. Adamu, "THD analysis for slider bearing with roughness: special reference to load generation in parallel sliders," *Acta Mechanica*, vol. 207, no. 1-2, pp. 11–27, 2009.

[19] G. Adamu and P. Sinha, "Thermal and roughness effects in a tilted pad slider bearing considering heat conduction through the pad and slider," *Proceedings of the National Academy of Sciences A: Physical Sciences*, vol. 82, no. 4, pp. 323–333, 2012.

[20] J. L. Neuringer and R. E. Rosensweig, "Magnetic fluids," *The Physics of Fluids*, vol. 7, no. 12, pp. 1927–1937, 1964.

[21] N. Tipei, "Theory of lubrication with ferrofluids: application to short bearings," *Journal of Lubrication Technology*, vol. 104, no. 4, pp. 510–515, 1982.

[22] V. K. Agrawal, "Magnetic-fluid-based porous inclined slider bearing," *Wear*, vol. 107, no. 2, pp. 133–139, 1986.

[23] R. C. Shah and M. V. Bhat, "Magnetic fluid based porous inclined slider bearing with velocity slip," *International Journal of Applied Mechanics and Engineering*, vol. 18, no. 2, pp. 331–336, 2003.

[24] G. S. Nada and T. A. Osman, "Static performance of finite hydrodynamic journal bearings lubricated by magnetic fluids with couple stresses," *Tribology Letters*, vol. 27, no. 3, pp. 261–268, 2007.

[25] N. S. Patel, D. P. Vakharia, and G. M. Deheri, "A study on the performance of a magnetic-fluid-based hydrodynamic short journal bearing," *ISRN Mechanical Engineering*, vol. 2012, Article ID 603460, 7 pages, 2012.

[26] R. C. Shah and M. V. Bhat, "Squeeze film based on magnetic fluid in curved porous rotating circular plates," *Journal of Magnetism and Magnetic Materials*, vol. 208, no. 1, pp. 115–119, 2000.

[27] G. M. Deheri and N. D. Abhangi, "Numerical modelling of a magnetic fluid-based squeeze film between rotating transversely rough curved circular plates," *International Journal of Computational Materials Science and Surface Engineering*, vol. 4, no. 3, pp. 185–204, 2011.

[28] J. T. Jenkins, "A theory of magnetic fluids," *Archive for Rational Mechanics and Analysis*, vol. 46, pp. 42–60, 1972.

[29] P. Ram and P. D. S. Verma, "Ferrofluid lubrication in porous inclined slider bearing," *Indian Journal of Pure and Applied Mathematics*, vol. 30, no. 12, pp. 1273–1281, 1999.

[30] R. C. Shah and M. V. Bhat, "Ferrofluid lubrication in porous inclined slider bearing with velocity slip," *International Journal of Mechanical Sciences*, vol. 44, no. 12, pp. 2495–2502, 2002.

[31] N. Ahmad and J. P. Singh, "Magnetic fluid lubrication of porous pivoted slider bearing with slip velocity," *Proceedings of the Institution of Mechanical Engineers Part J: Journal of Engineering Tribology*, vol. 221, no. 5, pp. 609–613, 2007.

[32] J. R. Patel and G. Deheri, "Combined effect of surface roughness and slip velocity on jenkins model based magnetic squeeze film in curved rough circular plates," *International Journal of Computational Mathematics*, vol. 2014, Article ID 367618, 9 pages, 2014.

[33] J. R. Patel and G. Deheri, "Shliomis model-based magnetic squeeze film in rotating rough curved circular plates: a comparison of two different porous structures," *International Journal of Computational Materials Science and Surface Engineering*, vol. 6, no. 1, pp. 29–49, 2014.

[34] J. R. Patel and G. M. Deheri, "Jenkins model based ferrofluid lubrication of a curved rough annular squeeze film with slip velocity," *Tribology in Industry*, vol. 37, no. 2, pp. 129–141, 2015.

[35] M. I. Shliomis, "Effective viscosity of magnetic suspensions," *Soviet Physics—Journal of Experimental and Theoretical Physics*, vol. 34, no. 6, pp. 1291–1294, 1972.

[36] D. Kumar, P. Sinha, and P. Chandra, "Ferrofluid squeeze film for spherical and conical bearings," *International Journal of Engineering Science*, vol. 30, no. 5, pp. 645–656, 1992.

[37] U. P. Singh and R. S. Gupta, "Dynamic performance characteristics of a curved slider bearing operating with ferrofluids," *Advances in Tribology*, vol. 2012, Article ID 278723, 6 pages, 2012.

[38] J.-R. Lin, "Fluid inertia effects in ferrofluid squeeze film between a sphere and a plate," *Applied Mathematical Modelling*, vol. 37, no. 7, pp. 5528–5535, 2013.

[39] B. L. Prajapati, *On certain theoretical studies in hydrodynamic and electro-magneto hydrodynamic lubrication [Ph.D. thesis]*, S.P. University, Vallabh Vidhyanagar, Undia, 1995.

[40] M. V. Bhat, *Lubrication with a Magnetic Fluid*, Team Spirit India, Ahmedabad, India, 2003.

[41] J. R. Patel and G. M. Deheri, "A comparison of porous structures on the performance of a magnetic fluid based rough short bearing," *Tribology in Industry*, vol. 35, no. 3, pp. 177–189, 2013.

10

A Novel Approach to Reduction of Frictional Losses in a Heavy-Duty Diesel Engine by Reducing the Hydrodynamic Frictional Losses

Staffan Johansson,[1] Per H. Nilsson,[2] Robert Ohlsson,[3] and Bengt-Göran Rosén[4]

[1] *Vehicle Engineering, Volvo Group Trucks Technology, 405 08 Gothenburg, Sweden*
[2] *Advanced Technology & Research, Volvo Group Trucks Technology, 412 88 Gothenburg, Sweden*
[3] *Powertrain Engineering, Volvo Group Trucks Technology, 405 08 Gothenburg, Sweden*
[4] *SET, Halmstad University, P.O. Box 823, 301 18 Halmstad, Sweden*

Correspondence should be addressed to Staffan Johansson; pusj@hotmail.com

Academic Editor: Yuansheng Jin

An important parameter in the reduction of fuel consumption of heavy-duty diesel engines is the Power Cylinder Unit (PCU); the PCU is the single largest contributor to engine frictional losses. Much attention, from both academia and industry, has been paid to reducing the frictional losses of the PCU in the boundary and mixed lubrication regime. However, previous studies have shown that a large portion of frictional losses in the PCU occur in the hydrodynamic lubrication regime. A novel texturing design with large types of surface features was experimentally analyzed using a tribometer setup. The experimental result shows a significant reduction of friction loss for the textured surfaces. Additionally, the textured surface did not exhibit wear. On the contrary, it was shown that the textured surfaces exhibited a smaller amount of abrasive scratches on the plateaus (compared to the reference plateau honed surface) due to entrapment of wear particles within the textures. The decrease in hydrodynamic friction for the textured surfaces relates to the relative increase of oil film thickness within the textures. A tentative example is given which describes a method of decreasing hydrodynamic frictional losses in the full-scale application.

1. Introduction

Tremendous demands are put on the combustion engines of today. A significant parameter, perhaps the most significant parameter within engine development, is the reduction of fuel consumption and CO_2 emission. Reducing frictional losses in the combustion engine means a reduction in CO_2 emission. The largest contributor to frictional losses in the engine is the Power Cylinder Unit (PCU); around 50% of the total frictional losses in the engine can be attributed to the PCU [1–4].

Much attention, from both academia and industry, has been paid to reducing the frictional losses in the boundary and mixed lubrication regime [5–8]; however, previous studies [9, 10] have shown that a large portion of the frictional losses in the engine occur in the hydrodynamic lubrication regime. There are several known techniques for reducing the hydrodynamic losses; however, although a reduction can be effected, there are also disadvantages:

(i) Viscosity of the engine oil is the dominating parameter in the hydrodynamic lubrication regime [11]. Reducing the viscosity of the engine oil decreases the hydrodynamic losses but could increase the frictional losses in the boundary and mixed lubrication regime [12], possibly causing an increase in wear.

(ii) Previous analysis has shown that, in the contact between the oil control ring and the cylinder liner, the main part of the surface significant for controlling friction, both hydrodynamic and mechanical [10], is the plateau part of the cylinder liner surface. The majority of the hydrodynamic friction losses occur at the midstroke region, because the highest piston velocity is achieved at this position [13]. An increase in the plateau roughness in the midstroke region would decrease hydrodynamic friction; however, the ever present smoothening of plateaus during running could decrease the initial effect of rougher plateaus; it is also possible that mechanical friction would increase with increased plateau roughness.

The aim of this study was to decrease the hydrodynamic losses of the PCU (Power Cylinder Unit) by means of surface texturing.

There are relatively few experimental studies analyzing the reduction of hydrodynamic friction losses for sliding line contacts that do not involve properties related to oil viscosity. It is possible to decrease the hydrodynamic frictional losses by applying surface texturing; a few examples of this are as follows:

(i) Qiu and Khonsari [14] analyzed the frictional behavior of elliptical and circular dimples in the hydrodynamic lubrication regime using an experimental tribometer setup. This experimental setup was meant to mimic the behavior of a laser textured piston ring. The result showed large frictional improvements with surface texturing; however, since the piston rings in the engine pass through all lubrication regimes, applying texturing on the piston rings might not prove ideal for a HD diesel engine. Also, the measured friction coefficient, up to 3 or even 4, does not mimic the friction in the engine. If the friction coefficient would prove to be of the suggested magnitude, much larger differences would be expected at midstroke compared to reversal zones [15].

(ii) Costa and Hutchings [16] analyzed the frictional behavior of several different types of surface textures in reciprocating tribometer experiments. Here, it was determined that surface textures with larger valley width than the axial Hertzian contact length gave smaller film thicknesses compared to surfaces without textures. However, only small frictional differences were measured between different texture types. In the experiments, a reciprocating frequency of 0,55 Hz was used, and the stroke length was 22 mm. This gives an average sliding speed of approximately 0,024 m/s which could be regarded as too small for an accurate representation of PCU components.

(iii) Podgornik et al. [17] completed an experimental study in which the frictional effects of different types and sizes of textures in full film lubrication were examined in tribometer experiments. The results from this study suggest that friction can be reduced in the hydrodynamic lubrication regime by the use of textured surfaces.

Previous analysis has shown that the main part of the surface significant for controlling friction is the plateau part [9, 10, 18, 19] and that the valley part of the surface plays no significant role. The result from the literature survey however suggests a possibility of using textures also for the control of the viscous friction losses and a hypothesis tested in this study was formulated based on the assumption that a texture with a valley geometry larger both laterally and vertically than previously reported traditional cylinder liner concepts would constitute a texture able to decrease the hydrodynamic friction losses.

The design parameters of the texturing were based on the following prerequisites:

(i) *Large Sized Features.* Since no correlation between surface parameters describing the valley and friction was found in previous studies [9, 10, 18], it was determined to analyze textures with a much larger geometry, in width, height, and depth, compared to the previously analyzed surface features (in comparison with the width, height, and depth of honing grooves; a schematic view of these geometries can be seen in Figures 3–5 in a previous study [10]).

(ii) *Closed Circular or Elliptical Textures.* The texturing would have to consist of elements that would extend from the plateau surface into the material, as closed void volumes. No extruding parts originating from the texture which could accelerate wear were desired. Closed void volumes as opposed to open volumes similar to traditional honing grooves were preferable to minimize the total oil transport by minimizing the oil transport within the textures [20], thus minimizing a possible increase in oil consumption caused by "open void channel" oil transport.

(iii) *A Significant Texture Fraction of the Total Area.* The area covered by the textures should be significant to obtain and test a significant difference in texture area coverage compared to the reference surface. Texturing area coverage of 30–35% was initially selected for this study. It is assumed that a significantly larger cover ratio would increase the effect of boundary lubrication and a significantly smaller cover ratio would not make the effect more difficult to detect.

The frictional properties of textured surfaces were evaluated using a reciprocating tribometer. Using a tribometer presents the opportunity to conduct isolated analysis of engine components and is a common procedure for analysis of frictional effects and the durability of PCU components [21, 22].

2. Experimental Methods

2.1. Manufacturing of Textured Samples. A five-axis digitally automated computer numerical control milling machine (CNC) was used to machine the texturing pattern directly on the cylinder liner samples. Samples were cut out from a production cylinder liner; the axial length of the samples was 50 mm and the tangential width was 10 mm. The milling operation in which a flat ended tool was used gave a sharp angle between texture and plateau surface. Two different protrusion depths were machined: 20 μm and 100 μm (termed T20 and T100 further on in the article); both textures had the same elliptical shape with the minor axis of 2 mm and the major axis of 3 mm.

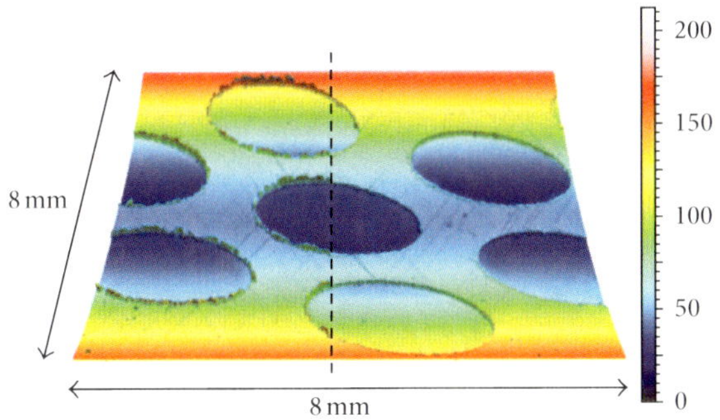

FIGURE 1: Illustration of the removal of the textures edge defects. The dotted line represents the reversal zone of the oil control ring; to the right of this line, the textures were removed in an initial experimental step.

TABLE 1: Overview of input parameters of test cycle steps.

Test cycle step	Time (min)	Temperature (°C)	Reciprocating frequency (Hz)	Load (N)
1	60	33	14	22
2	30	23	10	12
3	30	23	18	12
4	30	43	10	12
5	30	43	18	12
6	30	33	14	22
7	30	23	10	32
8	30	23	18	32
9	30	43	10	32
10	30	43	18	32
11	30	33	14	22

A previous study [23] has shown that it is possible to decrease hydrodynamic frictional losses using a textured pattern with a surface lay which is parallel to the sliding motion. However, using a surface with a surface lay which is parallel to the sliding direction will decrease the generation of hydrodynamic pressure and decrease the oil film thickness between the piston ring and the cylinder liner [24], and thus the orientation/lay of the textures of 20 degrees in reference to the direction of motion was selected in this analysis.

The milling operation causes sharp edges or "burr" at the boundary of each texture element. Since this defect causes additional wear particles, it was decided to remove the sharp edges before the experiments. By running each sample for five minutes using the experimental input parameters of the center point of the DoE setup, the burrs were effectively removed; this is exemplified in Figure 1. This running in stage was carried out using sample material (oil control rings and engine oil) that was not used in further experimentation. The running-in stage is performed on all samples, both textured and untextured.

2.2. Tribometer. A tribometer test setup was used to quantify the frictional properties of reference and textured surfaces; a schematic overview of the tribometer is shown in Figure 2. In the tribometer experiment, oil was continuously fed from the piston ring sample holder to the inner diameter of the oil control ring and into the gap between the two beams in the oil control ring. The oil was supplied using a peristaltic pump; 4,8 mL/min was continuously fed to the contact between the piston ring and the cylinder liner by injecting oil from the piston ring sample holder to the inner diameter at the position between the two beams of the piston ring. The continuous supply of oil with the said amount ensured fully flooded contact at all DoE cycle steps. The oil used was fully formulated 20W50 engine oil. The stroke length in the tribometer was set to 30 mm.

In the tribometer, the reference cylinder liner surface, REF, and two different textured surfaces, T20 and T100, were evaluated. The opposing surface was a coil spring loaded two-piece oil control ring with two beams and standard beam width; to ensure conformability, a complete piston ring was used in the experiments. The tribometer experiment was repeated four times for each surface. The input signals in the experiment were reciprocating frequency, temperature, and load; these signals were varied according to a Design of Experiment (DoE) setup (see Table 1) with high and low levels of all three input parameters. To verify the stability of the experiment over time, three parameter center points were added, as starting point, after half of the experimental points and at the end of the DoE setup. The measured output parameters were friction force and resistivity coefficient. Friction force was measured using piezoelectric transducers; resistivity coefficient was measured by quantifying the difference of the electrical contact resistivity between a resistor (with fixed level) and the electrical contact resistivity of the contact between the piston ring and the cylinder liner. The series resistor value was chosen to give the best discrimination of resistive coefficient comparing the value at the reversal and the value at midstroke for the reference material. In this study, the electrical resistance was measured using a voltage divider with the fixed resistor in series with the contact resistance. The resistive coefficient is not the electrical resistance but is actually the voltage drop over the tribocontact and is consequently inversely proportional to the electrical resistance. Electrical resistance is proportional to the area of contact [25], and the resistive coefficient is thus inversely proportional to the amount of contact. Comprehensive details of the experimental setup and quantification of input and output signals can be found in [9, 10].

2.3. 3D Profilometry. Surfaces were measured using CCP (Chromatic Confocal Probe) [26]. The complete surface of the cylinder liner sample, 48 mm $*$ 10 mm, was measured using point spacing of 10 μm, and the surface was measured before and after the experiment. To calculate geometric properties of surface features, it is common to remove the form prior to feature computation. Commonly, second-order polynomial form removal is sufficient for a cylinder liner surface; however, it was noticed that, for the textured surfaces, this type of form removal generated an undesired artificial

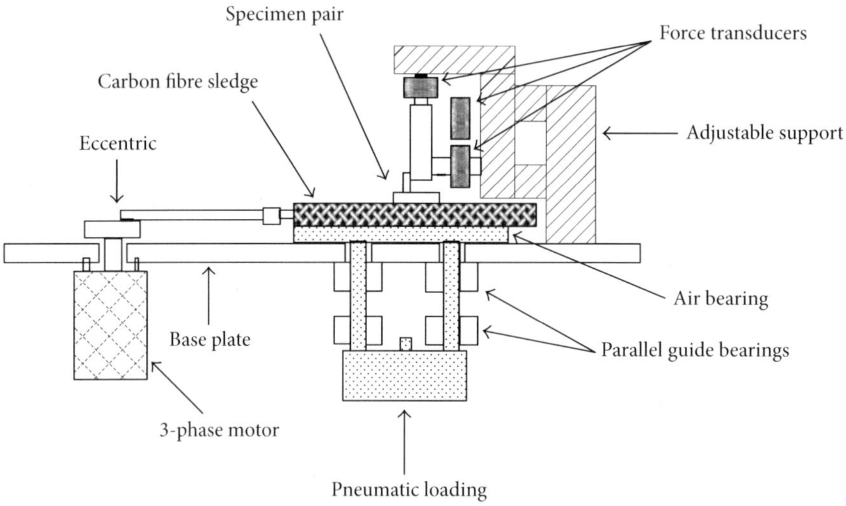

FIGURE 2: Schematic overview of the reciprocating eccentric tribometer.

form. To obtain a representative value of the dimensions of the textures, the surface form for the textured surfaces was removed with the following operations:

(1) Substitution of missing points by defined smooth shape (evaluation software: Mountains Map ver. 5.1, product of Digital Surf, Besançon, France).

(2) Second-order polynomial form removal from the original surface measurement.

(3) Edge detection technique (grain analysis modulus (evaluation software: Mountains Map ver. 5.1, product of Digital Surf, Besançon, France)) to define edges between the textures and the plateau surface.

(4) Extraction of grains; only grains belonging to the texture elements were selected.

(5) Masking of the texturing elements using the output of grain analysis. The texturing is thus removed from the surface (the datum of the texturing elements was replaced with missing points).

(6) Second polynomial form removal on the plateau surface (textures were removed using grain analysis in a previous step); the output from this step is the 2D form.

(7) Subtraction of the surface form generated in 6 with the surface obtained in 2.

Using the computational steps above, a surface without artificial form effects caused by the textures could be obtained; the geometry of each texturing element could thus be analyzed explicitly.

In order to quantify the wear depth, the surface measured before the experiment was subtracted from the surface measured after the experiment using the unworn part of the cylinder liner sample as reference for the absolute height datum.

3. Results

3.1. 3D Profilometry: Evaluation of Wear and Texture Geometry. The geometry of the texturing elements was evaluated using the grain analysis modulus in Mountains Map (see Figure 3). In the comparison between materials T20 and T100, it was found that the only difference in respect to texture geometry was obviously the depth of the textures; no other significant differences could be detected in the density of textures, the average maximum and minimum diameter, the average area density of the textures, the average texture orientation, and the average texture perimeter (for more information, see the Appendix). The angle at the boundary of the texture was difficult to quantify since the measurement technique is not capable of measuring sharp corners; a too high angle returns a missing point. After filling of the missing points, the angle was manually evaluated for several positions; the majority of the angular measurements were between 83° and 85°.

No wear was detected in the evaluation of wear depth (subtraction of surfaces before and after the experiment). As an additional analysis of wear, the surfaces were analyzed under a light optical microscope after the experiment. In this analysis, abrasive scratches were detected on the plateau part of the reference surface; however, virtually, no abrasive scratches were detected on the plateau part of the textures surfaces (see Figure 4). In closer inspection of the texture elements, it was seen that the texture itself contained significant amounts of wear particles (Figure 5).

3.2. Tribometer: Evaluation of the Stability of Input Signals. To gain representative values for each surface, the validity of the input signals was quantified; the said input quantified signal can be seen in Figure 6. To gain better representation of the input signals, these are recalculated; sliding speed is calculated from reciprocating frequency; oil dynamic viscosity is calculated from temperature; and contact pressure is calculated from load according to a previous study [18].

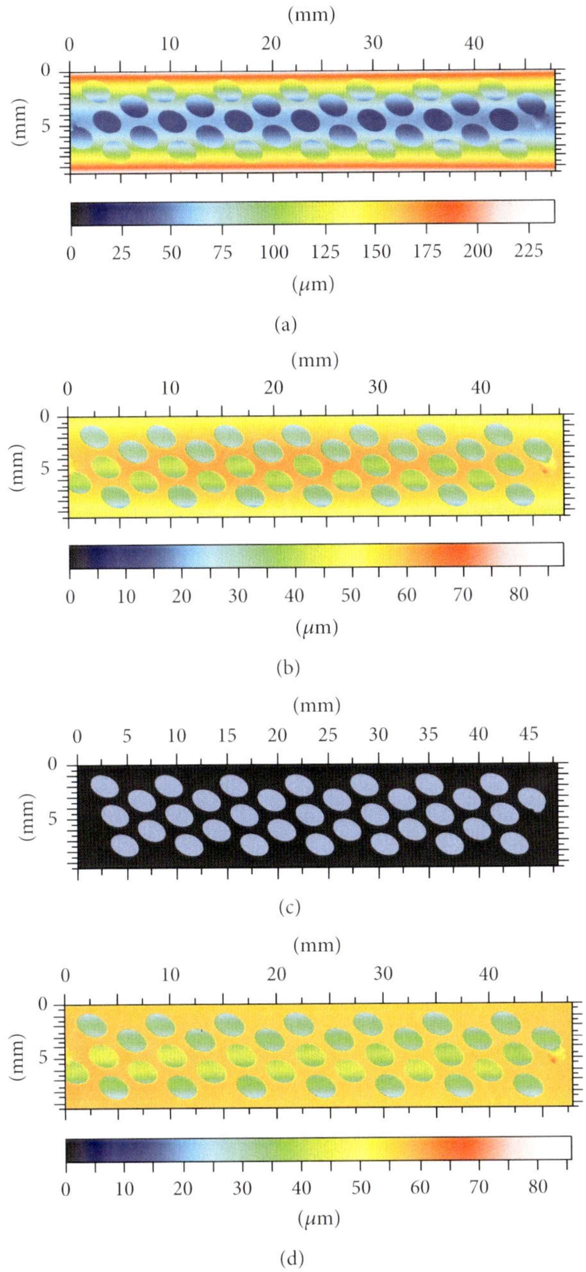

FIGURE 3: Example of form removal on a textured surface by removal of the significant texture prior to form removal and the consecutive adding of segmented features. (a) Original surface measurement. (b) Surface form removal by second-order polynomial extraction; it should here be noticed that this surface does not have a flat plateau area; much of the surface form remains after the first form removal step. (c) Grain analysis of textures in which texture elements are extracted. (d) Surface after second-order polynomial form removal based on grain analysis. It should here be noticed that the plateau part of the final surface does not have any residual form.

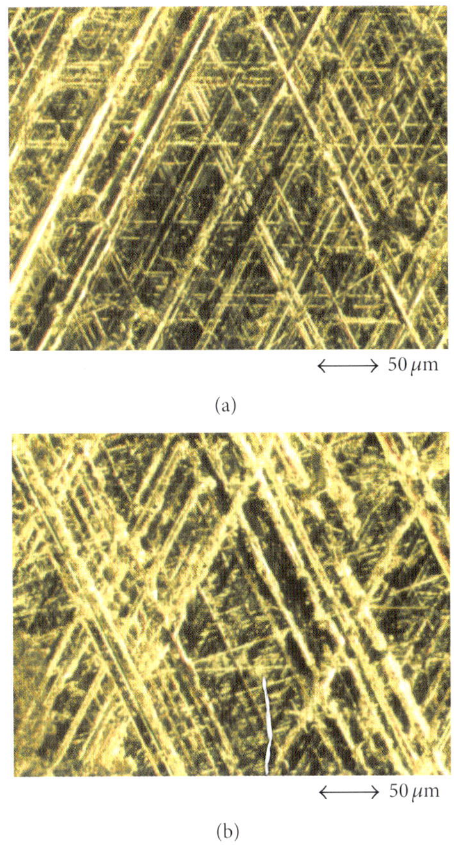

FIGURE 4: Microscope images after the tribometer experiment. (a) Reference material, example of axial abrasive scratches on plateaus. (b) T100: no axial scratches can be detected.

Performing the recalculation of input parameters also gives parameters which are independent of the test arrangement since sliding speed is dependent on stroke length, contact pressure is dependent on relative area of contact, and so forth. From the analysis of the input signals, it was detected that, for one of the experiments of T20 (T20-2), the dynamic viscosity is different from the other signals and for one of the experiments of T100 (T100-4) the contact pressure is different from the other measured signals; these two experiments are therefore removed in further evaluation.

3.3. Tribometer: Evaluation of Friction Coefficient and Resistive Coefficient. The experiments T20-2 and T100-4 were removed from this study; it should be noted that both of these experiments exhibited smaller values of friction coefficient compared to the average value for each surface texture. Figure 7 shows the measured friction coefficient for all experiments; in Figure 8, the average values of each surface are shown. Table 2 shows the standard deviation of the average friction coefficient as shown in Figure 8.

The resistive coefficient was measured in the tribometer experiment; Figure 9 shows the resistive coefficient for all experiments and in Figure 10 the average values of each surface are shown. Table 2 shows the standard deviation of the average resistive coefficient as shown in Figure 10.

3.4. Tribometer: Evaluation of Lubrication Regime and Experimental Output. For an illustrative analysis of different lubrication regimes, the cycle steps were plotted for each input

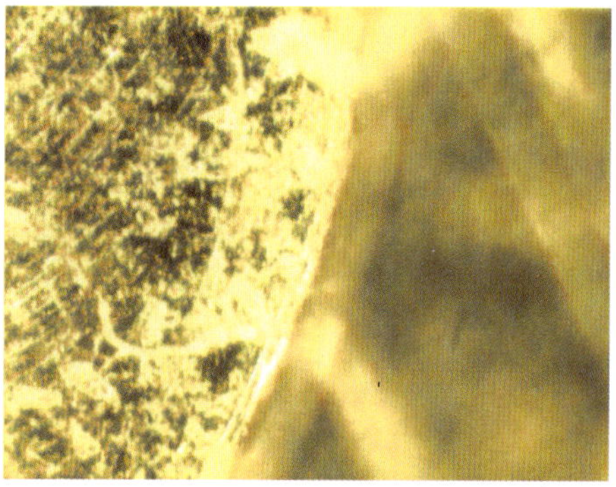

(a)

(b)

FIGURE 5: Microscope images after the experiment, T100. (a) Focus on the bottom part of the texture, wear particles trapped. (b) Focus on the plateau; this image shows the expected wear on the texturing boundary; however, no significant wear is detected on the neighbouring plateau.

TABLE 2: Average values of standard deviation of friction coefficient and resistive coefficient for experiments. T20 and T100 represent the values of standard deviation for the reduced set of experiments. T20∗ and T100∗ represent the values of standard deviation for all experiments, without reduction in the number of experiments.

	REF	T20	T20∗	T100	T100∗
Standard deviation of friction coefficient	0.0169	0.0043	0.0042	0.0109	0.0111
Standard deviation of resistive coefficient	0.5880	0.4190	0.3556	0.1321	0.1176

cycle step. What signifies the hydrodynamic lubrication regime is that friction increases for an increase in speed, an increase in oil viscosity, and a decrease in contact pressure. Each cycle step in the DoE setup is 30 minutes; to minimize the effect of transitions (e.g., thermal) between cycle steps, the values of friction coefficient are only calculated for the duration of 10–29 minutes within each cycle point. Each point in this statistical analysis represents the mean of all experiments for each surface.

On analyzing the lubrication regime transitions for the reference surface, it was shown that

(i) a shift towards the hydrodynamic regime is present for both increase in dynamic viscosity (as can be seen in Figure 11) and decrease in contact pressure (as can be seen in Figure 13);

(ii) a shift towards the hydrodynamic lubrication regime is present for low values of temperature (high level of viscosity), and a shift towards the boundary lubrication regime is present for high values of temperature (low level of viscosity).

On analyzing the lubrication regime transitions for the textured surfaces, it was shown that

(i) a shift towards the hydrodynamic regime is present for a decrease in contact pressure (as for the reference surface) (as can be seen in Figure 13);

(ii) a shift towards the hydrodynamic regime is present for an increase in dynamic viscosity for all cycle steps except for the cycle step with high level of load and low level of reciprocating frequency (similar to the reference surface, as can be seen in Figure 11);

(iii) a shift towards the hydrodynamic regime is present for an increase in sliding speed at low level of load and high level of reciprocating frequency (as can be seen in Figure 12); a shift towards the boundary lubrication regime is present for high values of temperature (as for the reference surface). For high level of load and low level of temperature, T20 and T100 show slightly different results where an increase in sliding speed shows a shift towards the boundary lubrication regime for T20 and a shift towards the hydrodynamic lubrication regime for T100.

The following conclusions can be drawn for the analysis of the frictional measurement and of the transitions of lubrication regime for different experimental points:

(i) The highest measured friction was found in experimental step number 3 which is a combination of high level of sliding speed, high level of dynamic viscosity, and low level of contact pressure; thus, the friction is the highest for contact with the most hydrodynamic lubrication condition.

(ii) The contact pressure at the investigated levels has the most significant impact on friction.

(iii) In general, the textured surfaces have the same frictional behavior as the reference surface; some

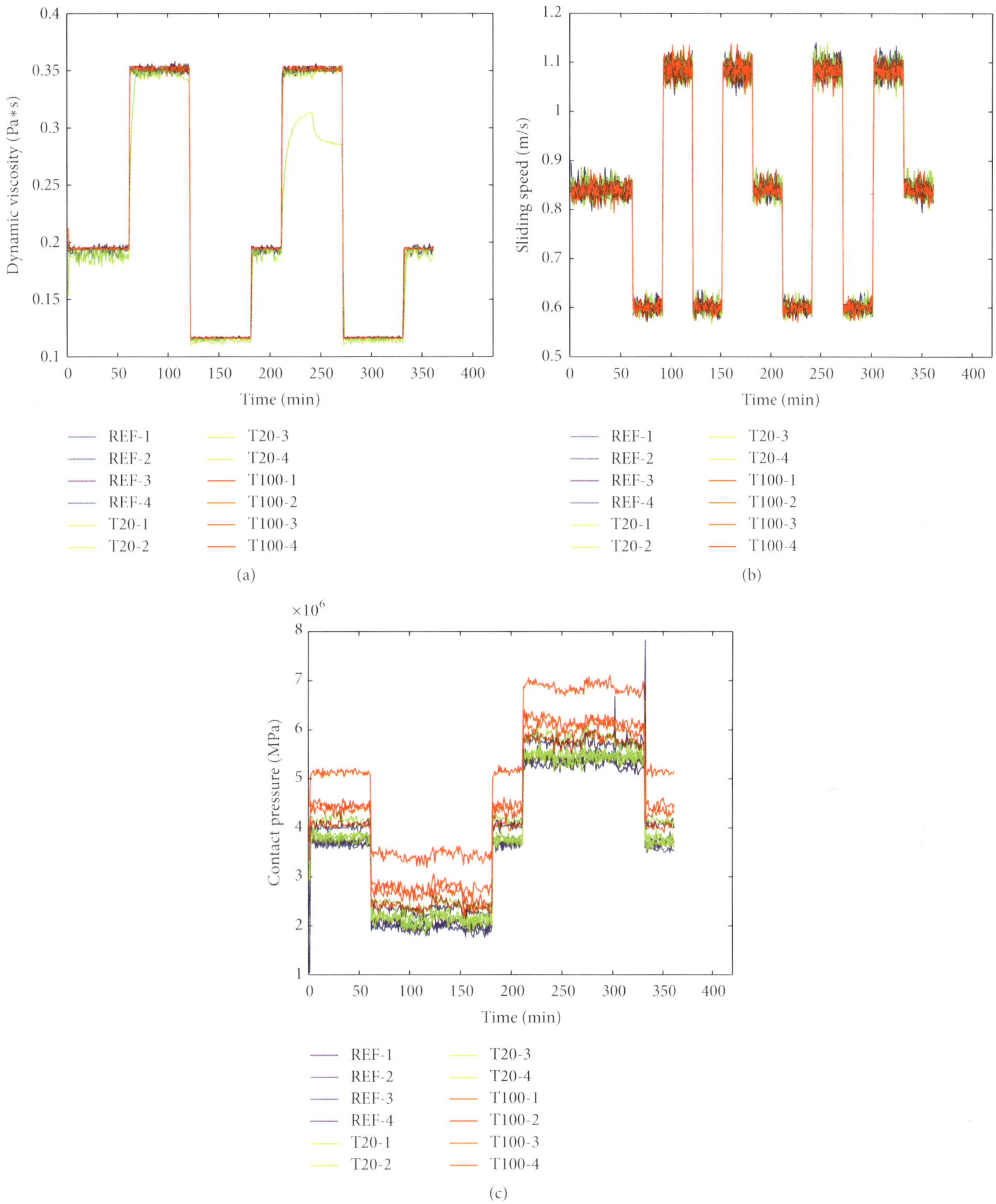

FIGURE 6: Measurement of input signals in experiments. (a) Dynamic viscosity. (b) Sliding speed. (c) Contact pressure.

differences are present for textured surfaces with shift towards the boundary lubrication regime; however, at low contact pressure and high viscosity, the friction increases with increased sliding speed for all investigated surfaces; thus, a shift towards the hydrodynamic lubrication regime is present for this contact condition for all surfaces, textured or untextured.

4. Discussion

4.1. Behavior of Oil Film Thickness and Textures. What causes the friction to decrease in sliding contact with textures even though a higher level of contact was measured for the textured surfaces? To visualize the interaction between two opposing surfaces in sliding motion in which one of these

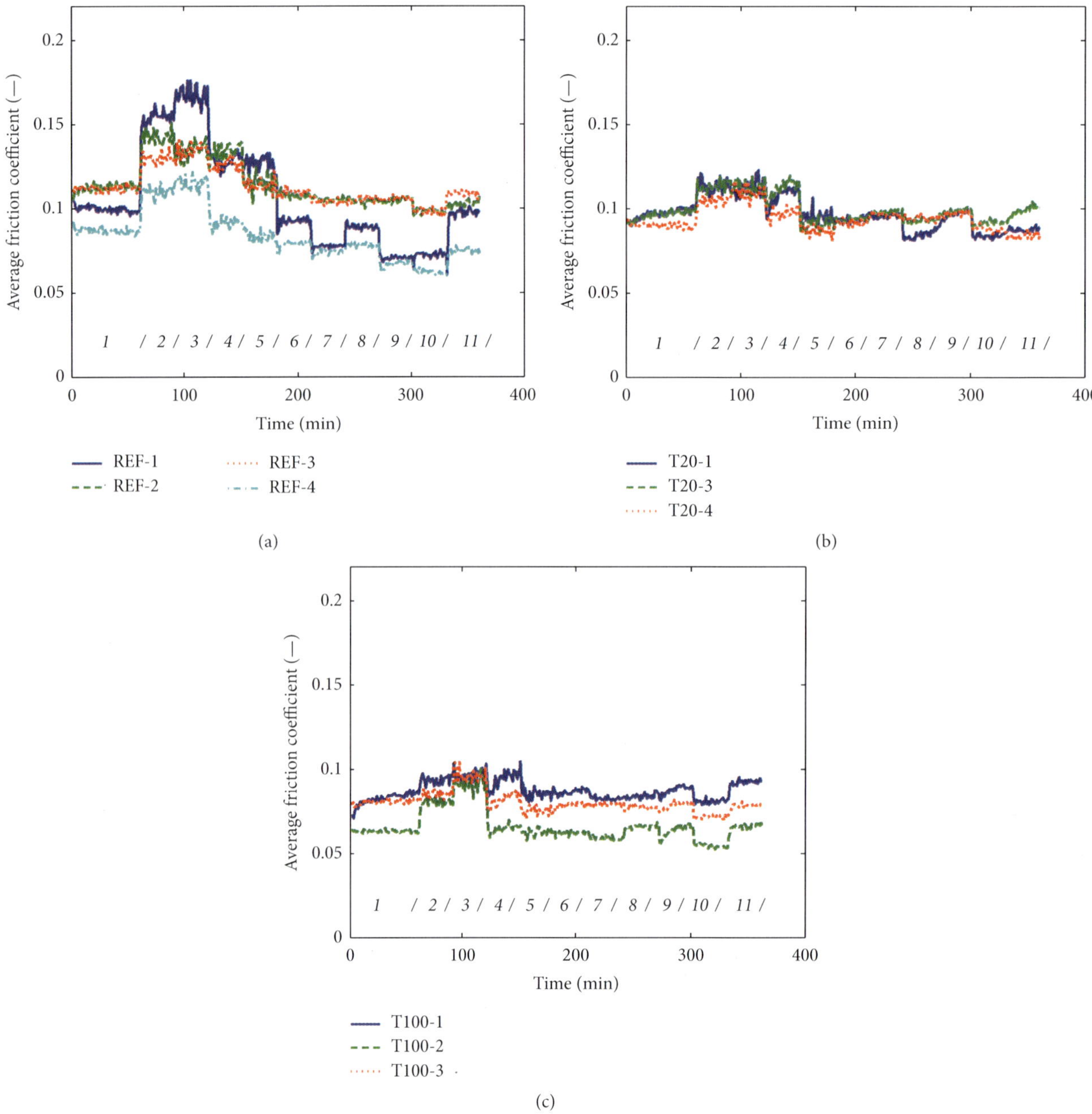

Figure 7: Measured friction coefficient for each experiment. The numbers above the x-axis *(italics)* indicate the duration of each experimental cycle step. (a) REF. (b) T20. (c) T100.

surfaces is textured, the problem has to be divided into two perspectives: either the contact is between the plateaus of the two surfaces (plateau of cylinder liner versus plateau of piston ring) or the contact is between the plateau part of the piston ring and the texture element of the textured surface (texture of cylinder liner versus plateau of piston ring). Equation (1) describes shear force, F_T, for two parallel planes fully separated by a Newtonian fluid. When a mating surface, for example, a piston ring, passes over a texture, the area, A, is also unaltered for the passage since the surface is not decreased or removed; there are still two parallel planes, although in the passage of a texture the planes are farther apart compared to the distance between the two plateaus of the mating surfaces. Hence,

$$F_T = \frac{\eta v A}{h} = \eta S A. \quad (1)$$

In an analysis of the oil viscosity, it is important to account for the non-Newtonian shear rate behavior of the engine oil. The shear rate is dependent on oil film thickness, h, and sliding velocity, v_0 (see (2)). The dynamic viscosity, η or μ, is dependent on the shear ratio; for low levels of shear rate, the value of viscosity is assumed to be that of zero shear, μ_0, and for high levels of shear rate the value of viscosity is assumed

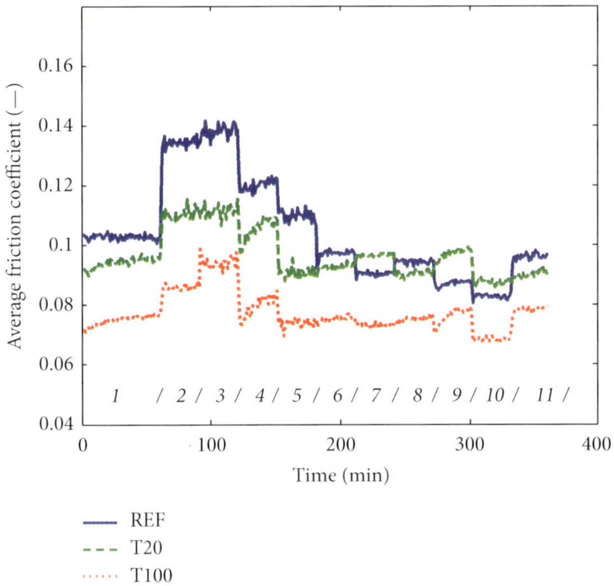

FIGURE 8: Measured friction coefficient, average of experiments for each surface type. The numbers above the x-axis *(italics)* indicate the duration of each experimental cycle step.

to be that of infinite shear, μ_∞ (see (3)). As was shown in the experimental study, the amount of contact increases for the textured surface compared to the reference surface. For passage of a texture, the oil film thickness increases and thus η or μ increases; however, the oil film thickness decreases for passage of a plateau and thus η or μ decreases. The effect of viscosity on shear force is however not believed to be highly significant since the decrease of dynamic viscosity for passage of plateau is partially cancelled out by the increase in dynamic viscosity for passage of a plateau. Therefore,

$$S = \frac{v_0}{h}, \quad (2)$$

$$\mu = \mu_\infty + \frac{\mu_0 - \mu_\infty}{1 + \gamma/\gamma_c}. \quad (3)$$

In essence, there is no alteration of the area and it is believed that there is no significant alteration of the dynamic viscosity for a textured surface compared to the reference surface; there is however a significant increase in the oil film thickness of passage of a texture if the oil film thickness is considered as the entire depth of the texturing element.

The resistive signal increases for the textured surfaces which indicates that the amount of metal-to-metal contact increased for the textured surface compared to the reference surface (see Table 3); this shows that there is generally a thicker oil film h_{ref} (see Figure 14) between the reference surface and the opposing surface compared to the oil film h_{oT} (oil film thickness outside of the texture) between the plateau of a textured surface and the opposing surface. However, for the textured surfaces and the passage of a texture, the oil film thickness is the same as the texture depth considering that the contact between the piston ring and the cylinder liner is fully flooded. The increase in metal-to-metal contact for the

TABLE 3: Average friction coefficient and average resistive coefficient for each surface type.

Surface type	REF	T20	T100
Average friction coefficient	0,105	0,097	0,078
Average resistive coefficient	1,209	1,399	1,575

textured surfaces is due to a decrease in the build-up of hydrodynamic pressure. There are two causes for loss of hydrodynamic pressure: firstly because of leakage of oil into the texture and secondly because less surface area is available for the generation of hydrodynamic pressure. The amount of metal-to-metal contact is greater for T100 compared to T20, since the areas of the textures are practically the same; this means that the leakage of oil into the texture is greater for the T100 textured surface. A summary of the average values of friction coefficient and the average values of resistive coefficient for all experiments and DoE cycle steps can be seen in Table 3.

In the tribometer experiment, fully flooded conditions were withheld for all experimental conditions; this gives an oil film thickness five times larger for passage over a texture of the T100 surface considering that the texture depth is five times larger for T100 (h_{iT100} in Figure 14, oil film thickness for passage of T100 texture) compared to T20 (h_{iT20} in Figure 14, oil film thickness for passage of T20 texture). In this calculation of the very specific oil film thickness within textures, the possible reduction in oil film thickness obtained due to the reduction in hydrodynamic pressure (for the two textured cases) is not taken into account. The calculated average oil film thickness mentioned in this paper for textured surface area is the average separation for the two components: average separation for the textured part of the surface times the texturing area density combined with the average separation

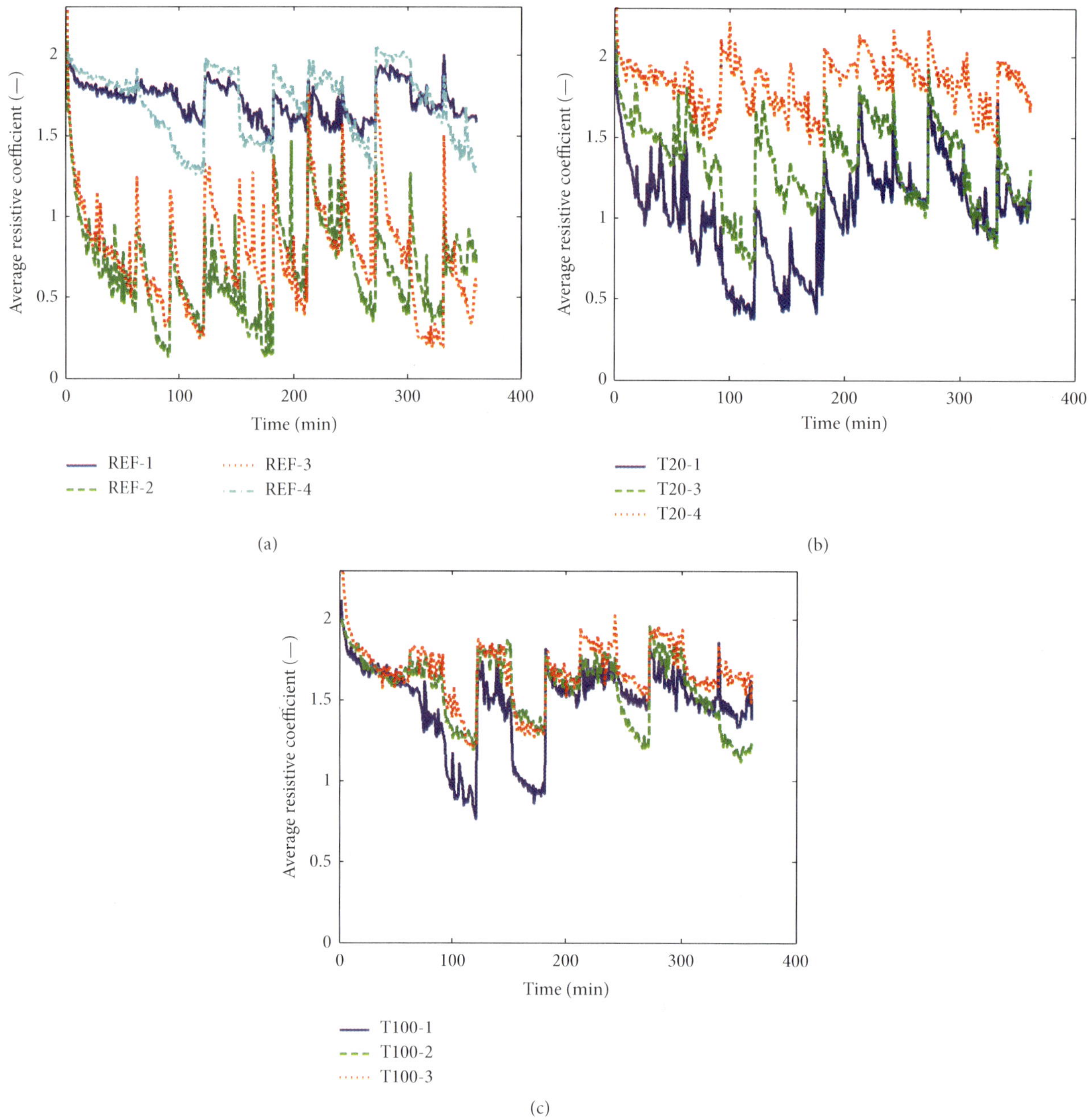

FIGURE 9: Measured resistive coefficient for each experiment. (a) REF. (b) T20. (c) T100.

for the plateau (nontextured) part of the surface times the plateau area density.

Since it is assumed that the oil transport from texture to plateau and from plateau to texture (in relation to the volume of the texture) will be small, the oil within the texture will be forced to recirculate; this is why there is a difference in the flow profiles for passage of textures compared to flow profiles for passage of plateau in Figure 14. The surface T20 exhibited a larger friction coefficient for some experimental cycle steps in which a high load was used (e.g., cycle step 9); the reason for this is believed to be that the increase in boundary friction for passage of a plateau was greater than the decrease in viscous losses for passage of texture, thus combined creating a higher friction than that of the reference surface.

4.2. Texture Properties Applicable to a Tribosystem in Hydrodynamic Lubrication. Textured surfaces with elements of similar geometry to the ones investigated in this paper can be applied on a tribosystem with reciprocating sliding with the aim of decreasing hydrodynamic friction. However, certain aspects should be considered. It is not optimal to put textures in the reversal zones of a tribosystem due to the following:

(i) The sliding speed is small at the reversal zones and thus the hydrodynamic friction losses are small.

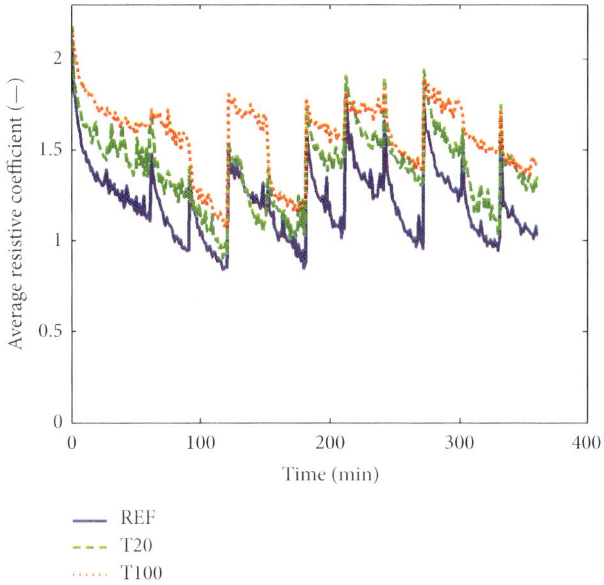

FIGURE 10: Measured resistive coefficient, average of experiments for each surface type.

(ii) If temperatures are high, the oil viscosity is low and thus the hydrodynamic friction losses are small.

(iii) If the contact pressure in the tribosystem is high, an addition of textures in severe tribological contact could increase wear [27]; thus, it is not regarded as beneficial to apply textures with similar dimensions as investigated in this study in the vicinity of the reversal zones.

The surface angle, which is the slope in the axial direction between the edge of the plateau surface and the bottom of the texture (for the textures shown in Figure 14(b), this angle is 90°), was high for the analyzed textures; this is here considered preferable since, with a high angle at the boundary, the oil film will be larger at a larger surface area. In perspective, this could be regarded as follows: either the counter body slides over a texture with high film thickness or it slides over the plateau surface to build up oil film between the two mating surfaces. The passage of texturing provides decreased hydrodynamic friction losses. The passage of a plateau provides oil film build-up between components in the tribosystem. To gain additional frictional improvement from a textured surface, it is important to produce a smooth surface on the plateaus to decrease the mechanical friction of the passage of a plateau.

The addition of the suggested surface textures increases the surface volume. An example of a sliding reciprocating tribosystem is the cylinder liner. In applying the knowledge to this system, it is important to analyze the effects on blow-by and oil consumption. Hegemier and Stewart [28] analyzed the effects on blow-by and oil consumption for different types of cylinder liner surfaces, although the analyzed surfaces were different from the suggested surface texturing in this study. Hegemier and Stewart found that different surface finishes had a little effect on blow-by and that the dominating factor that controlled oil consumption was the amplitude of the plateau roughness. Still, it could be necessary to optimize the geometries of the piston rings for efficient control of blow-by and oil consumption; one suggestion in this optimization is to analyze the effects on blow-by and oil consumption with different designs of gas tight top rings [29].

Seki et al. have experimentally analyzed the oil film thickness (OFT) [30] between piston rings and the cylinder liner; the result in this study shows that OFT increases with sliding speed. The following is a *tentative example* of how friction could be reduced by applying surface textures.

We start by assuming that oil film thickness increases linearly with sliding speed (this is a generalization despite being not that different from the study carried out by Seki et al.) for a reference plateau honed cylinder liner. We also assume that a varying area density of uniform texturing elements (with the geometry of T100) is applied on a textured cylinder liner surface, excluding the reversal zones (at the position of 0–20 crank angle degrees and at the position of 160–180 crank angle degrees). In this example, we do not define a location on the stroke for a specific component; this should be seen more as a general description. The area density of this texturing increases linearly from 21 to 90 crank angle degrees and decreases linearly from 90 to 159 crank angle degrees. On the untextured part of the textured cylinder liner, we assume that the oil film thickness between the plateau surface of the cylinder liner and the piston ring is the same as for the reference cylinder liner. We also assume that the oil film thickness is the same for the textured cylinder liner compared to reference cylinder liner for crank angle degrees that have a smaller value of oil film thickness compared to the constant value of oil film thickness. By controlling the area density of the texturing, the oil thickness between the plateau of the cylinder liner and the piston ring does not increase for crank angle 21–159 but is maintained at a constant value for the textured surface, which in this tentative example varies between 20% and 50% (see

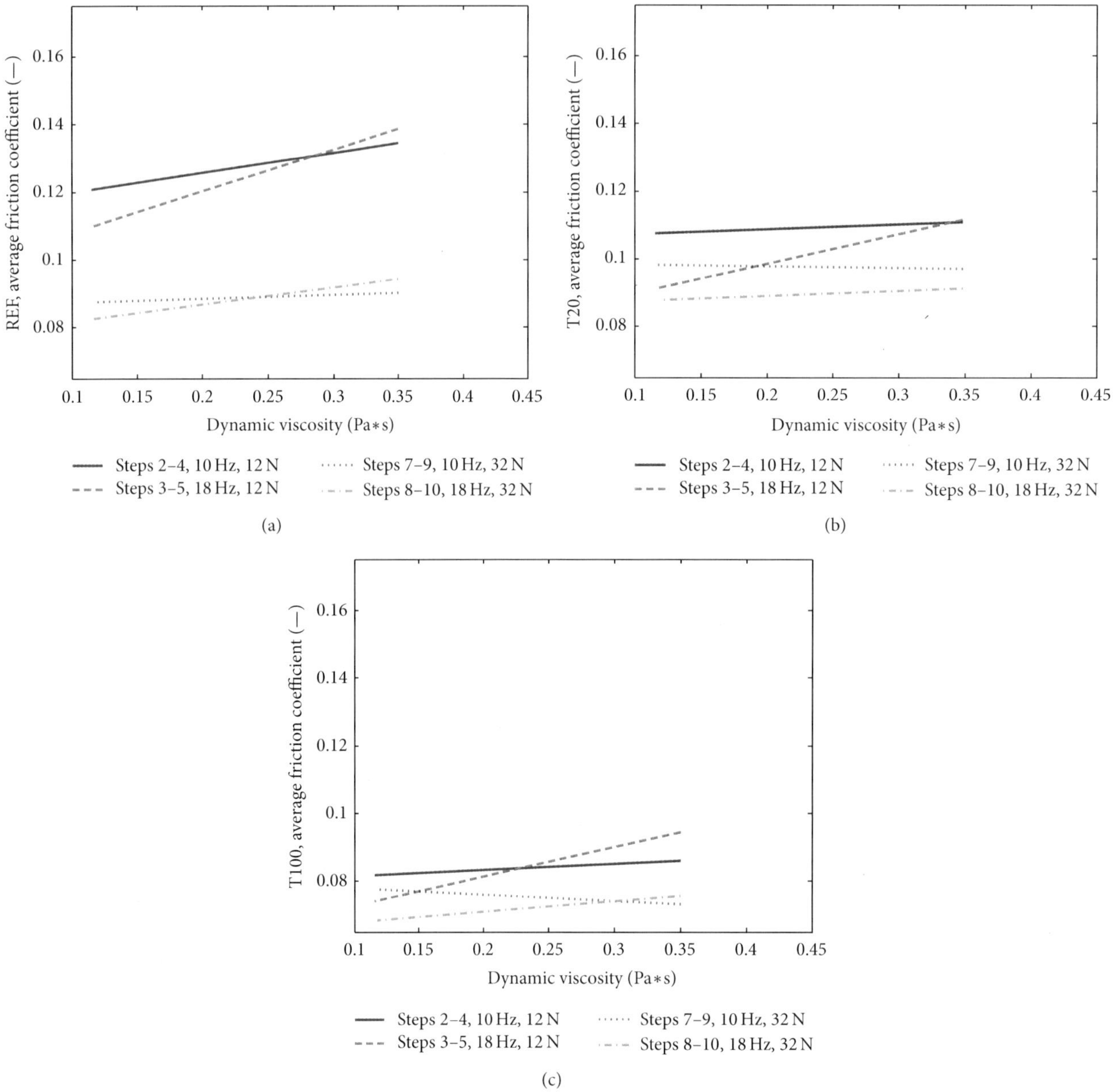

FIGURE 11: Evaluation of the different levels of DoE input parameters. The effect of dynamic viscosity on the friction coefficient. (a) REF. (b) T20. (c) T100.

Figure 15). The average oil film thickness for the textured surface including the oil film height within textures will thus be significantly higher than the oil film thickness for the reference surface; by this, it is possible to decrease the hydrodynamic friction.

In the example above, it is shown that the average oil film thickness with inclusion of the textures is greatly increased, and thus friction will decrease greatly according to (1). However, there is most likely a limit to which this equation is valid, assuming that viscous shear dependency on oil film thickness might in practice prove inaccurate for high values of oil film thickness. On the contrary, in the analysis of wet clutch engagement, the assumption that torque due to viscous shear is inversely proportional to film thickness is still considered valid for an oil film thickness of 0,001 inches *(25,4 μm)* [31] and it has been determined that if the contribution of viscous shear is neglected in a computational model the comparison to experimental data will be incorrect [32].

Also, it cannot be expected that, in the passage from plateau to texture, the flow profile of the oil is instantly fully developed down to the bottom of the texture. In this study, the axial length of the textures is far larger than the axial length of the piston ring and the sliding speed is smaller than that in the engine; thus, it is likely that the effect of time dependency and

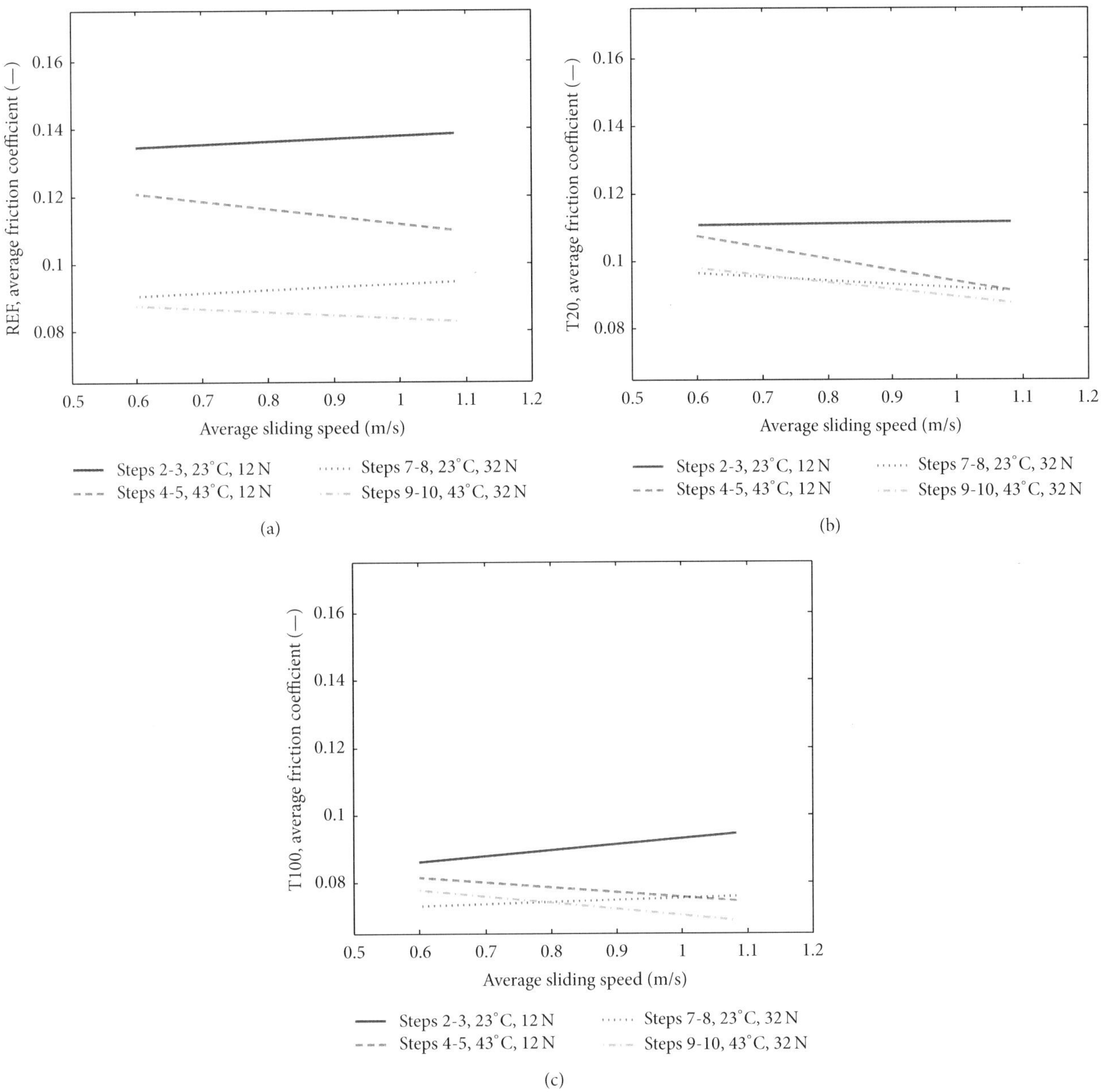

FIGURE 12: Evaluation of the different levels of DoE input parameters. The effect of sliding speed on the friction coefficient. (a) REF. (b) T20. (c) T100.

flow profile is small; however, for application in an engine, the flow profile dependency of texture geometry needs to be further investigated.

Manufacturing a consistent depth of 20 μm on T20 samples proved to be difficult; however, these is no reason to suppose from the experimental results that the distribution in depth is important.

4.3. Additional Discussion regarding Experimental Tribometer Setup. The tribometer used in this study was not originally constructed to perform tests with combined high sliding velocity and low levels of load; the result presented in this study is a representative quantification of the hydrodynamic lubrication regime, and, however, it is not possible to study more viscous contact using the presented tribometer without alteration of sample bodies. It is thus suggested that future experiments should be conducted in an experimental device capable of running higher sliding speeds. In an analysis of hydrodynamic friction losses, it could be possible to use an experimental device with constant sliding speed (as opposed to reciprocating, e.g., pin-on-disc, block-on-ring) although in an experimental setup it is important to maintain line contact between the surfaces.

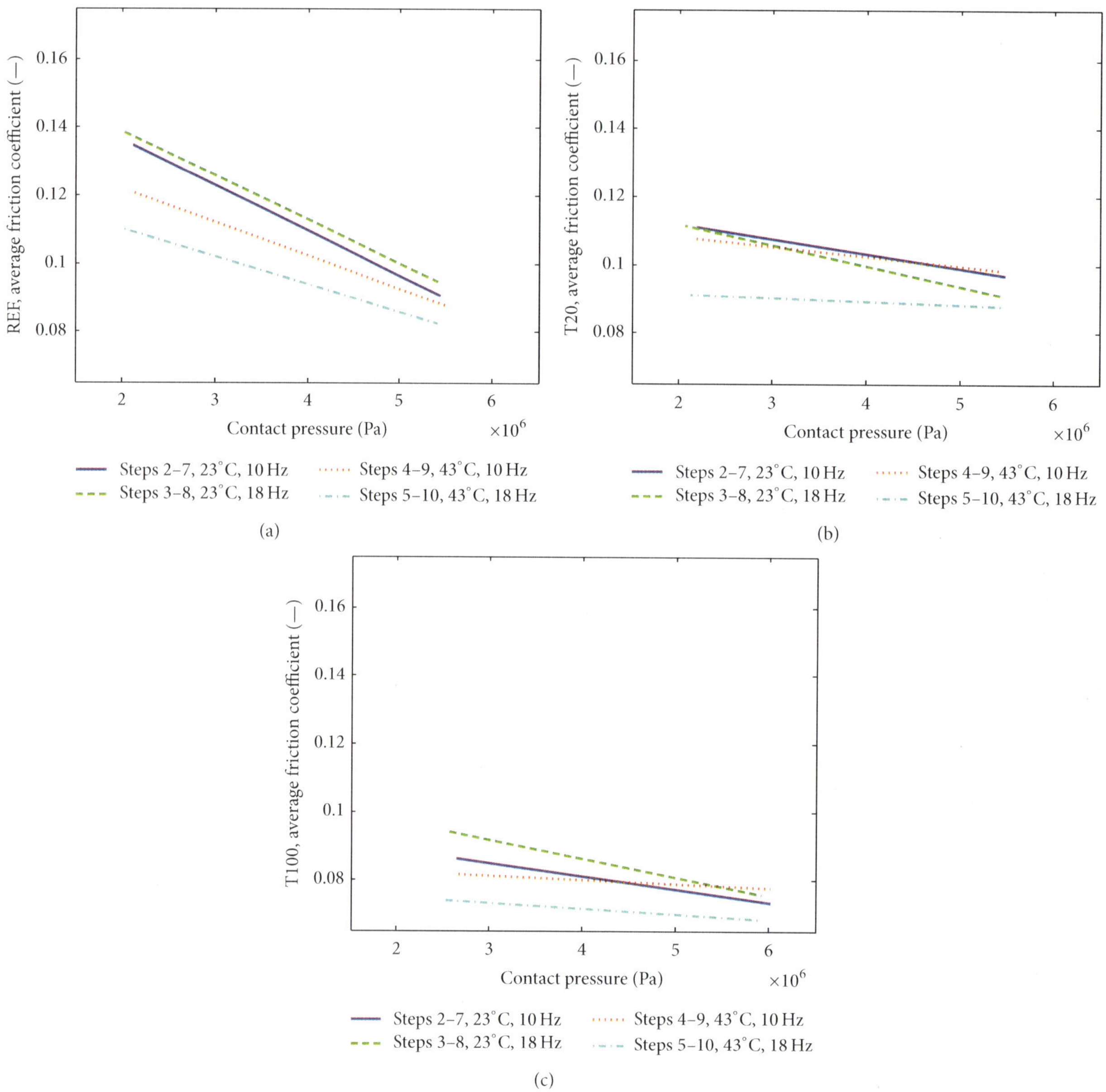

Figure 13: Evaluation of the different levels of DoE input parameters. The effect of contact pressure on the friction coefficient. (a) REF. (b) T20. (c) T100.

5. Conclusions

As a means of decreasing hydrodynamic friction losses, two different types of surface textures were analyzed in reciprocating tribometer experiments. The following conclusions can be drawn from this study:

(i) The two types of the investigated textured surfaces are quantified by 3D profilometry, by which it is shown that the only parameter that differs between the two types is the texture depth.

(ii) The novel type of surface texturing presented in the study decreases friction in the hydrodynamic regime in the contact between oil control ring and cylinder liner.

(iii) The analyzed textured surface does not increase wear; on the contrary, the textured surface exhibits smaller signs of abrasive wear in tribometer experiments; this was due to entrapment of wear particles in textures.

(iv) In further discussions, it is here suggested that the hydrodynamic frictional losses between the cylinder liner and the piston ring can be greatly reduced by the following design of a textured cylinder liner:

(a) The amplitude of the plateau surface on the complete axial length of a textured cylinder

Table 4: The table shows a complete overview of the measured dimensions of the textured samples.

(a)

	Density of grains (%)	Area (mm^2)	Area st. dev. (mm^2)	Perimeter (μm)	Perimeter st. dev. (μm)	Orientation (°)	Orientation st. dev. (°)
T20-1	32,2223	4,34707	0,0397027	8265,17	65,4723	159,4	1,25886
T20-2	32,8645	4,29443	0,147231	8277,15	358,444	158,529	6,6634
T20-3	33,8209	4,42867	0,0774429	8403,61	386,756	159,292	4,49348
T20-4	32,9664	4,42515	0,0374976	8304,15	62,9534	159,532	0,925686
T100-1	33,5814	4,39362	0,0402664	8303,7	67,3321	160,107	0,890755
T100-2	34,0799	4,44955	0,114732	8378,44	239,861	159,644	2,18038
T100-3	34,1404	4,44813	0,105916	8381,69	161,312	159,602	1,83101
T100-4	31,5613	4,42882	0,0626995	8332,02	99,071	159,907	1,58551

(b)

	Min. diameter (μm)	Min. diameter st. dev. (μm)	Max. diameter (μm)	Max. diameter st. dev. (μm)	Depth (μm)	Depth st. dev. (μm)
T20-1	1896,47	9,35645	2910	8,40168	19,59715882	6,337111655
T20-2	1872,57	11,0435	2917,71	121,281	18,16972941	4,905834966
T20-3	1912	5,75698	2947,43	81,393	22,64912941	7,889610536
T20-4	1915,29	6,52384	2926,18	7,67529	19,46122353	3,674416624
T100-1	1909,43	7,53766	2905,71	9,34596	94,58104412	13,81767236
T100-2	1919,71	6,08779	2930	103,178	87,75976176	11,92867327
T100-3	1919,14	6,03392	2928,29	73,4647	97,93557647	13,59051635
T100-4	1917,94	6,76471	2915,59	18,8166	87,18046471	11,65203976

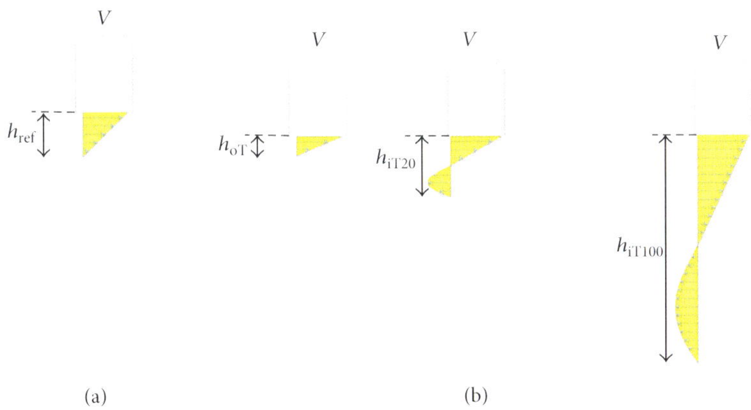

Figure 14: Schematic overview of the oil film build-up for the reference surface (a) and piston ring passage over the plateau of a textured surface; the piston ring passage over a 20 μm deep texture and the piston ring passage over a 100 μm deep texture (b).

liner is minimized to keep boundary friction at minimum.

(b) Texturing is applied on the part of the axial length on the cylinder liner below the upper reversal zone and above the lower reversal zone. The area density of texturing increases at midstroke position and decreases towards the reversal zones.

(c) The reversal zones remain untextured; there are two main reasons for this; firstly, the sliding speed is low in the vicinity of the reversal zones and thus the hydrodynamic friction is low; secondly, the introduction of textures with relatively large geometry might increase wear.

Appendix

See Tables 4(a) and 4(b).

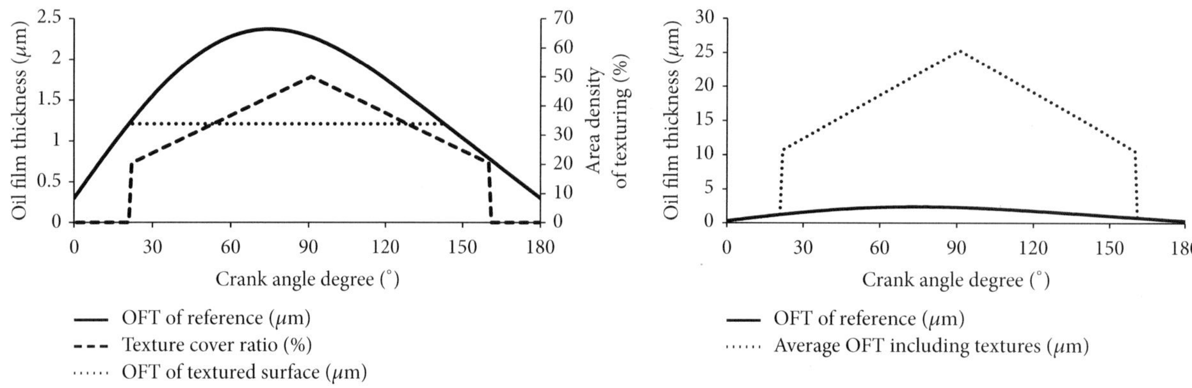

FIGURE 15: Tentative example of the oil film thickness with textured cylinder liner surfaces.

Conflicts of Interest

The authors declare that they have no conflicts of interest.

References

[1] J. B. Heywood, *Internal Combustion Engine Fundamentals*, McGraw-Hill, New York, NY, USA, 1988.

[2] R. I. Taylor, "Heavy duty diesel engine fuel economy: lubricant sensitivities," in *Proceedings of the International Spring Fuels & Lubricants & Exhibition*, Paris, France, June 2000.

[3] D. E. Richardson, "Review of power cylinder friction for diesel engines," *Journal of Engineering for Gas Turbines and Power*, vol. 122, no. 4, pp. 506–519, 2000.

[4] M. Nakada, "Trends in engine technology and tribology," *Tribology International*, vol. 27, no. 1, pp. 3–8, 1994.

[5] E. Tomanik, "Friction and wear bench tests of different engine liner surface finishes," *Tribology International*, vol. 41, no. 11, pp. 1032–1038, 2008.

[6] J. Li, D. Xiong, J. Dai, Z. Huang, and R. Tyagi, "Effect of surface laser texture on friction properties of nickel-based composite," *Tribology International*, vol. 43, no. 5-6, pp. 1193–1199, 2010.

[7] P. Andersson, J. Koskinen, S. Varjus et al., "Microlubrication effect by laser-textured steel surfaces," *Wear*, vol. 262, no. 3-4, pp. 369–379, 2007.

[8] S. Yuan, W. Huang, and X. Wang, "Orientation effects of micro-grooves on sliding surfaces," *Tribology International*, vol. 44, no. 9, pp. 1047–1054, 2011.

[9] S. Johansson, C. Frennfelt, A. Killinger, P. H. Nilsson, R. Ohlsson, and B. G. Rosén, "Frictional evaluation of thermally sprayed coatings applied on the cylinder liner of a heavy duty diesel engine: pilot tribometer analysis and full scale engine test," *Wear*, vol. 273, no. 1, pp. 82–92, 2011.

[10] S. Johansson, P. H. Nilsson, R. Ohlsson, and B.-G. Rosén, "Simulation and experimental analysis of the contact between oil control ring and cylinder liner in a heavy duty diesel engine," in *Proceedings of the 18th International Colloquium Tribology*, Stuttgart, Germany, January 2012.

[11] M. Priest and C. M. Taylor, "Automobile engine tribology—approaching the surface," *Wear*, vol. 241, no. 2, pp. 193–203, 2000.

[12] H. Moritani, H. Tokoro, M. Tohyama et al., "Challenge to the diesel engine lubrication with fuel," SAE Technical Paper 2007-01-1978, SAE International, 2007.

[13] Y. Wakuri, T. Hamatake, M. Soejima, and T. Kitahara, "Piston ring friction in internal combustion engines," *Tribology International*, vol. 25, no. 5, pp. 299–308, 1992.

[14] Y. Qiu and M. M. Khonsari, "Experimental investigation of tribological performance of laser textured stainless steel rings," *Tribology International*, vol. 44, no. 5, pp. 635–644, 2011.

[15] S.-W. Cho, S.-M. Choi, and C.-S. Bae, "Frictional modes of barrel shaped piston rings under flooded lubrication," *Tribology International*, vol. 33, no. 8, pp. 545–551, 2000.

[16] H. L. Costa and I. M. Hutchings, "Hydrodynamic lubrication of textured steel surfaces under reciprocating sliding conditions," *Tribology International*, vol. 40, no. 8, pp. 1227–1238, 2007.

[17] B. Podgornik, L. M. Vilhena, M. Sedlaček, M. R. Ripoll, Z. Rek, and I. Žun, "Effectiveness of surface texturing in different lubrication regimes," in *Proceedings of the 14th Nordic Symposium on Tribology (NORDTRIB '10)*, Storforsen, Sweden, 2010.

[18] S. Johansson, P. H. Nilsson, R. Ohlsson, and B.-G. Rosén, "Experimental friction evaluation of cylinder liner/piston ring contact," *Wear*, vol. 271, no. 3-4, pp. 625–633, 2011.

[19] S. Johansson, *A surface engineering approach to reduction of frictional losses of heavy duty diesel engines [Ph.D. thesis]*, Chalmers University of Technology, 2012.

[20] M.-S. Suh, Y.-H. Chae, S.-S. Kim, T. Hinoki, and A. Kohyama, "Effect of geometrical parameters in micro-grooved crosshatch pattern under lubricated sliding friction," *Tribology International*, vol. 43, no. 8, pp. 1508–1517, 2010.

[21] L. L. Ting, "Development of a reciprocating test rig for tribological studies of piston engine moving components—part II: measurements of piston ring friction coefficients and rig test confirmation," in *Proceedings of the International Congress & Exposition*, SAE 930686, Detroit, Mich, USA, March 1993.

[22] P. Andersson, "A new tribometer for piston ring friction measurements," *Tribological Research and Design for Engineering Systems*, vol. 41, pp. 501–510, 2003.

[23] H. Chen, J. Li, D. Chen, and J. Wang, "Study of the effect of transverse and longitudinal surface patterns on drag-reduction in shear-thinning fluid," *Proceedings of the Institution of Mechanical Engineers Part J: Journal of Engineering Tribology*, vol. 221, no. 2, pp. 87–94, 2007.

[24] M. Knopf, C. Eiglmeier, and G. P. Merker, "Calculation of unsteady hydrodynamic lubrication and surface contact at the piston-ring/cylinder-liner interface," in *Proceedings of the International Spring Fuels and Lubricants Meeting and Exposition*, Dearborn, Mich, USA, May 1998.

[25] J. A. Greenwood, "Constriction resistance and the real area of contact," *British Journal of Applied Physics*, vol. 17, no. 12, pp. 1621–1632, 1966.

[26] H.-J. Jordan, "Optical chromatic confocal probes," in *Proceedings of the 12th International Colloquium on Surfaces*, Chemnitz, Germany, January 2008.

[27] A. Kovalchenko, O. Ajayi, A. Erdemir, and G. Fenske, "Friction and wear behavior of laser textured surface under lubricated initial point contact," *Wear*, vol. 271, no. 9-10, pp. 1719–1725, 2011.

[28] T. Hegemier and M. Stewart, "Some effects of liner finish on diesel engine operating charateristics," in *Proceedings of the International Congress and Exposition*, Detroit, Mich, USA, March 1993.

[29] Y. Tateishi, "Tribological issues in reducing piston ring friction losses," *Tribology International*, vol. 27, no. 1, pp. 17–23, 1994.

[30] T. Seki, K. Nakayama, T. Yamada, A. Yoshida, and M. Takiguchi, "Study on variation in oil film thickness of a piston ring package: variation of oil film thickness in piston sliding direction," *JSAE review*, vol. 21, no. 3, pp. 315–320, 2000.

[31] H. Wu, "An analysis of the engagement of wet-clutch plates," *Wear*, vol. 24, no. 1, pp. 23–33, 1973.

[32] P. Payvar, "Laminar heat transfer in the oil groove of a wet clutch," *International Journal of Heat and Mass Transfer*, vol. 34, no. 7, pp. 1791–1798, 1991.

11

A New Integrated Approach for the Prediction of the Load Independent Power Losses of Gears: Development of a Mesh-Handling Algorithm to Reduce the CFD Simulation Time

F. Concli,[1] A. Della Torre,[2] C. Gorla,[3] and G. Montenegro[2]

[1] *R&D Department, Bonfiglioli Mechatronic Research, Via Fortunato Zeni 8, 36068 Rovereto, Italy*
[2] *Department of Mechanical Engineering, Politecnico di Milano, Via La Masa 1, 20156 Milano, Italy*
[3] *Department of Energy, Politecnico di Milano, Via La Masa 1, 20156 Milano, Italy*

Correspondence should be addressed to F. Concli; franco.concli@mail.polimi.it

Academic Editor: Michael M. Khonsari

To improve the efficiency of geared transmissions, prediction models are required. Literature provides only simplified models that often do not take into account the influence of many parameters on the power losses. Recently some works based on CFD simulations have been presented. The drawback of this technique is the time demand needed for the computation. In this work a less time-consuming numerical calculation method based on some specific mesh-handling techniques was extensively applied. With this approach the windage phenomena were simulated and compared with experimental data in terms of power loss. The comparison shows the capability of the numerical approach to capture the phenomena that can be observed experimentally. The powerful capabilities of this approach in terms of both prediction accuracy and computational effort efficiency make it a potential tool for an advanced design of gearboxes as well as a powerful tool for further comprehension of the physics behind the gearbox lubrication.

1. Introduction

More and more severe requirements in terms of efficiency have encouraged the gearbox manufacturers to increase the investments for power dissipation reduction. An energy efficient solution, besides a pure power consumption reduction, shows lower operating temperatures and therefore has a higher reliability. The losses are dissipated in form of heat through the housing and shafts. While the exchange area/volume is often limited by external constrains, the internal design of a gearbox is completely in charge of the gearbox manufacturer. Considering that in industrial gearboxes the power losses are more or less equally shared between the bearings, the gear meshing, and the interaction with the oil, it is clear that at least this main contribution (there are other contributions related, e.g., to the seals) should be considered when a new gearbox is developed. Being able to correctly model the power losses enables the capability to predict also the operating temperatures under each operating condition and consequently the so-called thermal limit. While for bearings and for the gear meshing losses accurate models already exist [1–5], for the oil churning/sloshing power losses only basic research was carried out. All the experimentally derived models results are accurate only as long as the operating conditions and geometrical configurations are similar to those used in the original experiments. One of the first works on this topic was carried out by Mauz [6]. Mauz tested several geometries and operating conditions. From the authors experience [7] this model is not capable of taking into account very important parameters like the helix angle. Other studies in this field were carried out by other researchers such as Dawson [8] who concentrated on windage, Seetharaman and Kahraman [9] who concentrated on churning, and several other authors [10–14]. The authors maintain that a deeper understanding of the physical phenomena can only be achieved by means of numerical techniques. Different contributions have already shown the goodness of the CFD approach for such purposes [15, 16]. In the past the authors approached the problem using commercial software [17–22]. The main limitation to the wide diffusion of such approaches is related to the computational effort, the complexity of setting up a model with a general-purpose software, and the costs for the licenses. In this scenario the authors started studying a simplified configuration [23, 24] with the open-source code OpenFOAM [25]. This approach allows both to overcome the license costs (GNU license) but also, thanks to the specificity of the developed algorithm, to reduce the setup and calculation times. In this paper a mesh generation and handling algorithm that enables the simulation of the gear engagement is presented. The developed code was tested and the results show good agreement compared with experimental data [17].

2. Power Loss Classification

According to [5] the power losses of gears can be subdivided into load dependent and load independent losses and again according to the mechanical component that is responsible for their generation

$$P_L = P_{LG} + P_{LG0} + P_{LB} + P_{LB0} + P_{LS} + P_{LX}. \quad (1)$$

The subindexes G, B, S, and X stand for gears, bearings, seals, and other generic components (like clutches or synchronizers). The subindex 0 indicates the load independent losses. The focus of this research is on the load independent gearing power losses P_{LG0}. It results from the interaction of the rotating (or in general moving) mechanical components and the lubricant. This load independent contribution of the gears can further be subdivided into windage, churning, and squeezing/pocketing effects. Windage and churning as in external aerodynamics concern the main interaction of the components with the oil/air or with the oil-air mixture. The pocketing effects instead are related to the volume variation of the gap between the teeth in the mating region and the additional fluxes that take place.

3. Objective of the Study

The main objective of this study is to find a reliable method in order to calculate the load independent losses of gears. Beside the accuracy of the results the models should also be manageable in terms of computational times so as to enable systematic studies in reasonable times, a requirement that is particularly stringent for industrial applications.

4. Numerical Modeling

4.1. Conservation Equations. For a generic volume it is possible to write five conservation equations: the averaged mass conservation equation for no-stationary incompressible flows and the averaged momentum equation (and eventually the energy conservation equation):

$$\begin{aligned}
&\frac{\partial \langle \rho u_i \rangle}{\partial x_i} = 0, \\
&\frac{\partial}{\partial t}\left(\rho \langle u_i \rangle\right) + \frac{\partial}{\partial x_j}\left(\rho \langle u_i \rangle \langle u_i \rangle\right) \\
&\quad = -\frac{\partial}{\partial t}\langle p \rangle \\
&\qquad + \frac{\partial}{\partial x_j}\left[\mu\left(\frac{\partial \langle u_i \rangle}{\partial x_j} + \frac{\partial \langle u_j \rangle}{\partial x_i}\right) - \frac{\partial}{\partial x_j}\tau_{ij}\right],
\end{aligned} \quad (2)$$

where $x_{i,j}$ represent the Cartesian coordinate and $u_{i,j}$ is the velocity component, p the pressure, ρ the density, and τ_{ij} the Reynolds term.

Solving these equations numerically it is possible to describe the fluid behavior in terms of macroscopic properties such as pressure, velocity, and their space and time derivatives. Simulations were run adopting a transient incompressible pressure-velocity coupled solver. In particular, the implementation of the pimpleDyMFoam solver of OpenFOAM (PIMPLE solver, a flexible implementation of a transient solver that allows operation in both PISO [26] and transient SIMPLE mode) was adopted. In this work, the temperature of the lubricant was assumed as uniform in the domain and a priori known; therefore the energy conservation equation was not considered. In a similar way the density and the viscosity values were set on the basis of the temperature.

4.2. Boundary Condition. In order to correctly move the gear boundaries according to the kinematic laws, a new boundary condition was developed. The relation on which the boundary condition is based is

$$\psi^n = \psi^{n-1} + \omega \cdot dt \quad (3)$$

in which ψ^n represents the angular position of the gear at the actual time step, ψ^{n-1} the angular position at the previous time step, ω the rotational speed of the gear, and dt the time step. In this manner it was possible to move the gear boundaries according to the kinematic law.

4.3. Mesh. OpenFOAM is capable of handling the domain deformations by moving the mesh. The objective of the mesh motion is to accommodate externally prescribed boundary deformations by changing the positions of mesh points (Figure 2). During the motion the mesh must remain geometrically valid. The motion $\mathbf{u}$ is calculated considering the Laplace smoothing equation (4) and the pseudo-solid equation (5) (linearization of the motion equations for small deformations)

$$\nabla \cdot (\gamma \nabla \mathbf{u}) = 0, \quad (4)$$

$$x^{n+1} = x^n + \Delta t \mathbf{u}, \quad (5)$$

where γ represents the diffusivity, x the position of the nodes of the mesh, and t the actual time.

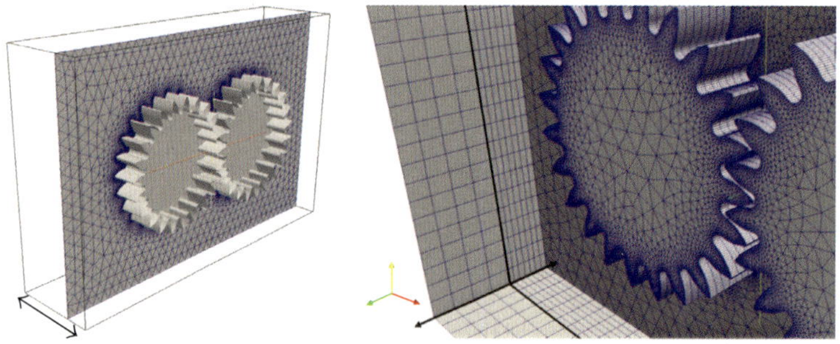

(a) (b)

FIGURE 1: Discretization of the internal planar faces: this mesh will then be extruded in both directions.

This approach is effective only as far as the deformation does not affect too much the quality of the mesh which should be updated in this case. Different from many commercial codes, OpenFOAM does not allow local mesh regeneration to circumvent the mesh quality degeneration.

In order to simulate the complete rotation and meshing of the gears, an approach which employs a multiple number of meshes was applied [27, 28]. In this framework, each mesh is generated for one specific rotational angle and has a certain angular validity. Once a mesh is created, motion is imposed at the boundaries, whereas the inner points of the grid are moved according to the solution of the Laplace equation in order to adapt to the boundary motion. Usually, when the mesh is deformed, its quality in terms of skewness, nonorthogonality and geometrical/topological validity decreases. As a consequence, when the deformation becomes excessive a new mesh must be created and the solution mapped from the old mesh onto the new one resorting to field interpolation techniques. In this case, the interpolation of the computed flow field from one mesh to the next one is performed by means of a second-order, inverse distance weighting method. The conservativeness of the approach was tested in order to verify that the errors induced by interpolation were negligible both in terms of integral qualities (fluid mass) and velocity profiles.

In OpenFOAM two main meshing tools are available. The first one, blockMesh, is capable of creating parametric meshes with grading and curved edges; blockMesh is however not suitable to handle complex geometries. SnappyHexMesh, another utility of OpenFOAM, generates 3-dimensional meshes containing hexahedra and split-hexahedra automatically from triangulated surface geometries in stereolithography format. The mesh approximately conforms to the surface by iteratively refining a starting mesh (generated with blockMesh) and morphing the resulting split-hex mesh to the surface. The surface handling is robust with a prespecified final mesh quality but the procedure is also very time consuming and not suitable for the purposes of this study. For this reason, a new and more efficient approach was tested. The geometry is generated with SALOME [29]. A python script allows the parametrization of the geometrical entities that are defined through an analytical formulation. In a second step, the domain is meshed. In order to reduce the computational time a special strategy was implemented. The domain (a box with two mating gears) is cut by a plane that lies on the gear sides. The generated internal faces (Figure 1) are meshed with Netgen [30]. Netgen follows a top down strategy. It starts by computing the corner points. Then the edges are defined and meshed into segments. Next the faces are discretized by an advancing front surface mesh generator. A fast Delaunay algorithm generates most of the elements, but sometimes it fails for the last elements, so a slower back-tracking rule-base algorithm takes over. Basically the Delaunay algorithm subdivides the edges into segments according to the local seed prescribed element size (Figure 3(a)). Then, the surfaces are seeded with internal points (Figure 3(b)). These were connected (triangulation) together. The results of such procedure is often a surface (or volume) mesh that does not match the boundary mesh (Figure 3(c)) so the outer and the intersecting elements are removed (Figure 3(d)).

The internal two-dimensional mesh is successively extruded in both directions generating a 2.5D discretization. This procedure reduces significantly the mesh generation times: even if the computational domain is three-dimensional, the meshing algorithms are applied only to two-dimensional surfaces and the resulting meshes copied to create swept elements.

The elements belonging to each boundary can be grouped taking advantage of the analytical geometrical entities. The mesh is finally exported and converted into the OpenFOAM format.

For this test case the whole geometry and mesh generation takes approximately 30 seconds on a single core 12.8 GFLOPS machine. On the previously tested commercial code [17], meshing the same geometry (with tetrahedrons) takes approximately 6 minutes (same amount of cells—13.2 GFLOPS machine). The algorithm is so effective that also the local mesh regeneration, performed by the commercial code [17] after each iteration to regenerate the collapsed cells, takes more time than the actual procedure takes for the complete remeshing of the domain.

5. Torque Calculation

The total resistant torque around a specified axis is computed by summing the cross products of the pressure $\mathbf{F}_p$ and viscous

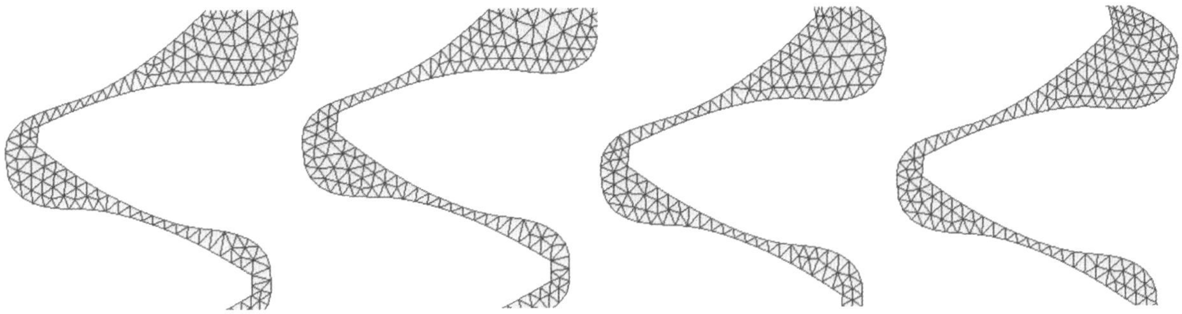

FIGURE 2: Detail of the first 4 meshes (symmetry plane); before mapping the data, the $(i-1)$th mesh is deformed so that its boundary will coincide with the boundaries of the ith mesh.

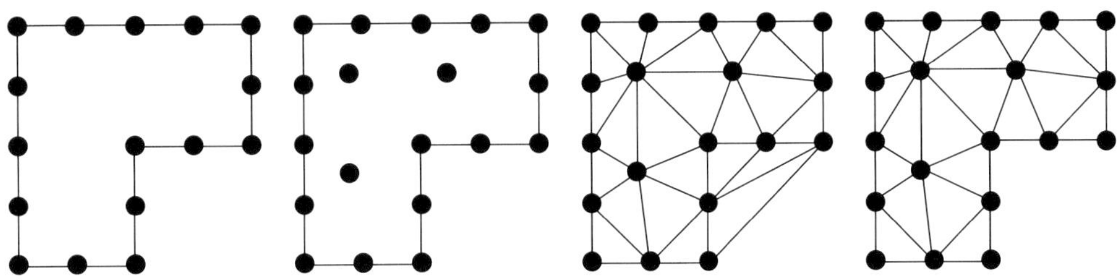

(a) (b) (c) (d)

FIGURE 3: Steps performed by the Delaunay algorithm.

force vectors $\mathbf{F}_\tau$ for each face and the vector $\mathbf{r}_{AB}$, defining the distance of each face center from the specified axis

$$\mathbf{M}_A = \mathbf{r}_{AB}\mathbf{F}_P + \mathbf{r}_{AB}\mathbf{F}_\tau. \quad (6)$$

The pressure and viscosity force vectors are calculated as

$$\begin{aligned} \mathbf{F}_P &= pA_i, \\ \mathbf{F}_\tau &= \mu\frac{\partial \mathbf{u}}{\partial \mathbf{x}}A_i, \end{aligned} \quad (7)$$

where A_i represents the area of the i-face of the mesh.

6. Model Validation

The presented model was validated on the basis of some already available experimental results [7, 17]. The tests were performed in 2012 at FZG [31]. Two gears, whose properties are listed in Table 1, are set into rotation with an electric motor whose rotational speed and torque are measured by sensors.

Both gears are mounted on cantilevered shafts each one supported by two rolling bearings. The primary shaft is connected to the motor while the second rotates freely. The power supplied by the motor is therefore completely dissipated by the gears (and shafts), by the bearings, and by the seals. Dedicated tests without the gears were performed in order to quantify the power losses due to bearings, shafts, and seals. In this manner it was possible to separate the pure gear loss contribution.

The dimensions of the test rig housing are 290 × 145 × 185 mm. The volume has been completely filled with an FVA2 lubricant and pressurized to 6 bars. The lubricant properties are listed in Table 2.

TABLE 1: Gear parameters.

Width b [mm]	40
Tip diameter d_a [mm]	102.5
Helix angle β [°]	0
Pressure angle α [°]	20
Normal module m_n [mm]	4
Center distance a [mm]	91.5
Number of teeth z [ ]	23
Chordal thickness s_j [mm]	5.35

TABLE 2: Fluid properties.

ρ_{15} [Kg/m^3]	ν_{100} [mm^2/s]	ν_{100} [mm^2/s]
871	29.8	5.2

Measurements were performed up to 4000 rpm (22 m/s of tangential velocity on the pitch diameter). The measured values are the sum of the contributions given by the 4 rolling bearings, the two seals, the shafts interaction with the lubricant, and the gear interaction with the lubricant.

This configuration was simulated with the above described approach. Despite the cantilevered shafts, the model was considered symmetric and only half of the domain was modeled.

7. Results

Figure 4 shows a comparison between the numerical results of present work (OpenFOAM), the experimental ones, and

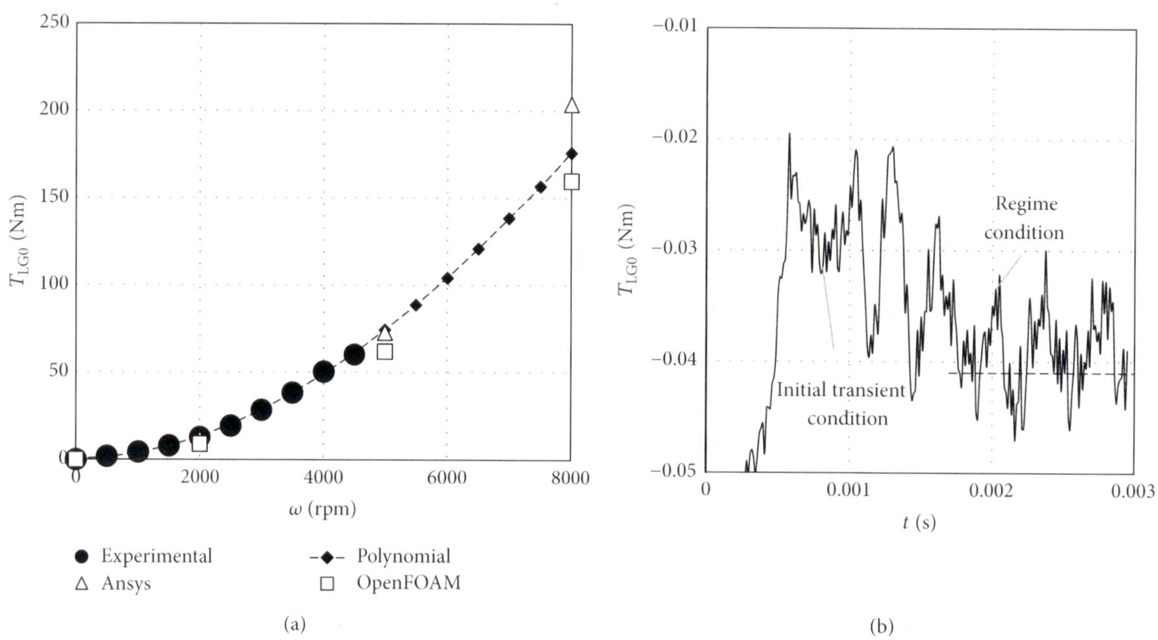

FIGURE 4: (a) Gear load independent resistant torque: experimental versus numerical data and (b) convergence monitor.

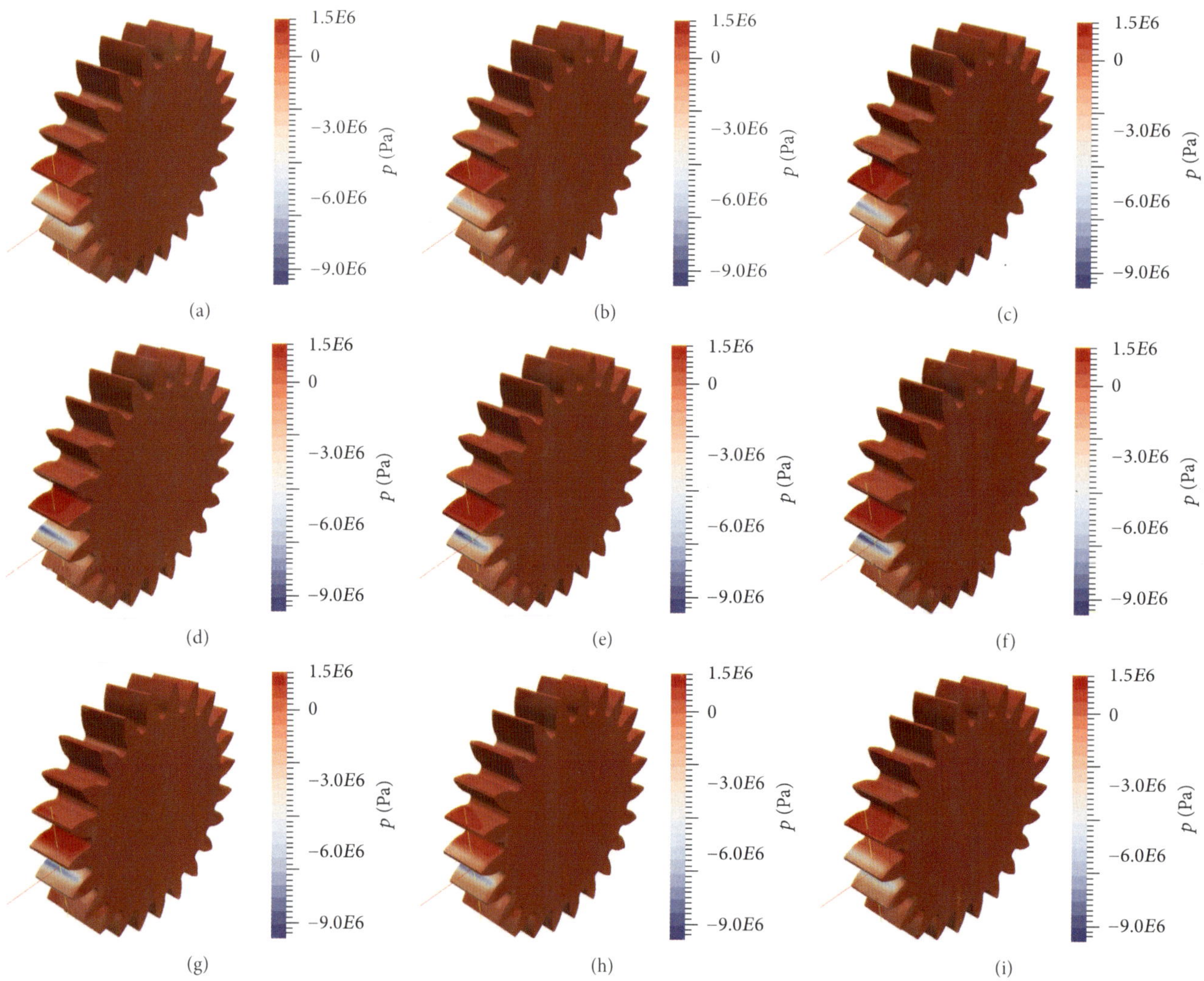

FIGURE 5: Relative pressure variation in meshing positions [Pa].

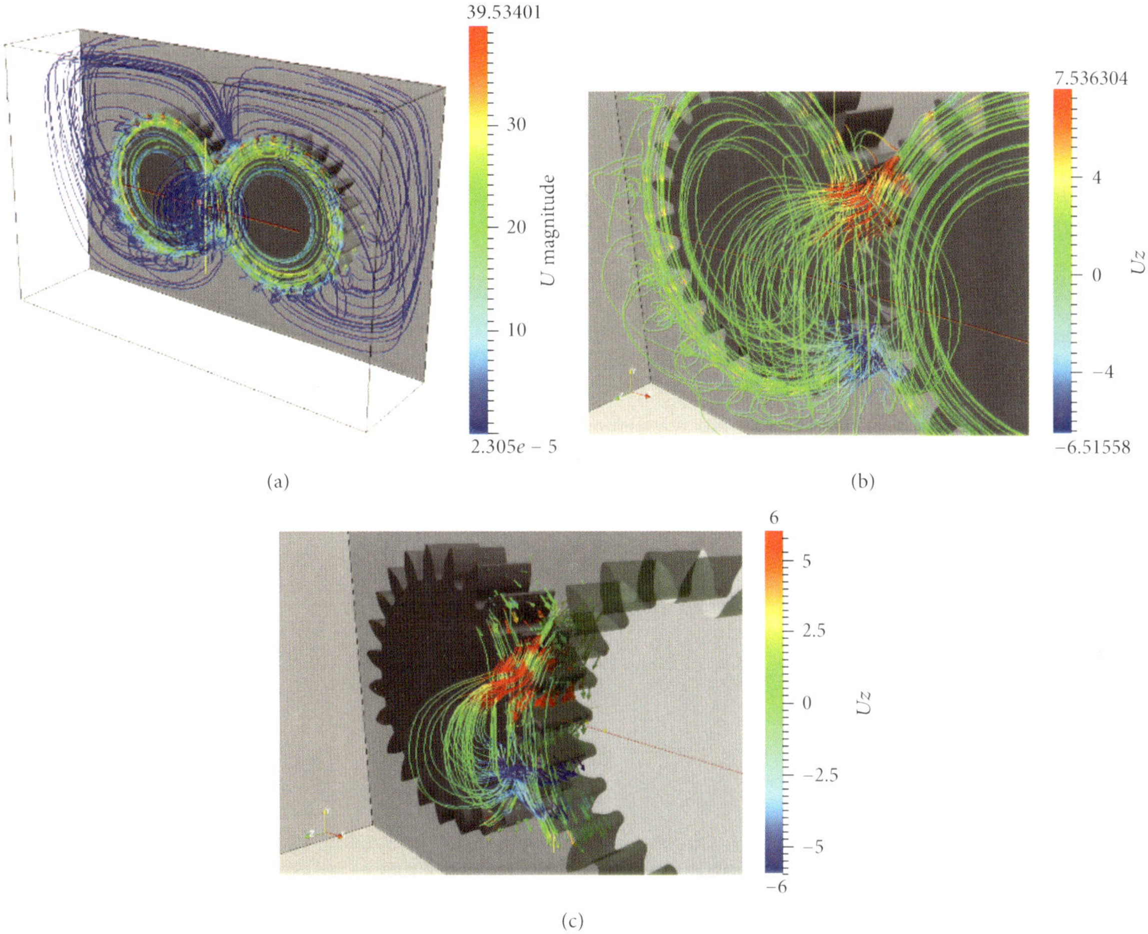

FIGURE 6: (a) Path-lines (U [m/s]) and (b, c) detail (Uz [m/s]).

those obtained by the authors with the commercial general-purpose software in a previous research [7, 17]. The results are presented in terms of resistant torque versus rotational speed. The stripes represent a 3rd-order polynomial extrapolation of the numerical results.

The trend of the curve is well reproduced by the simulations even if the absolute values result is underestimated. This can be explained considering the geometrical simplification (the domain was modeled as symmetric despite the cantilevered shafts which are not modeled).

Figure 5 shows the pressure evolution on the flanks during the engagement. The pressure on the mating flanks increases gradually when the gap between the teeth reduces the volume and squeezes the lubricant generating an axial flow (squeezing/pocketing effect). After the contact point passed through the pitch point, the volume increases again generating a negative pressure and a suction effect that is visible in Figure 6(b) looking to the axial velocity Uz.

These effects that can be evaluated by looking to the path-lines (Figure 7) can also be quantified in terms of torque. In a previous study [23] the authors have used the different CFD approaches to calculate the power losses generated by a single rotating gear. The aim of that study was to validate different approaches to reduce the computational effort. The

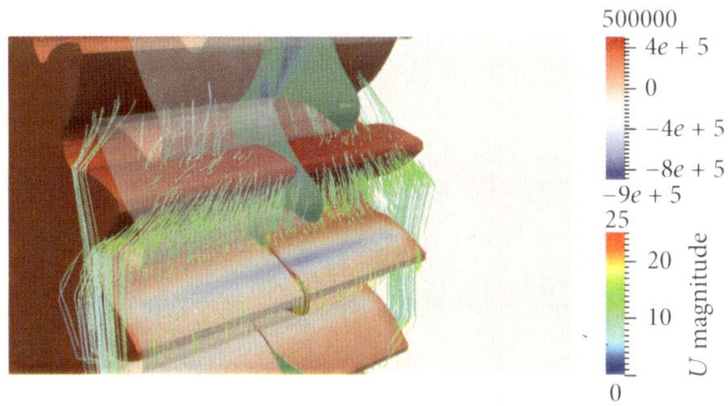

FIGURE 7: Detail of the contact.

results of that study can be used for a comparison between the losses generated by 2 mating gears and a single rotating gear multiplied by 2.

Figure 8 represents the differences between the torque generated by two mating gears and that generated by one single rotating gear (×2) that is attributable to the squeezing effects. Considering the measurement accuracy and the big simplifications introduced in the numerical model, even if differences in the squeezing torque can be seen with respect to the measured values, the presented approach seems to be

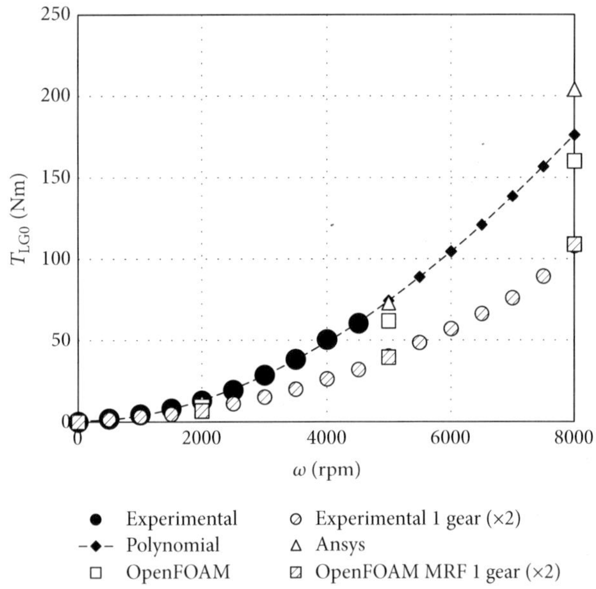

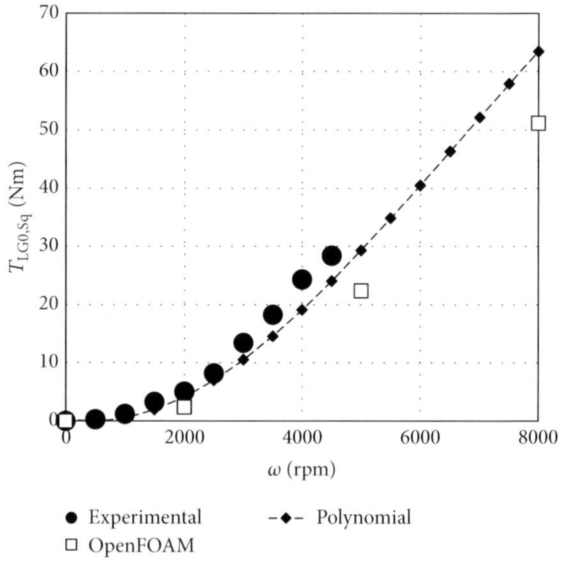

FIGURE 8: (a) Resistant torque generated by 2 mating gears and by 1 gear (×2): experimental and numerical values and (b) squeezing torque evaluated as the difference between the torque generated by 2 engaging gears and a single rotating gear (×2): experimental and numerical values.

able to predict also this phenomena with good accuracy and therefore can be elected as suitable for the prediction of each source of load independent losses of gears.

Regarding computational effort the solution time of each simulation was reduced from approximately 20 h on 16 3.3 GHz CPUs (211 GFLOPS) needed in a general-purpose commercial environment, to 21 h on a single 3.2 GHz CPU (12.8 GFLOPS) with a net reduction of the computational time of 93.5% (considering a perfect scalability). This is mainly due to two aspects. The first one is the reduced time for the mesh-handling and regeneration. The solution of the mass and momentum conservation equations represent just a small part of the total computational time and the presented approach takes less than 1/12th of the time needed by the commercial solver for the mesh generation. The second can be found in the capability of OpenFOAM to reach, with this specific mesh-handling technique/mesh type, convergence also for bigger time steps ensuring a significant reduction of the number of time steps needed to reach the regime condition. The commercial code remeshing algorithm substitutes the original distorted cells with tetrahedrons that can have also very small sizes affecting the maximum possible time step (considering to limit the Co to 1) leading to meshes that are bigger in size with respect to the initial one.

8. Conclusions

The results of the experiments confirm that CFD represents a valid method to predict power losses. The error in the predictions with respect to the measured ones is less than 10%. This little difference can be explained with the fact that the geometry was simplified. Simulations with and without the engagement show the capability of the model to predict also the squeezing effects.

The main limitation of the wide diffusion of such an approach is traditionally the big computational effort both for modeling and for solving and the license costs. These drawbacks were overcome with the development of specific algorithms capable of handling the geometry generation, the boundary motion, and effective mesh generation and handling. This ensured a reduction of the computational effort of about 93.5%. In other words, with this approach it is possible to perform 15 times more simulations with the same computational effort making such technology suitable for industrial applications.

A future work will be the systematic application of this approach for a better understanding of the power loss phenomena and the influence of the different operating conditions.

The method can be of course generalized. The first simulations on a planetary configuration [21, 22] have shown good agreement with the experimental evidences and a net computational time reduction of approximately 93% with respect to the simulation performed by the authors with commercial software. For this reason the author maintains that this approach has a big potential and can be successfully extended to each gear geometry and possible gearbox configuration showing the same benefits highlighted with the presented example.

Competing Interests

The authors declare that they have no competing interests.

References

[1] INA/FAG, Rolling Bearings, Schaeffler Technologies GmbH & Co. KG, 2007.

[2] SKF, Catalogo Generale, 6000 IT, 2006.

[3] D. Talbot, S. Li, and A. Kahraman, "Prediction of mechanical power loss of planet gear roller bearings under combined radial and moment loading," *Journal of Mechanical Design*, vol. 135, no. 12, Article ID 121007, 2013.

[4] S. Li and A. Kahraman, "Prediction of spur gear mechanical power losses using a transient elastohydrodynamic lubrication model," *Tribology Transactions*, vol. 53, no. 4, pp. 554–563, 2010.

[5] G. Niemann and H. Winter, *Maschinenelemente—Band 2: Getriebe Allgemein, Zahnradgetriebe—Grundlagen, Stirnradgetriebe—2. Auflage*, Springer, Berlin, Germany, 2003.

[6] W. Mauz, *Hydraulische Verluste von Stirnradgetrieben bei Umfangsgeschwindigkeiten bis 60 m/s*, Bericht des Institutes für Maschinenkonstruktion und Getriebebau nr. 159, Universität Stuttgart, Stuttgart, Germany, 1987.

[7] C. Gorla, F. Concli, K. Stahl et al., "CFD simulations of splash losses of a gearbox," *Advances in Tribology*, vol. 2012, Article ID 616923, 10 pages, 2012.

[8] P. H. Dawson, "Windage loss in larger high-speed gears," *Proceedings of the Institution of Mechanical Engineers*, vol. 198, no. 1, pp. 51–59, 1984.

[9] S. Seetharaman and A. Kahraman, "Load-independent spin power losses of a spur gear pair: model formulation," *Journal of Tribology*, vol. 131, no. 2, Article ID 02220111, 2009.

[10] Y. Ariura and T. Ueno, "Lubricant churning loss and its behavior in gear box in cylindrical gear systems," *Journal of Japan Society of Lubrication Engineers*, vol. 20, no. 3, pp. 167–175, 1975.

[11] A. S. Terekhov, "Hydraulic losses in gearboxes with oil immersion," *Russian Engineering Journal*, vol. 55, no. 5, pp. 7–11, 1975.

[12] R. J. Boness, "Churning losses of discs and gears running partially submerged in oil," in *Proceedings of the International Power Transmission and Gearing Conference: New Technologies for Power Transmissions of the 90's*, pp. 355–359, Design Engineering Division, ASME, Chicago, Ill, USA, April 1989.

[13] R. B. Höhn, K. Michaelis, and T. Vollmer, *Thermal Rating of Gear Drives Balance between Power Losses and Heat Dissipation*, AGMA FTM, 1996.

[14] P. Luke and A. V. Olver, "A study of churning losses in dip-lubricated spur gears," *Proceedings of the Institution of Mechanical Engineers, Part G: Journal of Aerospace Engineering*, vol. 213, no. 5, pp. 337–346, 1999.

[15] Y. Marchesse, C. Changenet, F. Ville, and P. Velex, "Investigations on CFD simulations for predicting windage power losses in spur gears," *Journal of Mechanical Design*, vol. 133, no. 2, Article ID 024501, 2011.

[16] M. J. Hill, R. F. Kunz, R. B. Medvitz et al., "CFD analysis of gear windage losses: validation and parametric aerodynamic studies," *Journal of Fluids Engineering*, vol. 133, no. 3, Article ID 031103, 2011.

[17] C. Gorla, F. Concli, K. Stahl et al., "Hydraulic losses of a gearbox: CFD analysis and experiments," *Tribology International*, vol. 66, pp. 337–344, 2013.

[18] C. Gorla, F. Concli, K. Stahl et al., "CFD simulations of splash losses of a gearbox," *Advances in Tribology*, vol. 2012, Article ID 616923, 10 pages, 2012.

[19] F. Concli and C. Gorla, "Analysis of the oil squeezing power losses of a spur gear pair by mean of CFD simulations," in *Proceedings of the ASME 11th Biennial Conference on Engineering Systems Design and Analysis (ESDA '12)*, pp. 177–184, Nantes, France, July 2012.

[20] F. Concli and C. Gorla, "Oil squeezing power losses in gears: a CFD analysis," *WIT Transactions on Engineering Sciences*, vol. 74, pp. 37–48, 2012.

[21] F. Concli, C. Gorla, and E. Conrado, "Analysis of power losses in an industrial planetary speed reducer: measurements and computational fluid dynamics calculations," *Proceedings of the Institution of Mechanical Engineers Part J: Journal of Engineering Tribology*, vol. 228, no. 1, pp. 11–21, 2014.

[22] F. Concli and C. Gorla, "Computational and experimental analysis of the churning power losses in an industrial planetary speed reducer," *WIT Transactions on Engineering Sciences*, vol. 74, pp. 287–298, 2012.

[23] F. Concli, C. Gorla, A. Della Torre, and G. Montenegro, "Windage power losses of ordinary gears: different CFD approaches aimed to the reduction of the computational effort," *Lubricants*, vol. 2, no. 4, pp. 162–176, 2014.

[24] F. Concli, C. Gorla, A. Della Torre, and G. Montenegro, "Churning power losses of ordinary gears: a new approach based on the internal fluid dynamics simulations," *Lubrication Science*, vol. 27, no. 5, pp. 313–326, 2015.

[25] OpenFOAM, http://www.openfoam.com.

[26] H. K. Versteeg and W. Malalasekera, *An Introduction to Computational Fluid Dynamics—The Finite Volume Method*, Longman Group, London, UK, 1995.

[27] G. Montenegro, A. Della Torre, A. Onorati et al., "CFD simulation of a sliding vane expander operating inside a small scale ORC for low temperature waste heat recovery," SAE Technical Paper 2014-01-0645, 2014.

[28] T. Lucchini, A. Della Torre, G. D'Errico, G. Montenegro, M. Fiocco, and A. Maghbouli, "Automatic mesh generation for CFD simulations of direct-injection engines," SAE Technical Paper 2015-01-0376, 2015.

[29] SALOME, http://www.salome-platform.org.

[30] Netgen, https://gitlab.asc.tuwien.ac.at/jschoeberl/ngsolve-docu/wikis/home.

[31] B.-R. Höhn, K. Michaelis, and H.-P. Otto, "Influences on no-load gear losses," *Ecotrib*, vol. 2, pp. 639–644, 2011.

12

Friction Reduction Capabilities of Silicate Compounds used in an Engine Lubricant on Worn Surfaces

Devendra Singh, G. D. Thakre, L. N. Sivakumar Konathala, and V. V. D. N. Prasad

CSIR-Indian Institute of Petroleum, Mohkampur, Haridwar Road, Dehradun 248005, India

Correspondence should be addressed to Devendra Singh; devendra@iip.res.in

Academic Editor: Yuansheng Jin

Effects of magnesium silicate and alumina dispersed in engine lubricant on friction, wear, and tribosurface characteristics are studied under boundary and mixed lubrication conditions. Magnesium silicate and alumina, henceforth called as friction reducing compounds (FRC), were dispersed in engine lubricant in very low concentration of 0.01% weight/volume. Four-ball wear test rig was used to assess friction coefficient and wear scar diameter of balls lubricated with and without FRC based engine lubricant. Scanning electron microscopy (SEM) equipped with Energy Dispersive X-ray (EDX) was used to analyse the tribosurface properties and elemental distributions on worn surfaces of the balls. Test results revealed that FRC based engine lubricant increases friction coefficient but marginally reduces wear scar diameter of new balls, whereas, test on the worn-out balls running on FRC based engine lubricants shows 46% reduction in friction coefficient compared to the new balls running on engine lubricants without FRC. Investigations on tribosurfaces with respect to morphology and elemental distribution showed the presence of Si and O elements in micropores of the worn surfaces of the balls, indicating role of FRC in friction coefficient reduction and antiwear properties. These FRC based engine lubricants may be used in the in-use engines.

1. Introduction

In the recent past a number of original equipment manufacturers (OEMs) recommended a lower viscosity grade engine lubricant to improve fuel economy of their vehicles. Effect of engine lubricant viscosity on fuel consumption was studied by many researchers and it has been reported that lower viscosity grade engine lubricants result in reduction in engine fuel consumption [1, 2]. But there are some concerns associated with the use of a low viscosity grade engine lubricant in terms of wear characteristics and durability of an engine. Most vulnerable parts of an engine prone to wear, associated with lower viscosity grade engine lubricants, are the surfaces operating in boundary lubrication regime such as cam and follower of valve train system, top dead centre (TDC), and bottom dead centre (BDC) of a cylinder liner. Therefore, these low viscosity grade engine lubricants require some sort of surface modifiers which must act on the tribosurfaces operating in boundary lubrication regime via physical or chemical adsorption and forming a protective tribofilm, able to reduce friction coefficient and preventing the excessive wear of the engine parts during engine operation.

During the last decade, many researchers have studied serpentine mineral, very similar to the FRC used in this study, as a potential surface modifier possessing excellent tribological characteristics. Serpentine mineral, basically a clay material, is stoichiometrically represented by the chemical formula of magnesium silicate hydroxide, $Mg_6Si_4O_{10}(OH)_8$. Serpentine-group minerals are composed of chrysotile, lizardite, and antigorite, and based on the analysis it emerged that these are either members of a complex solid solution series or are separate chemical species,= but are not simply polymorphs [3]. Yuansheng et al. [4] had put some light on the basic structure of $Mg_6Si_4O_{10}(OH)_8$; it was shown that Si–O tetrahedron plane and the Mg–O/OH octahedron plane form the layered structure via cross-linkage between them. A concept of *in situ* mechanochemical reconditioning of worn tribosurfaces of a locomotive diesel engine using serpentine powder added in engine oil in the ratio of 5 mg/mL, with particle size not larger than 2 μm, was also explained. Nanohardness of protective layer formed was reported to be twice as high as that of the substrate and surface roughness of 0.0694 μm, coefficient of friction between real piston ring and cylinder bore, was stabilized to 0.005. Yu et al. [5] studied the compositions of a thin tribofilm generated on worn surface by using 1.5 weight% serpentine powders, with an average size of 1 μm, dispersed in SN 500 mineral base oil. It was reported that a nanocrystalline tribofilm was generated on worn surface with a thickness of 500–600 nm, mainly composed of Fe_3O_4, FeSi, SiO_2, AlFe, and Fe–C compounds (Fe_3C). Zhang et al. [6] investigated the tribological characteristics and self-repairing effect of 5% hydroxy-magnesium silicate (HMS) with average diameter about 0.5 μm, dispersed in gasoline engine lubricant SJ 10W40 on steel-to-steel friction pairs "plane on plane" configuration with various surface roughness; it was stated that wear resistance of HMS self-repairing material has a relationship with the surface roughness of friction pairs and the average diameter of self-repairing material particle. When the powder average diameter is close to the surface roughness value of friction pairs, the HMS self-repairing material shows friction reduction, antiwear, and self-repairing properties. Nan et al. [7] also used 0.5 wt% as an optimum concentration of ultrafine magnesium aluminum silicate powders, as lubricant additive in SRV oscillating friction and wear tester and found that ultrafine powder was very efficient in reducing friction and wear at different loads (10 N–100 N). Tribological characteristics of serpentine powder used in engine oil or base oil with respect to formation of tribofilm, surface modifying capabilities, reduction of friction coefficient, and wear was extensively studied by many other researchers [8–13]. Similar results on reduction of friction coefficients by 24.63% and diameters of friction spots by 41.88% as compared to the base oil were reported by Zhao et al. [14]. Lyubimov et al. [15] reported the merits of Kaolin based powder over serpentine powder in terms of tribological characteristics.

In contrast to other researchers, an alternate mechanism of the self-repairing function was studied by Yue et al. [16]. According to his work, aluminum and silicon elements of silicate particles were not found in the repaired layer, hypothesizing that the silicate particles did not participate in the formation of the repaired layer rather it act as a catalyst to promote a series of complex tribochemical reactions to form a regenerated layer with amorphous carbon structure on the worn surface under high-friction temperature and pressure in the friction and wear process.

Qi et al. [17, 18] proposed generation mechanism of tribofilm formed by nanoscale serpentine powder (average size < 100 nm) at a high temperature of 400°C. Effect of thermal activation on tribological characteristics of serpentine ultrafine powder added in liquid paraffin was studied by Yu et al. [19]; it was found that a temperature range from 300°C to 600 °C increases the film forming ability of the serpentine and at higher temperature beyond 850°C the layer structure is destroyed and aggravated the friction and wear.

Based on the literature review it may be inferred that serpentine mineral was added in engine lubricant or base oil by researchers in concentration ranging from 0.025% to 5%, normal load applied in different tribometers was varying from 50 N to 200 N, and experiments were conducted, either at very high temperature ranging from 300°C to 600 °C or at room temperatures. In this paper a four-ball wear test rig was used to investigate the lubricity properties of FRC based engine lubricant, applied load on balls was around 490 N, and engine lubricant temperature was controlled at 90°C, simulating the boundary lubrication condition of a typical cam and follower of valve train system of an internal combustion engine. A very low concentration (0.01%) weight/volume of FRC (a mixture of Mg silicates hydroxide and alumina) was added in an engine lubricant SAE 5W-30, API SL/CF. SEM-EDX was used to examine and analyse the role of FRC in influencing surface properties of worn surfaces.

2. Experimental Set Up

2.1. Engine Lubricants Sample Preparation. Friction reducing compound was synthesized by crushing and mixing Mg silicates, alumina, and some catalyst together in pestle mortar. Chemically the main constituent of the FRC powder belongs to the serpentine family, which is stoichiometrically represented by the chemical formula of magnesium silicate hydroxide, $Mg_6Si_4O_{10}(OH)_8$. The average particle size of FRC was around 9 μm. Engine lubricant sample for the experiments was prepared by adding 1 mg of FRC in 10 mL of engine lubricant, SAE 5W-30, API CF/SL. Engine lubricant sample comprising of FRC was prepared by first stirring it using a magnetic stirrer at 2000 rpm followed by an ultrasonic bath treatment for one hour, keeping bath temperature at 50°C. Following engine lubricants samples were used for the study:

Oil A, SAE 5W-30,

Oil A1, 0.1 mg of FRC added in 1 mL of SAE 5W-30.

It was also observed that by adding FRC in low concentrations of 0.1 mg/mL, it does not affect physicochemical characteristic of the engine lubricants adversely.

2.2. Rolling Contact Fatigue Four Ball Test Rig. The tribological performance of lubricants was carried out on a four-ball tribotester shown in Figure 1. The tribotester utilizes four-ball geometry in a trapezoidal form. The top ball is fixed into the spindle and rotates at the predefined speed, that is, 1200 rpm. The bottom three balls are fixed in a ball cradle filled with lubricant. The four balls make three-point contacts. The test conditions employed are given in Table 1. These operating conditions were slightly different from that prescribed in ASTM D 4172, applied load was kept as 50 kg instead of 40 kg, and oil temperature was kept as 90°C in place of 54°C, to adequately represent the boundary lubrication conditions of a typical cam and follower of valve train system of an internal combustion engine. Friction encountered within the contact is continuously monitored and recorded using the data acquisition software.

2.3. SEM-EDX. FESEM (Field Emission Scanning Electron Microscope) with EDX/EDS (Energy Dispersive X-ray/Spectroscopy) system was used to examine and analyse the morphology and elemental distributions of worn surfaces of the balls.

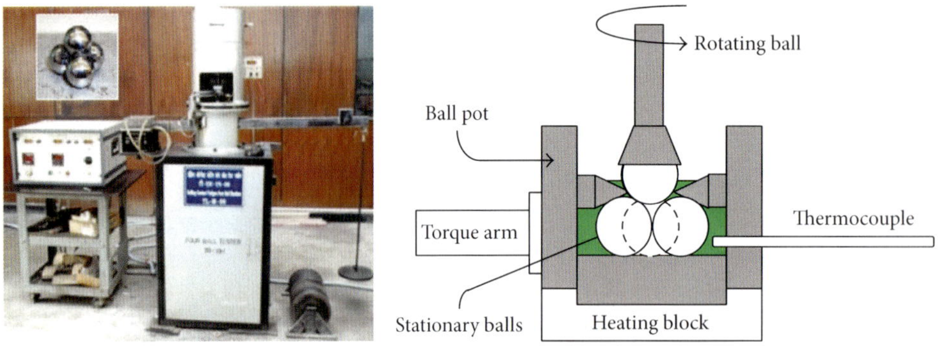

FIGURE 1: Four-ball tribotester.

TABLE 1: Test operating conditions of four-ball test rig.

1	Load (kgf)	50
2	Ball specimen diameter (mm)	12.7
3	Speed (rpm)	1200
4	Oil temperature (°C)	90
5	Ball type	EN 31 steel
6	Ball roughness R_a (microns)	0.6
7	Composition of EN 31 steel	C: 1%, Mn: 0.50%, Cr: 1.40%, Si: 0.2%

TABLE 2: Test matrix of the experiment.

Test number	Duration in min	Type of balls	Engine lubricants
1	60	New	Oil A
2	60	New	Oil A1
3	120	New	Oil A
4	120	New	Oil A1
5a	60	New	Oil A
5b	75	Balls used in test 5a	Oil A1

3. Test Matrix and Methodology

A comparative performance of the engine lubricants without FRC (Oil A) and with FRC (Oil A1) was conducted on a four-ball test rig with respect to friction coefficient, wear scar diameter, and surface properties. Test matrix of the experimental study is shown in Table 2. A total of five tests were conducted, first two tests were carried out on new balls using Oil A and Oil A1 each of one hour' duration, next two tests were conducted on new balls for two hours each to study the effect of extended duration on friction coefficient and wear, and last test was conducted in two phases, initially with Oil A (test number 5a) for one hour, then draining Oil A without disturbing the balls positions and refilling it with Oil A1 for carrying out the test 5b on same used balls till friction coefficient gets stabilized.

Bench marking of test rig in terms of friction coefficient and wear scar diameter of balls was done in test number 1 using Oil A on new balls for one hour' duration, followed by test number 2 using Oil A1 on new balls for one hour. These tests were conducted in order to examine the efficacy of FRC, added in engine lubricant in very low concentration, on friction coefficient, wear scar diameter, and surface properties of new balls specimens.

In order to investigate the effect of test duration on friction coefficient, wear scar diameter, and surface properties, two more experiments (experiments numbers 3 and 4) were conducted on new balls using Oil A and Oil A1 for extended test duration of two hours each.

Finally, test number 5a was conducted on new balls using Oil A for one hour followed by test number 5b on same set of used balls (worn-out balls without changing the tribosurfaces of interacting balls) that were undertaken using Oil A1 until stabilization of friction coefficient. Friction coefficient stabilization was reached within 1 hour and 15 min (4500 sec). This test was carried out to investigate the effect of FRC added in engine lubricant, that is, Oil A1, on worn-out balls with respect to friction coefficient, wear scar, and surface properties.

Friction coefficient for each test was obtained from the measured value of friction torque of the contact surface geometry of the balls by a data logging system.

Wear scar diameter (WSD) in mm, for each oil sample, was calculated by taking the average worn scars of all three stationary balls. Morphology and elemental analysis of the worn surfaces of the balls was done by selecting a ball among the three stationary balls randomly followed by the SEM and EDX analysis on it, to understand the role of FRC on the surface properties.

4. Result and Discussion

4.1. Friction Coefficient and Wear Characteristics. Friction coefficients of the ball surfaces along with the standard deviation and wear scar diameter (WSD) results of all tests on rolling contact fatigue four-ball test rig are shown in Table 3.

4.1.1. Effect of FRC on Friction Coefficient and WSD of New Balls. It is observed from Table 3 that the addition of FRC in engine lubricant, even in very low concentration (0.01%), can influence friction coefficient of tribosurfaces. Wear scar diameters of balls run on both engine lubricants (Oil A and Oil A1) were almost similar during the one hour' test.

TABLE 3: Stabilized friction coefficient and wear scar diameter.

Test number	Duration in min	Types of balls	Engine lubricants	Stabilized friction coefficient	Standard deviations	Wear scar diameter (WSD), mm
1	60	New	Oil A	0.109	0.015	0.525
2	60	New	Oil A1	0.142	0.015	0.525
3	120	New	Oil A	0.098	0.009	0.650
4	120	New	Oil A1	0.131	0.015	0.600
5a	60	New	Oil A	0.109	0.012	0.660
5b	75	Balls used in test 5a	Oil A1	0.076	0.020	

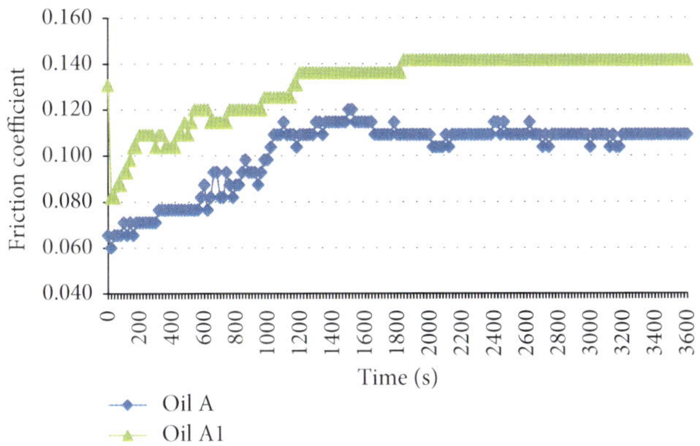

FIGURE 2: Friction coefficient variation with time using new balls (tests 1 and 2).

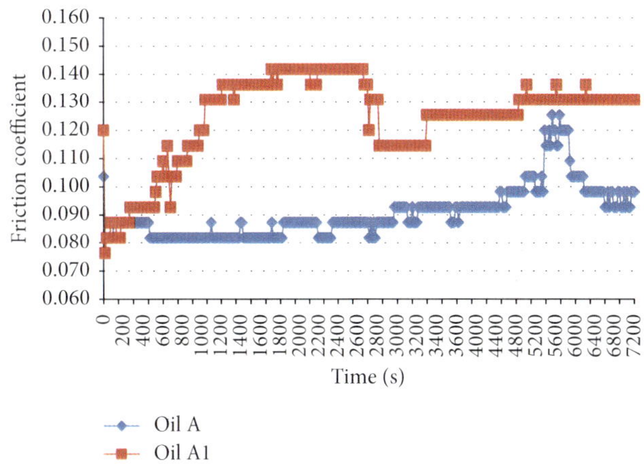

FIGURE 3: Friction coefficient variation with time using new balls (tests 3 and 4).

Figure 2 shows the comparative results of friction coefficient of new balls running on Oil A and Oil A1 each for one hour' duration. Comparative test results revealed that FRC based engine lubricant that is Oil A1 aggravates the friction between the interacting balls, resulting in higher values of friction coefficient as compared to Oil A, which showed lower values of friction coefficient, throughout the test. This may be explained with the fact that Mg silicate particles may act as abrasives on smooth surfaces of new balls having surface roughness values of $R_a = 0.6\,\mu$m, which is much lower than the average size of FRC particles, that is, 9 μm. Bigger diameter FRC particle may bring certain degree of grain abrasion during the test, which enhances frictional force between interacting surface of balls resulting in increase in friction coefficients. Hence, surface roughness and FRC particle size play a vital role in influencing friction; this finding is consistent with the study done by Zhang et al. [6]. Friction coefficient values varied from 0.082 to 0.142 for Oil A1 and for Oil A; it varied from 0.065 to 0.109 and according to Stribeck curve [20], friction pairs with friction coefficient $\mu \geq 0.10$ were considered as operating in boundary lubrication conditions, and those with $0.001 < \mu < 0.10$ were assumed to operate in mixed lubrication regime. So inferences may be drawn from these results that the lubrication regime of interacting balls transformed from mixed to boundary for Oil A1 owing to presence of FRC, whereas lubrication regime for Oil A in test 1 predominantly remained in mixed lubrication only.

4.1.2. Effect of Extended Test Duration on Friction Coefficient and WSD of New Balls. Another set of test was conducted on new balls for extended time duration of two hours, lubricated with Oil A and Oil A1 (tests numbers 3 and 4). These tests were conducted to investigate the role of FRC in formation of protective film on tribosurfaces if any, which reduces the friction coefficient. It is observed from Figure 3 that the friction coefficient values for Oil A1 keep on increasing from 0.076 to 0.131, during the whole test and stabilized at 0.131 which was very close to the stabilized value of 0.142 obtained for Oil A1 during the one hour' test (test number 2). This increasing trend of friction coefficient value negated the idea of friction reduction capabilities of FRC. On the contrary it may be inferred from the results of tests numbers 2 and 4, shown in Figures 2 and 3, that FRC based engine lubricant enhances the friction coefficient of new balls under the prescribed operating conditions irrespective of test duration. Wear performance of FRC in terms of wear scar diameter (WSD) shows marginal reduction in WSD values, that is, from 0.650 for Oil A to 0.600 for Oil A1, as indicated in Table 3.

4.1.3. Effect of FRC on Friction Coefficient and WSD of Used Balls. Test 5 was carried out in two phases, 5a and 5b; during first phase, friction coefficient value which varied from 0.082 to 0.109 for Oil A as shown in Figure 4 and

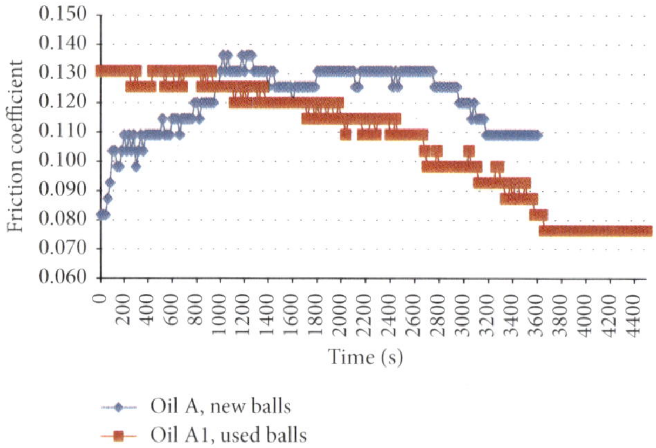

FIGURE 4: Friction coefficient variation with time using worn-out balls (tests 5a and 5b).

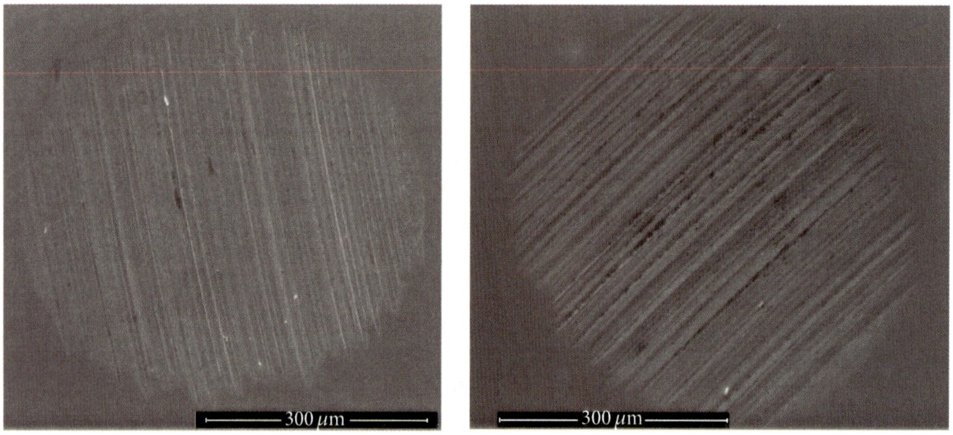

(a) Oil A (b) Oil A1

FIGURE 5: SEM images of ball after 1 hr test (a) ball run on Oil A and (b) ball run on Oil A1.

when Oil A1 was used in second phase of test on the same set used balls (worn surfaces), and friction coefficient values were continuously decreasing from 0.131 to 0.076 and stabilized at 0.076 after 75 minutes. By this test it was demonstrated that FRC based engine lubricants, that is, Oil A1, performed as an excellent friction reducing compound for the worn surfaces. Lubrication regime also transformed from boundary condition to mixed lubrication condition. It may be depicted from these results that FRC may acts as friction reducing compound when used on worn surfaces and significantly reduces the friction coefficient. Further it is hypothesized that on worn surfaces, FRC particles may get adsorbed and embedded into the micropits or furrows and showed some reaction activity on interacting surfaces culminating into much smoother surfaces and ultimately helps in friction reduction. It is assumed that some sort of friction reducing film may be formed on the interacting ball surfaces which help in friction coefficient reduction.

4.2. Surface Morphology and Elemental Composition

4.2.1. Effect of FRC on Surface Morphology. SEM images of worn surface of balls run on Oil A and Oil A1 are shown in Figures 5(a) and 5(b). Certain degrees of furrow were observed in the rubbing directions, but these rubbing marks are more pronounced and prominent in case of ball run on Oil A1. This may be explained with the fact that FRC particles, mainly silicates, may act as an abrasive on a smooth surface of EN31 steel ball under high load of 490 N which results in higher value of friction coefficient and large number of typical grain abrasion and fatigue wear.

Elemental composition of the highlighted surface shown in Figures 6(a) and 6(b) demonstrates the presence of various base elements of EN31 alloy steel balls like Fe, C, S, P, and Cr, whereas presence of Ca, O, and Zn on worn surface of ball may be originated from engine lubricant additives (antioxidant, detergent). Figure 6(a) shows the elemental composition of the ball run on Oil A and absence of Si from the worn surface indicates that the balls were run on engine lubricant (Oil A) without FRC. Presence of Si, O elements on the worn surface of the ball run on Oil A1, shown in Figure 6(b), confirms the role of FRC in influencing the new ball surface properties in terms of deterioration and abrasion of smooth surfaces.

4.2.2. Effect of FRC on Surface Morphology during Extended Hour Test. Figures 7(a) and 7(b) shows micrographs of worn

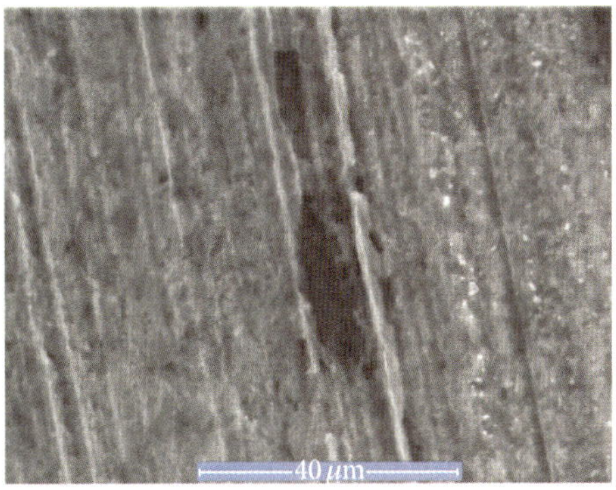

Element	Weight%
C K	13.39
O K	13.95
P K	05.01
S K	01.03
CaK	05.49
CrK	01.28
FeK	58.31
ZnK	01.53

(a)

Element	Weight%
C K	13.89
O K	14.75
ZnL	5.81
SiK	0.36
P K	2.12
S K	2.65
CaK	2.48
CrK	1.54
FeK	56.41

(b)

FIGURE 6: Elemental composition of the ball surfaces run on (a) Oil A and (b) Oil A1.

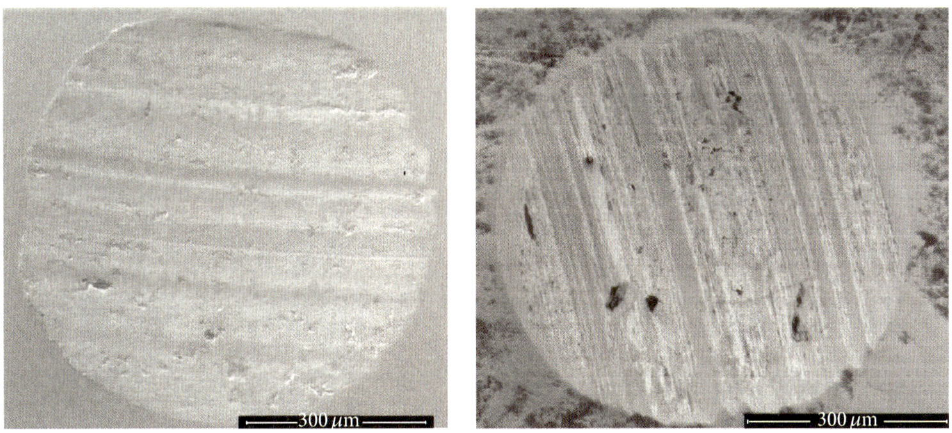

(a) (b)

FIGURE 7: SEM images of ball after 2 hr test (a) ball run on Oil A and (b) ball run on Oil A1.

surface of the tests run on Oil A and Oil A1, respectively, for two hours each. Many deep scratches and adhesion phenomenon budding on worn surfaces of steel ball were observed for Oil A1, indicating the worn mechanisms mainly as abrasive wear and adhesive wear. On the contrary, lesser scratch marks were observed in the micrograph of ball surfaces run on Oil A, shown in Figure 7(b). This may be explained with the fact that FRC was absent in Oil A, resulting in lesser abrasive wear on the worn surface.

4.2.3. Surface Morphology of Balls Used in Tests Numbers 5a and 5b. Figure 8 shows the micrographs of the worn surface, which are initially lubricated with Oil A using new balls for one hour followed by test on same set of balls (worn-out balls

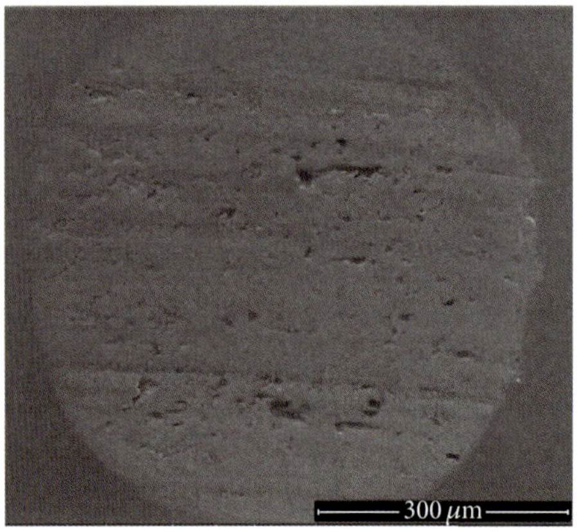

Figure 8: SEM image of a ball which initially run on Oil A for 60 min and then on Oil A1 without changing the surface contact for 75 min.

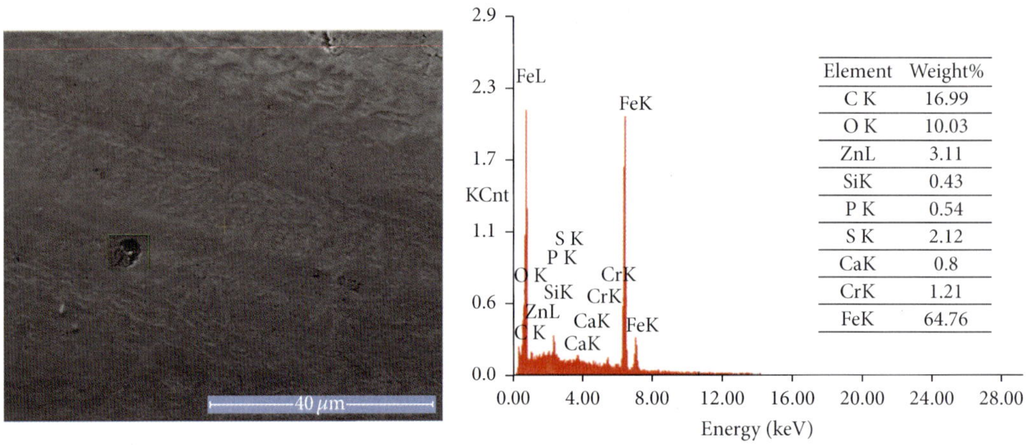

Element	Weight%
C K	16.99
O K	10.03
ZnL	3.11
SiK	0.43
P K	0.54
S K	2.12
CaK	0.8
CrK	1.21
FeK	64.76

Figure 9: SEM image and elemental composition of surface of a ball which initially run on Oil A for 60 min and then on Oil A1 using same balls for 75 min.

without changing the tribosurfaces of interacting balls) using Oil A1 until stabilization of friction coefficient. The surface exhibits much smoother morphologies but a large number of micropits were also seen on the worn surface.

Figure 9 illustrated a typical elemental composition of one of the micropits on the worn surface. Presence of Si and O, in micropits, corroborates the role of FRC in friction reduction of interacting surfaces, consistent with the previous studies, highlighting the benefits of these micropits on the friction reducing mechanisms of surface texture [19, 21]. Micropits on friction surface can disarm abrasive dusts by entrapping them, thereby suppressing grain abrasion and adhesion. Besides, it is obvious that the micropits can act as oil reservoirs, which transport or retain oil to be released in emergency situations. The grinding action between tribosurface due to foreign particles removed from worn surface were captured by the micropits during the sliding and shearing of the worn surfaces, resulting in the reduction in abrasive wear of the contact surface.

Figures 10(a) and 10(b) describe the elemental distribution of the worn surface of the ball used for test run 5a and 5b and ball run on Oil A1 for one hour' test. Presence of Fe Si and O on the worn surfaces of both balls accounted for the possibility of some sort of ferrosilicate layer on tribosurfaces which may help in friction reduction.

5. Conclusions

This study revealed some of the important facts about the performance of FRC based engine lubricant with respect to friction, wear, and surface properties using a four ball tribometer. The following points may be concluded:

(i) Engine lubricant with very low concentration of FRC influenced the friction coefficient.

(ii) FRC based engine lubricant aggravates the friction between the interacting surfaces of new balls and lubrication regime transformed from mixed to boundary conditions during the test. SEM micrographs demonstrated it with the deep scratch marks in the rubbing direction when the FRC based engine lubricant was used.

(iii) Friction reducing performance of FRC based engine lubricant on worn surfaces is phenomenal and during

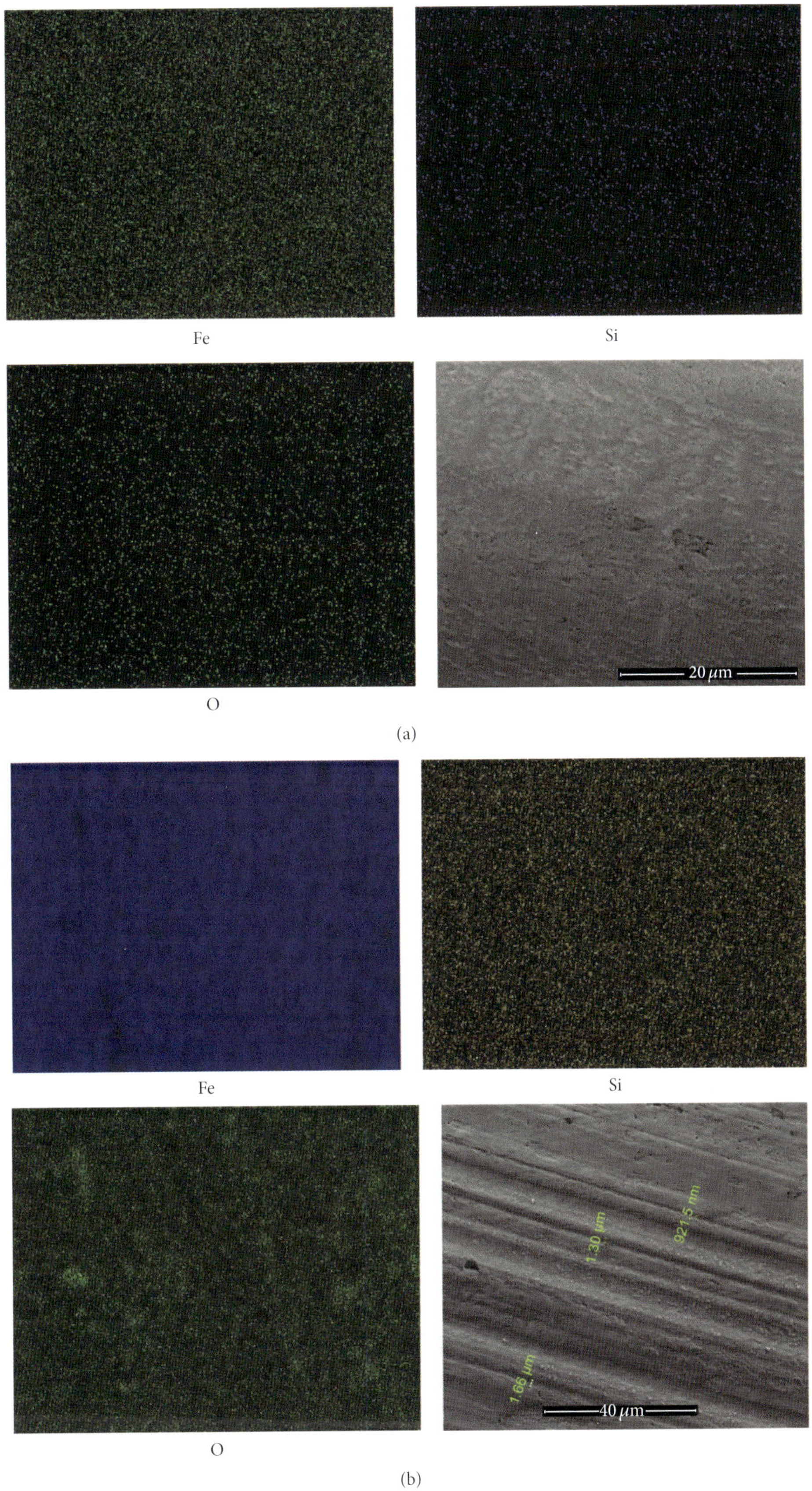

FIGURE 10: Worn surface mapping of the balls used in (a) tests numbers 5a, 5b and (b) test number 2 demonstrating the presence of Fe, Si, and O.

the test on worn surfaces, lubrication regime transformed from boundary condition to mixed lubrication condition.

(iv) Presence of Fe, Si, and O on the worn surfaces of the used balls may account for the ferrosilicate layer on tribosurfaces.

Conflict of Interests

The authors declare that there is no conflict of interests regarding the publication of this paper.

References

[1] R. I. Taylor, "Engine friction: the influence of lubricant rheology," *Proceedings of the Institution of Mechanical Engineers. Part J: Journal of Engineering Tribology*, vol. 211, no. 3, pp. 235–246, 1997.

[2] D. Singh, J. D. Fieldhouse, A. K. Jain, M. R. Tyagi, and S. K. Singal, "Investigating the effect of operating variables and engine lubricant viscosity on engine friction—a DOE approach," SAE Technical Paper 2011-28-0052, 2011.

[3] J. P. Norman, "Chemical differences among the serpentine polymorphs," *American Mineralogist*, vol. 53, pp. 201–215, 1968.

[4] J. Yuansheng, L. Shenghua, Z. Zhengye, Y. He, and W. Feng, "In situ mechanochemical reconditioning of worn ferrous surfaces," *Tribology International*, vol. 37, no. 7, pp. 561–567, 2004.

[5] H. Yu, Y. Xu, P. Shi et al., "Microstructure, mechanical properties and tribological behavior of tribofilm generated from natural serpentine mineral powders as lubricant additive," *Wear*, vol. 297, no. 1-2, pp. 802–810, 2013.

[6] B. Zhang, B.-S. Xu, Y. Xu, and B.-S. Zhang, "Tribological characteristics and self-repairing effect of hydroxy-magnesium silicate on various surface roughness friction pairs," *Journal of Central South University of Technology*, vol. 18, no. 5, pp. 1326–1333, 2011.

[7] F. Nan, Y. Xu, B. Xu, F. Gao, Y. Wu, and Z. Li, "Tribological behaviors and wear mechanisms of ultrafine magnesium aluminum silicate powders as lubricant additive," *Tribology International*, vol. 81, pp. 199–208, 2015.

[8] W. Chen, Y. Gao, and H. Zhang, "XPS and SEM analyses of self-repairing film formed by mineral particles as lubricant additives on the metal friction Pairs," in *Advanced Tribology: Proceedings of CIST2008 & ITS-IFToMM2008*, pp. 660–664, Springer, Berlin, Germany, 2010.

[9] K. N. Dolgopolov, D. N. Lyubimov, A. G. Ponomarenko, G. G. Chigarenko, and M. V. Boiko, "The structure of lubricating layers appearing during friction in the presence of additives of mineral friction modifiers," *Journal of Friction and Wear*, vol. 30, no. 5, pp. 377–380, 2009.

[10] L. I. Pogodaev, I. A. Buyanovskii, E. Y. Kryukov, V. N. Kuz'min, and V. V. Usachev, "The mechanism of interaction between natural laminar hydrosilicates and friction surfaces," *Journal of Machinery Manufacture and Reliability*, vol. 38, no. 5, pp. 476–484, 2009.

[11] P. Wang, J. Lv, L. Wang, Q. Ma, and X. Zhu, "Research on tribological properties of serpentine particles as lubricating oil additives," *Advanced Materials Research*, vol. 284–286, pp. 1001–1005, 2011.

[12] D. N. Lyubimov, K. N. Dolgopolov, A. T. Kozakov, A. V. Nikolskii, and E. A. Glazunova, "Tribochemical aspects of interactions between high-dispersed serpentine particles and metal friction surface," *Journal of Friction and Wear*, vol. 33, no. 2, pp. 108–114, 2012.

[13] J. Zhang, B. Tian, and C. Wang, "Long-term surface restoration effect introduced by advanced silicate based lubricant additive," *Tribology International*, vol. 57, pp. 31–37, 2013.

[14] F. Zhao, Z. Bai, Y. Fu, D. Zhao, and C. Yan, "Tribological properties of serpentine, $La(OH)_3$ and their composite particles as lubricant additives," *Wear*, vol. 288, pp. 72–77, 2012.

[15] D. N. Lyubimov, K. N. Dolgopolov, A. T. Kozakov, and A. V. Nikolskii, "Improvement of performance of lubricating materials with additives of clayey minerals," *Journal of Friction and Wear*, vol. 32, no. 6, pp. 442–451, 2011.

[16] W. Yue, C. Wang, Y. Liu, H. Huang, Q. Wen, and J. Liu, "Study of the regenerated layer on the worn surface of a cylinder liner lubricated by a novel silicate additive in lubricating oil," *Tribology Transactions*, vol. 53, no. 2, pp. 288–295, 2010.

[17] X. W. Qi, Z. N. Jia, H. Chen, Y. L. Yang, and Z. Wu, "Self-repairing characteristics of serpentine mineral powder as an additive on steel-chromium plating pair under high temperature," *Tribology Transactions*, vol. 56, no. 3, pp. 516–520, 2013.

[18] X. W. Qi, Z. N. Jia, Y. L. Yang, and B. L. Fan, "Characterization and auto-restoration mechanism of nanoscale serpentine powder as lubricating oil additive under high temperature," *Tribology International*, vol. 44, no. 7-8, pp. 805–810, 2011.

[19] H. L. Yu, Y. Xu, P. J. Shi, H. M. Wang, W. Zhang, and B. S. Xu, "Effect of thermal activation on the tribological behaviours of serpentine ultrafine powders as an additive in liquid paraffin," *Tribology International*, vol. 44, no. 12, pp. 1736–1741, 2011.

[20] J. B. Heywood, *Fundametals of Internal Combustion Engine*, McGraw-Hill, New York, NY, USA, 1988.

[21] H. L. Yu, Y. Xu, P. J. Shi et al., "Tribological behaviors of surface-coated serpentine ultrafine powders as lubricant additive," *Tribology International*, vol. 43, no. 3, pp. 667–675, 2010.

13

Theoretical Analysis of the Influence of Asperity's Dimensions Affected by a Scale Factor on the Mixed Lubrication between Parallel Surfaces

F. Robbe-Valloire,[1] R. Progri,[1,2] and T. Da Silva Botelho[1]

[1] *QUARTZ (EA 7393), Supmeca, 3 rue Fernand Hainaut, 93407 Saint-Ouen, France*
[2] *Retired Supmeca Professor, France*

Correspondence should be addressed to F. Robbe-Valloire; francois.robbe-valloire@supmeca.fr

Academic Editor: Michael M. Khonsari

Mixed lubrication between a given pair of surfaces is directly related to the parameter $\eta u/p$. Any change in microgeometry produces a shift in the transition between the boundary and lubricated regimes. Using an asperity based model including five families of asperities (2 for full fluid lubrication and 3 for local boundary lubrication), we simulate mixed lubricated behavior. Our theoretical results confirm the relation between the mixed regime and the $\eta u/p$ parameter. All homothetic changes in microgeometry affecting the vertical and horizontal directions by the same scaling factor Sc induce a shift in the transition. The most interesting result is that this shift is exactly the same if speed u is scaled by 1/Sc with the initial microgeometry. This particular behavior, which is in good agreement with experimental results proposed in the literature, comes from the fact that behavior at each asperity can be written using dimensionless parameters. Most of these parameters are independent to any scaling of the microgeometry and only one, the speed parameter, needs to be artificially scaled in order to remain unchanged by the modification of the microgeometry.

1. Introduction

Mixed lubrication is a specific regime of lubrication characterized by an increasing friction coefficient (and wear) compared to that classically obtained when the fluid film properly separates both surfaces in contact (full lubrication regime). This increasing friction coefficient is related to the existence of part of contact where local film thickness can be so thin (on the order of few molecules of oil) that speed accommodation through the film cannot occur. As a consequence of this behavior, one can observe increasing friction (and wear). This situation takes place at the summit of the highest asperities, and, because of the classical range of microgeometry amplitude, it may be possible to observe a local continuous film on the other part of the contact pair, particularly at the lowest asperities. This physical aspect clearly shows the link between mixed lubrication and microgeometry, but despite the large amount of literature devoted to this relationship, it is not fully understood, in particular for large flat contacts. The main cause for this gap in understanding appears to be the number of diverse geometric configurations of lubricated contacts and the specificities of calculation associated with each one of them.

Local point and concentrated line contacts are probably the most studied typologies of lubricated rough contact. Theoretical studies about such contact are generally based on the well-known deterministic method using a mesh to model the microgeometry. The limited area of contact permits a deterministic description of the microgeometry and makes this numerical method well suited for studies of fluid perturbation in the case of a noncontact situation and/or contact under more severe conditions. A bibliography synthesis proposed by Hooke [1] in 2005 gives illustration of the number of publications using this numerical method. More recently, some works try to associate numerical Reynold's based analysis for fluid part of the interface with analytical statistical analysis of direct contact [2]. Concentrated contacts, such as that used in rolling contact between ball bearings or gears, generally working in rolling / sliding cinematic condition, are not considered in the present work. In this study, we only consider large flat contact.

Concerning large conformal contact, there is a lack of research on general relationship between mixed lubrication and microgeometry. Many studies are primarily devoted to evaluating the modification of the hydrodynamic effect induced by microgeometry. Only a few studies have also undertaken analysis of direct contact and the corresponding friction enhancement. The classical deterministic method cannot be directly used to model this situation because of the large number of points along the dimension of contact and the resolution needed for a microgeometry description. A number of studies have addressed the specific case of journal bearing using the deterministic method and have proposed evolutions of the method. Dobrica [3] suggests numerical simplification in order to reduce the number of numerical operations. Finite Element Method analysis, for both fluid and solid analysis, using a commercial finite element analysis software, is proposed by Albers [4], but only on a small representative part of the bearing. A coupled analysis using statistical analysis for direct contact and average flow Reynolds formulation is proposed by Zhang [5] for studying tribological effect of texturing in mixed lubrication. Multiscale analysis of defects was proposed by Fricke [6]. In this study, influence of surface form defect is calculated on the whole bearing, while microgeometry effect is quantifying using a representative small area of the bearing.

For mixed lubrication induced between parallel flat surfaces, publications including experimental results are very rare, particularly concerning the influence of microgeometry. Key articles on the subject are proposed by Emmens [7] and Felder [8], who present experimental based studies of the influence of microgeometry in the context of parallel surfaces. Both authors perform tribological tests of flat contact in the deep drawing of steel sheets where several microgeometries are present. Both authors confirm a mixed lubrication transition directly related to the microgeometry of the contacting surfaces. For the same set of contact parameters, the friction coefficient is reduced by a decrease in the height of microgeometry. Quantitatively, they suggest new parameter associations that include surface roughness based on the classical parameter $\eta u/p$. On the basis of their experiments, the use of these parameters gives a better localization of the points near a single curve (or master curve) than the use of the classical parameter $\eta u/p$. Emmens [7] proposes $H_e=\eta u/pR_p^2$ and adds the R_p parameter (altitude of the mean line of the profile from the lowest point). This new parameter H_e is homogeneous of a length. Emmens suggests that this length could be related to the periodicity of the microgeometry defects. Felder [8] proposed the dimensionless parameter $H_f= 3\eta lu/pR_a^2$, where R_a is the arithmetic mean value and l is the space between two consecutive asperities. He proposes this expression in order to obtain a parameter in the form of p_l/p, where p_l is the theoretical pressure induced by the pouches generated by the microgeometry. This theoretic pressure is obtained assuming a perfect regular microgeometry and Rayleigh's pad shape. More recently, Nogueira [9] completed these experimental studies, comparing the mixed lubrication behavior of various materials (steels, copper alloys) that have different microgeometries. In addition to the microgeometry effect, she observes a significant influence exerted by the mechanical properties of materials in contact. The best correlation of the experimental results is obtained using Emmens' parameter with a term that includes the elastic properties of the contacting materials $H_n= H_e / Eq^{2/3}$, where Eq is the equivalent Young's modulus of the contacting materials.

Existing theoretical approaches proposed for mixed lubrication in parallel flat surfaces mostly concern the case of externally pressurized contact used in the field of dynamic faced sealing. Lebeck [10] offers a wide-ranging survey of possible mechanisms and analyses effects such as undesirable pad tilting or thermal deformation or the squeeze effect coupling with dynamic behavior of the supports. But the most promising approach appears to be the asperity lubrication effect known as the microhydrodynamic effect. More recently, authors [11] attempt to integrate the statistical description of microgeometry, and the contact behavior is studied using the Greenwood-Williamson theory. All contacting asperities undergo plastic deformation while behavior for the noncontacting area is obtained using the classic Reynold's equation, including Patir and Cheng's flow factor, in order to take into account the effect of roughness on fluid flow.

Theoretical approach for parallel flat surfaces in contact without external pressurization is mainly devoted to the field of sheet metal forming. Methodology coupled numerical average flow Reynolds based analysis for fluid and statistical analysis for deformed asperities. In this case deformation regime of asperities is also only plastic [12]. Specific evolution takes into account flattening of workpiece asperities and ploughing induced by tool asperities [13].

For this situation of parallel lubricated contact without oil pressurization, we seek to construct the behavior of the contact also using the statistical analysis but by association of the local behavior of each asperity [14], in order to take into account contact situations where elastic to plastic deformation occurs. Contacting asperities can be assigned to one of three regimes (elastic, elastoplastic, or full plastic regimes) depending on the level of indentation [15]. Noncontacting asperities work with hydrodynamic lubrication or elastohydrodynamic lubrication depending on the level of elastic deformation. In the following, we aim to determine whether theoretical analysis can confirm the influence of microgeometry on mixed lubrication between parallel surfaces. This paper deals only with effects induced by a scaling of the microgeometry. This kind of modification affects by similar amplification both the vertical and lateral directions of the microgeometry. In our methodology, we study a specific case and simulate the scaling of microgeometries tested in the same condition (Section 3). The surprisingly consistent shift of the mixed lubrication transition, obtained with the performed scaling of the microgeometry, suggests a theoretical explication based on a geometrical dimensionless formulation detailed in Section 2.

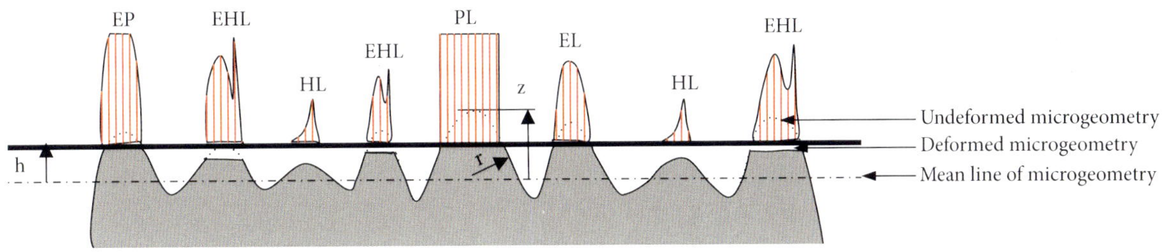

FIGURE 1: The five lubrication regimes at the asperity scale.

2. Description of the Model

The mixed lubrication model proposed in past articles by the authors has not undergone any modification [14]. Accordingly, in this section we only provide a brief outline of the model, discussing its behavior in order to highlight the effect of a scaled microgeometry on mixed lubrication.

The basis of this mixed lubrication model is to extend Greenwood's theory initially proposed for dry contact between nominally flat rough surfaces, to lubricated rough contact. This is made by association of the individual normal load calculated at each asperity for a given separation (or mean oil thickness) h between the surfaces. A test and trial method, based on the h separation variable, can identify the separation able to generate a given normal load at the point of contact. Concerning the normal load at each asperity, we have detected for a given contact the possibility of 5 families of asperities (Figure 1) depending on the local geometry of the asperity. The specific mechanism for each of the 5 families depends on the value of the local indentation of the asperity (z-h value). For each family, a normal analytical expression is proposed. The aim of each paragraph in this section is to present this expression and to discuss the possibility for its use in a geometrically dimensionless formula.

The different families are the following:

(i) Elastic deformation (*EL*): this situation concerns asperities without a local fluid effect subjected to a normal load level, keeping all the materials in the linear reversible elastic deformation mechanism.

(ii) Elastoplastic deformation (*EP*): owing to the indentation effect, local stress can exceed the yield stress of the material. The contact mechanism is called elastoplastic deformation because both elastic and plastic deformation take place. The degree of plastic deformation increases with the degree of indentation.

(iii) Full plastic deformation (*PL*): this mechanism occurs when plastic deformation is generalized across the whole area of contact.

(iv) Isoviscosity and piezoviscosity hydrodynamic lubrication (*HL*): pressure in the fluid film is low and insufficient to cause elastic deformation of the asperity. For large film thickness, the calculation is based on the assumption of rigid bodies with isoviscous fluid. On decreasing the film thickness, the fluid pressure increases and the initial effect is a change of viscosity, which is described by the classical exponential Barus' law of piezoviscosity.

The association of elementary normal load is a simple summation and, as a consequence, neglected interaction between neighboring asperities is a necessary assumption of the calculation. Of course due to the very large number of asperities present in the contact, simple summation can be advantageously replaced by probability analysis using a statistical description of the microgeometry. The variables used for the statistical description of microgeometry must be the same as that used in the expression of individual normal load: z, the altitude of asperity. If $F_{regime}(z)$ is the expression of the normal load on an asperity working with the considered regime (EL, EP, PL, HL, or EHL regimes), we suggest to adopt the well-known load sharing concept, firstly proposed by Johnson [16], where total load is sharing between contacting asperities and lubricating film. We only extend it to the five families of asperity we have identified. As a consequence, the total normal load N acting on all these asperities is expressed as

$$N = n_0 \int_{z_{min}}^{z_{max}} f(z) * F_{regime}(z)\, dz$$
$$= N_{EL} + N_{EP} + N_{PL} + N_{HL} + N_{EHL} = N_b + N_l \quad (1)$$

In this relation, n_0 is the total number of asperities in the considered area of contact A_0 ($n_0 = \beta A_0$ with β density of asperities), and $f(z)$ is the probability to have an asperity at the altitude z. N_{regime} (subscript *regime* can be *EL, EP, PL, HL, and EHL*) are normal loads supported by asperities in the considered regime. $N_b = N_{EL}+N_{EP}+N_{PL}$ is normal load supported by all asperities in the local boundary regime and $N_f = N_{HL}+N_{EHL}$ is the normal load for asperities in the local fluid regime.

Friction force in the considered contact is induced by both contributions: $T = T + T_f$, but it is important to note that T_f can be generally neglected due to the very low friction at full fluid sites compared to that produced at boundary lubricated sites in a mixed lubrication regime. Assuming a mean friction coefficient f_b for boundary lubrication, friction force at the contact can be expressed as

$$T \cong f_b N_b \quad (2)$$

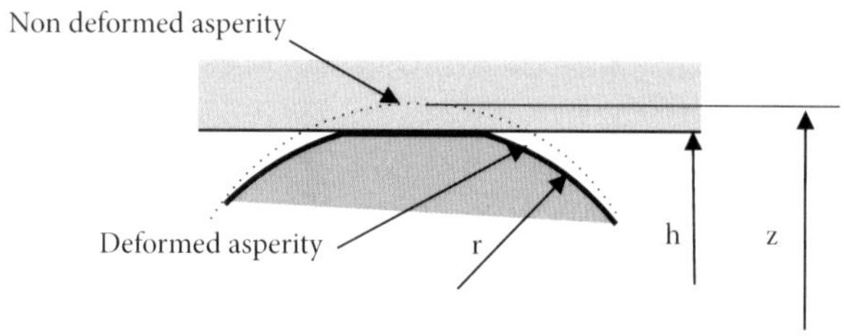

FIGURE 2: Local boundary contact.

Consequently, an approximation of the friction coefficient in mixed lubrication is

$$f \cong f_b \frac{N_b}{N} = f_b \phi \tag{3}$$

$\phi = N_b/N = (N_{EL} + N_{EP} + N_{PL})/N$ is the ratio of the normal load transmitted by the local boundary lubrication. This ratio appears as a governing factor for friction coefficient evaluation in a mixed lubrication regime. Numerical values of this ratio, obtained in a given situation of scaled micro-geometries, will be studied in the case detailed in Section 3. First, however, the analytical expression of individual normal load at asperities will be presented using a geometrically dimensionless expression.

2.1. Elastic Deformation Regime. The local boundary situation refers to asperities in direct contact with the antagonist. A necessary condition is that the altitude of the undeformed asperity z is greater than the altitude h of the antagonist (Figure 2). In this condition, if we neglect the boundary film thickness, we obtain $\delta = z - h$ as the indentation value.

The elastic deformation regime is a first local boundary situation and corresponds to the lowest values for the indentation δ and, in this case, the behavior is based on the Hertz theory for a ball-on-plan contact. The relationship between the normal load and the indentation is expressed as

$$F_{EL} = \frac{2E_q}{3} r^{1/2} \delta^{3/2} \tag{4}$$

$E_q = 2/[(1-\nu_1^2)/E_1 + (1-\nu_2^2)/E_2]$ corresponds to the equivalent Young modulus of both materials in contact and r is the radius of asperity. In order to facilitate a general discussion of mixed lubrication, we suggest the use of dimensionless parameters such as $\Delta=\delta/r$. We obtain

$$F_{EL} = \frac{2E_q r^2}{3} \Delta^{3/2} \tag{5}$$

Using a dimensionless parameter for the normal load, $W_{EL} = F_{EL}/E_q r^2$, yields

$$W_{EL} = \frac{2}{3} \Delta^{3/2} \tag{6}$$

As suggested by authors [17, 18], the limit of the elastic deformation regime corresponds to the appearance of plasticity in the contact, and assuming a Tresca criterion, we obtain for the critical indentation δ_{Elim}

$$\frac{\delta_{ELlim}}{r} = \Delta_{ELlim} = \left(\frac{0.5\pi}{0.3}\frac{Rp_e}{Eq}\right)^2 \approx 27.4 \frac{Rp_e^2}{Eq^2} \tag{7}$$

We can observe that, in elastic regime, the normal load and limit of this regime can be easily expressed using geometrical dimensionless parameter. If the indentation is greater than this limit (asperity could have a greater altitude or a smaller radius), elasticity and plasticity coexist in the contact. Equation (4) ceases to be applicable and the regime becomes elastoplastic.

2.2. Elastoplastic Deformation Regime (EP). In this case, elastic and plastic deformations coexist in the contact. The ratio between the two kinds of deformation depends on the contact conditions. We use theoretical results first proposed by Hill and reported by Johnson [17]. These results refer to plastic deformation caused by a uniform pressure imposed by a spherical indenter. The corresponding equation for the normal load is

$$F_{EP} = \frac{2\pi}{3} Rp_e a^2 \left[1.80 + \ln\left(\frac{1}{6}\frac{E_q}{Rp_e}\frac{a}{r}\right)\right] \tag{8}$$

Chang et al. [18] propose a link between the radius a of the contact area and the indentation δ. This link is based on the classical assumption of matter conservation during plastic deformation. Redistribution of matter is localized near the contact and preserves a rounded shape:

$$a^2 = r\left[2\delta - \delta_{ELlim}\right] \tag{9}$$

This elastoplastic behavior can also be formulated using the dimensionless parameters W_{EP} (resp., Δ) for normal load (resp., indentation):

$$W_{EP} = \frac{2\pi}{3}\frac{Rp_e}{E_q}\left(2\Delta - \Delta_{ELlim}\right) \cdot \left[1.80 + \ln\left(\frac{E_q}{Rp_e}\frac{\sqrt{2\Delta - \Delta_{ELlim}}}{6}\right)\right] \tag{10}$$

This mechanism continues to be applicable until plasticity occupies all the contact area to produce the so-called plasticity regime.

The use of a finite element model [19, 20] for this situation shows that the assumption of a constant pressure is not really adapted. Uniform pressure underestimates finite element results by up to 38% for intermediate interference. Another interesting result of these finite element studies [19, 20] is that solution can be expressed with empirical curve fitting including 2 stages of power laws expressions using dimensionless parameters deduced from both parameters we use in relation (10): Δ and Δ_{ELlim}. Such results indicate that a way of improvement of quantitative results can be performed with new law but conclusion of this work, directly based on the existence of laws based on dimensionless parameters, as explained in Section 3.4, could not change by using new set of dimensionless relation.

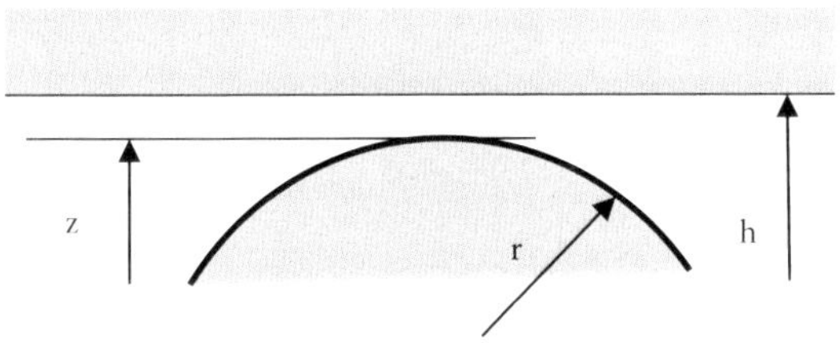

FIGURE 3: Local boundary contact.

2.3. Full Plastic Deformation Regime (PL). The plastic deformation mechanism corresponds to a generalized plastic deformation in the contact. Thus the mean contact pressure p_m remains constant at $p_m = 3Rp_e$. As regards elastoplastic deformation, we use a matter conservation assumption based on Chang's equation relating indentation and the radius of the contact area. The results can be also written using the dimensionless parameters:

$$W_{PL} = 3\pi \frac{Rp_e}{Eq} \left(2\Delta - \Delta_{ELlim}\right) \tag{11}$$

Assuming continuity with the elastoplastic domain, full plastic deformation begins at the interference:

$$\Delta_{EPlim} = \frac{\delta_{EPlim}}{r} \cong 3980 \left(\frac{Rp_e}{E_q}\right)^2 \tag{12}$$

2.4. Fluid Lubricated Hydrodynamic Regime (HL). Fluid lubricated asperity concerns asperity where a continuous film separates both surfaces. $h_0 = (h - z)$ is the film thickness of this contact (Figure 3) and by coherence with the notations used in Sections 2.1–2.3, we suggest the use of the geometrical dimensionless parameter $H_0 = (h - z)/r$.

This first domain corresponds to a fluid contact situation and is characterized by the highest local values of the fluid film thickness. In order to obtain a local fluid film inside the contact, a necessary condition is that the summit of the considered asperity is below the altitude of the antagonist. This first domain also corresponds to the furthest asperities from the counterpart. The assumptions made are the following:

(i) The solids are rigid.

(ii) The viscosity of the fluid follows Barus' law: $\eta = \eta_0 \exp[\alpha * p]$, where α is the piezoviscosity coefficient and p the local pressure in the lubricant fluid.

As for the previous regimes, an analytical solution can be proposed using a two-step calculation. Firstly, calculation is made using assumption of a constant viscosity for the lubricant (IVR isoviscous regime). The solution [21] is expressed as

$$W_{IVR} = 8.1 \frac{U}{H_0^{\ 0.5}} \tag{13}$$

In this relation, U is the dimensionless speed parameter: $U = \eta u / 2E_q r$ and $W_{IVR} = F_{IVR}/E_q r^2$.

The application of Barus' law in Reynold's equation assumes a simple relationship between local pressures in the piezoviscous (HL) and the isoviscous regimes (IVR). As a consequence, the dimensionless normal load, using piezoviscous regime, can be approached [14] by

$$W_{HL} = W_{IVR} \left[1 - 0.194 \ln\left(1 - \frac{1.05GU}{H_o^{3/2}}\right)\right] \tag{14}$$

In this relation, G is the dimensionless material parameter $G = \alpha\, E_q$. Fortunately, in this new regime, relation uses the dimensionless parameters. Quantitative analysis of the evolution of the normal load shows a rapid increase at a given value H_{oHLlim}. In this case, the assumption of rigid solids becomes unacceptable and the elastohydrodynamic regime begins. Accordingly, we choose to stop using the piezoviscous hydrodynamic regime near this value. The validity domain for the hydrodynamic situation is

$$H_o = \frac{h - z}{r} \geq H_{oHLlim} = 1.126\,(GU)^{2/3} \tag{15}$$

2.5. Fluid Lubricated Elastohydrodynamic Regime (EHL). If the dimensionless film thickness is lower than the previously proposed value, the large increase in pressure induced by the rigid asperity suggests that the regime becomes elastohydrodynamic. Due to the spherical shape of asperity, the formula proposed by Hamrock and Dowson [22] seems applicable; however, the problem with a normal load evaluation is to identify the value of the film thickness because it should be influenced by elastic deformation of the asperity. If we neglect the existence of the final convergent at the end of the EHL contact, and if we suppose a constant film thickness corresponding to the central film thickness deduced from the Hamrock and Dowson equation for circular contact, we obtain the following equations:

$$H_c = 1.916 U^{0.670} G^{0.53} W_{EHL}^{-0.067} \tag{16}$$

$$H_c = \Delta + H_o \tag{17}$$

As proposed by Grubin, we suggest the determination of the indentation Δ using the Hertz equation deduced from dry elastic contact ($W_{EHL} \cong W_{EL}$). The corresponding assumption is that the film thickness is low compared to the indentation. Δ and H_c can be eliminated in (17), using (6) and (16). We obtain

$$1.916 U^{0.67} G^{0.53} W_{EHL}^{-0.067} = \left(\frac{3W_{EHL}}{2}\right)^{2/3} + H_o \tag{18}$$

This relation immediately yields the expression of H_0 (position of the asperity) as a function of W_{EHL} (normal load at the considered asperity). But unfortunately, this equation cannot be inversed in an analytical way. A second-order approximation for the solution of this equation gives

$$W_{EHL} = \frac{2}{3} \left[H_P \left(1 - 0.922 \frac{H_0}{H_P} + 0.007365 \left(\frac{H_0}{H_P}\right)^2\right)\right]^{3/2} \tag{19}$$

with $H_p = 1.81(G^{0.48} U^{0.61})$ and $H_o = (h - z)/r$.

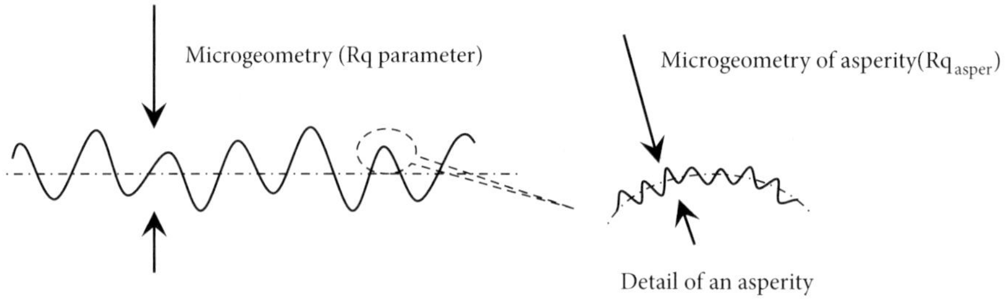

FIGURE 4: Definition of the microgeometry of asperities.

The limit of this elastohydrodynamic regime corresponds to a film thickness too thin compared to the height of roughness [14]. If we name Rq_{asper} the roughness of the asperities (Figure 4), consequently, this criterion is expressed as $h_{cEHL\lim} = 1.5Rq_{asper}$.

If $h_c \geq 1.5Rq_{asper}$, the mean film thickness can be sufficient to maintain a continuous fluid film. We obtain

$$H_{oEHL\lim} = -\left(\frac{1.97U^{0.67}G^{0.53}}{Rq_{asper}/r}\right)^{9.95} \tag{20}$$

3. Theoretical Results and Discussion

In this section, we seek to illustrate the effect of microgeometry using an example. We compare several lubricated contacts with the same characteristics (listed in Table 1) except for their microgeometry.

Note that all parameters represent a fixed value except the speed parameter u (in the range of 0.1 to 2 m/s) which is used as variable for characterizing the mixed lubrication domain.

The studied microgeometry is presented in Figure 5. This isotropic microgeometry is measured on a 2.5 mm x 2.5 mm area of a specimen manufactured by shoot preening. Roughness parameter for this surface is Rq ~ 2.8 μm.

In the following, we consider scaled microgeometries deduced from the reference microgeometry (presented in Figure 5) by a homothetic factor S_c in the vertical direction and also in the horizontal direction (Figure 6). As a consequence, the vertical roughness parameters are scaled by the same factor Sc. The same scaling is observed for horizontal parameters, including mean periodicity between asperities, and, finally, for parameters such as the radius of asperities.

3.1. Statistical Description of the Microgeometry. As proposed by Greenwood and Williamson [23], we adopt a normal distribution for the peak heights z and a constant radius r for all asperities. Peak distribution uses the mean value m_z and the standard deviation s_z to express the peak altitude.

$$f_1(z) = \frac{1}{s_z\sqrt{2\pi}}\exp\left[-\frac{1}{2}\left(\frac{z-m_z}{s_z}\right)^2\right] \tag{21}$$

A key issue is the determination of both parameters for this normal distribution. The method for determining m_z, s_z, and r is explained in another article [24]. By applying this method, we can obtain the parameters presented in Table 2.

TABLE 1: Parameters of the studied contact.

Parameters	Nominal
Contact	Smooth on rough (30 mm^2 area)
Lubricant	Newtonian fluid
	$\eta = 0.100$ *Pas* and $\alpha = 2.0$ 10-8 Pa^{-1}
Speed	0.1 to 10 m/s
Microgeometry	Rough on Smooth
Mean apparent pressure	1 MPa
Materials	Steel on Brass
Steel	$E_1 = 2\ 10^{11}$; $\nu_1 = 0.3$
Brass	$E_2 = 1\ 10^{11}$; $\nu_2 = 0.3$
	$Rp_e = 500$ MPa
Boundary friction coefficient	f_b=0.12

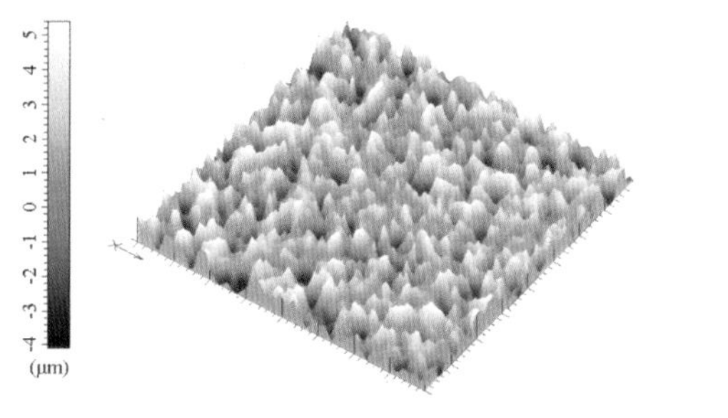

FIGURE 5: The reference microgeometry.

Finally, as proposed in the theoretical part, the microgeometry description must be completed by a description of the surface relief of the asperities. To achieve a quantitative evaluation of this surface relief for each asperity, we use a filtering technique adjusted to suppress the shape of the asperity. As presented in [14], the crucial point is the determination of the high pass cut-off. We propose to quantify this microrelief using the root mean square (RMS) standardized parameter Rq. This value ($Rq_{asperity} = 1.4\ \mu m$ for the proposed microgeometry) will be used in the next section as data for the description of the transition from local boundary lubrication to local continuous fluid lubrication.

3.2. Stribeck Curve. Figure 7 presents 3 Stribeck curves (friction coefficient versus $\eta u/p$ parameter) calculated using

TABLE 2: Parameters of the statistical description.

Parameters	Definition	Value
m_z	Mean altitude for summit from reference line	4.25 μm
s_z	Standard deviation for asperity summit	1.80 μm
r	Radius of asperities	400 μm
β	Density of asperities	18 mm^{-2}

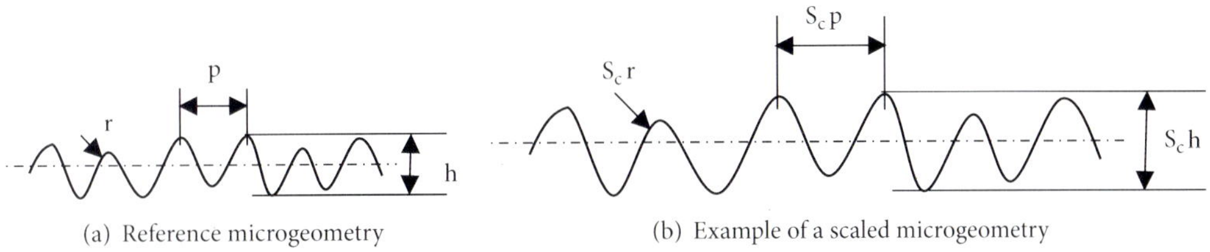

FIGURE 6: Principle of a scaled microgeometry.

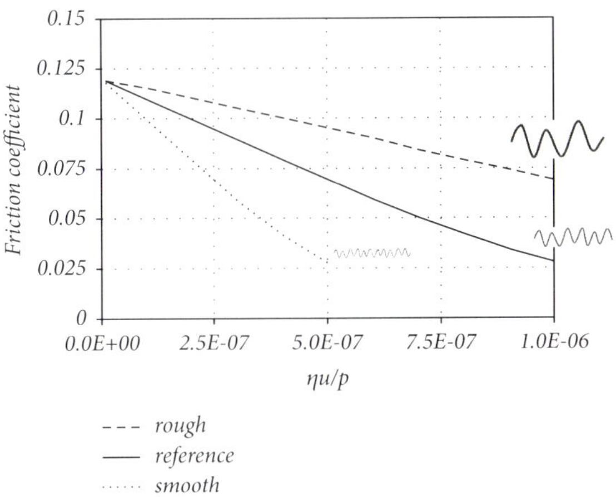

FIGURE 7: Stribeck curves obtained for the reference condition and influence of the microgeometry.

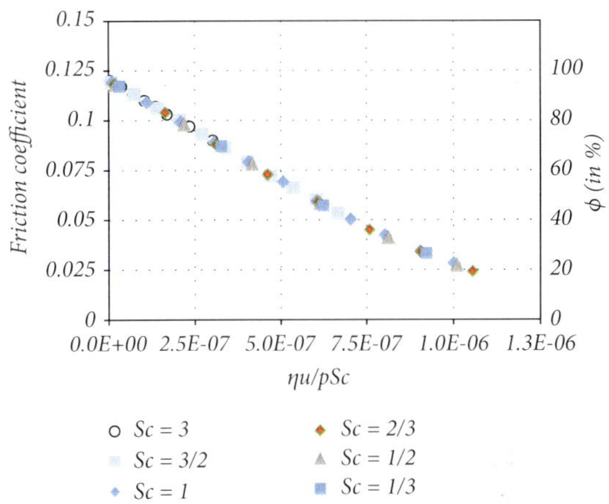

FIGURE 8: Stribeck curve obtained in the reference condition showing the influence of the microgeometry.

the methodology proposed in Section 2 and obtained with the same condition of contact (detailed in Table 1), but with three different microgeometries. The reference microgeometry corresponds to the microgeometry presented in Figure 5. Other curves, labeled rough or smooth, correspond to homothetic microgeometries. The rough and smooth microgeometries are deduced from the reference using a magnification or scale factor of 2 (for the rough geometry) and 1/2 (for the smooth geometry). Consequently, the rough microgeometry shows asperities with a double height and a double periodicity compared to the reference's asperities, whereas the asperities on the smooth surface are half the height and half the thickness of the asperities in the reference microgeometry.

According to the assumption used in (2), calculations are only performed for the domain where boundary lubrication remains significant compared to full fluid lubrication (friction coefficient ranging from 0,125 to 0,025).

As can be observed, the model correctly predicts the shape of the evolution in the friction coefficient with the $\eta u/p$ parameter and suggests that a large influence is exerted by the microgeometry of the surface. As expected, in a mixed lubrication regime, an increasing $\eta u/p$ parameter decreases the friction coefficient. Decreasing the amplitude of the microgeometry also has this effect. More precisely, with respect to the microgeometry, we notice that the scale factor S_c plays a significant role because a factor of 2 (or 1/2) is sufficient to completely modify the Stribeck curve. In addition, a smooth microgeometry reduces the domain of mixed lubrication.

In order to compare the impact of the microgeometry scale factor S_c compared to that of the Stribeck parameter $\eta u/p$, we plot the calculated evolution of the friction coefficient in mixed lubrication using a new parameter $\eta u/pS_c$, which tries to include the microgeometry effect with the scale factor S_c (Figure 8).

From a quantitative perspective, it is interesting to note that the use of the scale factor in the denominator of the Stribeck parameter induces perfect superposition of all points on a single curve and consequently suggests an envelope curve for the shape of the considered microgeometry. Two remarks result from this perfect organization over a wide range of scale parameters (1/3 to 3) and the Stribeck parameter:

(i) A comparison with experimental results proposed in the literature

(ii) A theoretical analysis of this perfect single curve obtained with microgeometries with asperities modified by a scale factor

3.3. Confrontation with Existing Experimental Results. This envelope curve for the theoretical results can be compared to the experimental results obtained by Emmens [7] and Felder [8]. Both authors perform experimental study using various microgeometries and propose the inclusion of the microgeometry effect in a modified Stribeck parameter, with slight differences from one author to the other: Emmens proposes $H_e=3\eta u/pR_p^{\ 2}$, where R_p is the altitude of the mean line of the profile from the lowest point. However, in the final discussion, the author explains that something is missing in this equation because this parameter has a dimension of length. Emmens suggests that the missing entity could be a parameter describing the periodicity of the asperity. As a consequence, the microgeometry parameter proposed by Emmens is proportional to l/Rp^2. Application of a scale factor Sc to the microgeometry affects the numerator by a factor Sc (coming from l) and the denominator by Sc^2 (coming from Rp^2). Equivalent parameter for microgeometry deduced from such experimental works is $1/Sc$ in the case of scaled microgeometries, so the same we have obtained in our theoretical developments.

Meanwhile, Felder proposes $H_f = 3\eta lu/pR_a^{\ 2}$, where R_a is the arithmetic mean and l describes the space between two consecutive asperities. The proposed parameter is dimensionless, and if we applied the scale effect to this parameter, we can observe that H_f is affected by S_c at the numerator (due to the l parameter) and $S_c^{\ 2}$ at the denominator due to the term Ra^2. It is important to note that the global effect ($1/S_c$) also exactly corresponds to the effect shown by the model we have developed.

3.4. Theoretical Aspects. The second remark is an explanation for the perfect shape of the envelope curve obtained with homothetic microgeometries. In fact, as proposed in Section 2, all theoretical expressions (normal load, variables, limits, etc.) are formulated using dimensionless parameters. The examination of these parameters shows that most of them are insensitive to the scale effect in the microgeometry:

(1) $Z = (z - m)/s_z$ is the dimensionless variable describing the altitude of summits; m_z and s_z are proportional to the scale factor of the microgeometry and, consequently, if the variable z is chosen proportional to the scale factor, the dimensionless parameter Z is independent of S_c.

(2) Both the indentation parameter $\Delta = (z-h)/r$ and the local film thickness parameter $H_0 = (h-z)/r$ include z and r variables that are proportional to the scale factor S_c. If we impose a h value proportional to the scale factor, both of these dimensionless parameters are independent of the microgeometry scale factor.

(3) Only one dimensionless parameter is scale-sensitive: the speed parameter $U = \eta\, u/2\, r\, E_q$ depends on the scale factor only for the term r. If we artificially modify another term r (for example, by changing the viscosity from η to ηS_c), then this parameter, as all other parameters, remains independent of the scale factor.

Consequently, the resulting calculation of individual normal loads at asperities only depends on factor terms used to transform the dimensionless term into dimensional quantities. The proposed calculation for the friction coefficient is only based on an evaluation of normal loads (3). It is interesting to note that all normal loads on asperities are scaled by the square of the radius value and, consequently, by the square of the scale factor. The friction coefficient, as a ratio of loads, is not influenced by the scale parameter of the microgeometry. As a synthesis, if friction using the modified viscosity (viscosity η is replaced by a new viscosity ηS_c) is not modified by the scale factor, the result corresponding to the homothetic microgeometry and the initial viscosity is obtained with the $\eta u/pS_c$ value for the Stribeck parameter. Finally, if we adopt this modified Stribeck term, the friction coefficient is independent of the scale parameter, and it is logical to obtain the same curve for all homothetic geometries.

Because all expressions in the theoretical approach can be expressed with a dimensional parameters, existence of a master curve is not limited to friction coefficient evolution but also for fraction of force transmitted by each of the 5 families of asperity (Figure 9).

Figure 9 gives the fraction of normal load supported by each family of asperity as a function of the modified Stribeck parameter: $\eta u/pS_c$. As evidence, because one asperity must be working with one of the 5 identified mechanisms, the summation of these fraction has to be always equal to 100%. But repartition depends on the situation and in particular with the modified parameter $\eta u/pS_c$. The considered situations in this study show that sharing of normal load is not uniform into all five mechanisms. Depending on lubrication condition expressed by the modified Stribeck parameter, one can observe 2 major mechanisms (Figure 9(a)):

(i) Elastoplastic mechanism for direct contact (ranging from 0 to 96 %)

(ii) Hydrodynamic lubrication for lubricated contact (ranging from 0 to 100 %)

As explained in [14], main reason for the existence of only 2 from the five possible mechanisms is the large-scale between transitions from one mechanism to the next compared to the variability of situation due to the range of altitude for asperities. Other families mechanisms are identified on details of the global curve:

(i) Elastohydrodynamic lubrication (ranging from 0 to 18 %)

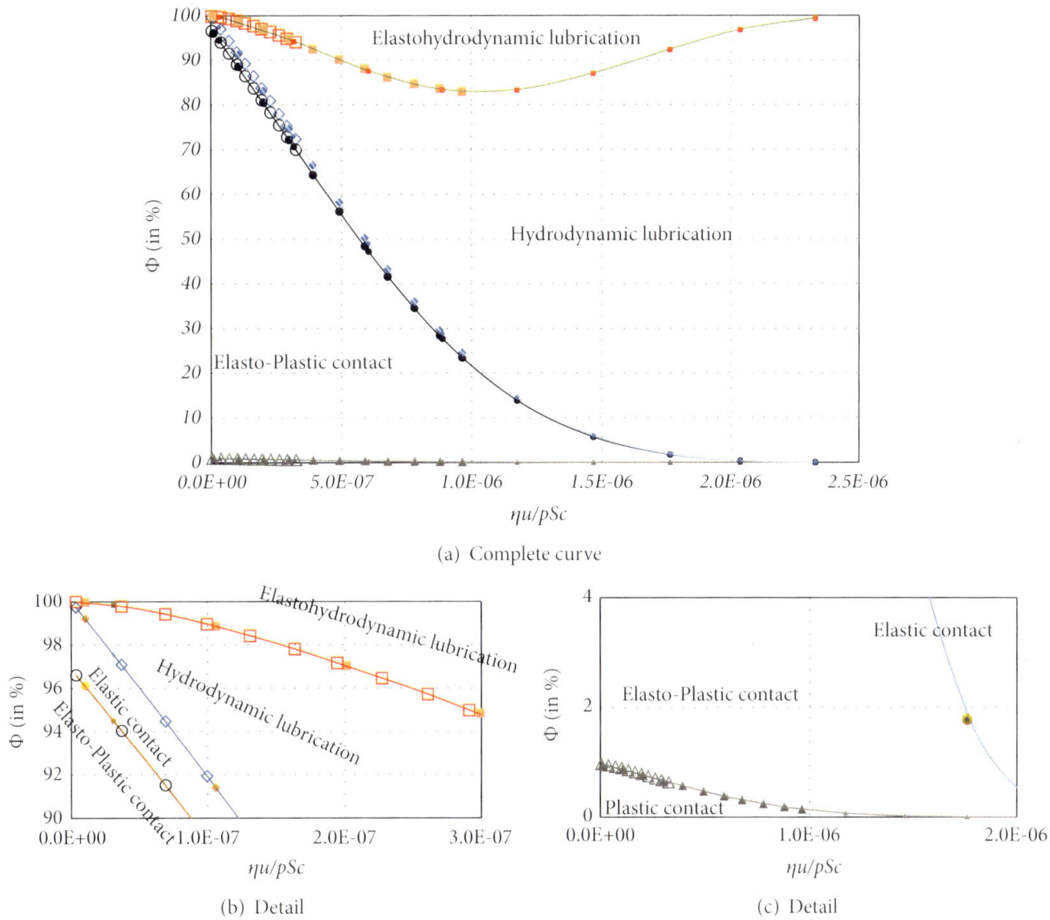

FIGURE 9: Fraction of force transmitted by each of the 5 families of asperity ((a) global curve, (b) and (c) details).

(ii) Elastic mechanism (less than 3% in detail (b) of Figure 9)

(iii) Plastic deformation (less than 1% in detail (c) of Figure 9)

It is also important to note that Figure 9 regroups information obtained using 3 different microgeometries:

(i) Reference microgeometry represented by intermediate filled marks

(ii) Scaled microgeometry with a factor of 1/3 represented by small filled marks

(iii) Scaled microgeometry with a factor of 3 represented by big marks

The use of the modified Stribeck parameter gives an excellent continuity for fraction of asperities working in the same condition. This confirms the existence, for all families of asperity, of a dimensionless formulation adapted to the modified parameter.

4. Conclusions

This article presented theoretical predictions for friction evolution under variation of contact parameters and microgeometry. The applied model was based on the decomposition of the contact in neighboring asperities. Depending on the contact parameters and the geometry of the considered asperities, the behavior was assigned to 5 possible regimes. The behavior of the contact was obtained by associating the elementary behavior determined for each asperity.

From a qualitative perspective, simulations using this model exhibit the classical decreasing Stribeck curve between the friction coefficient and the well-known parameter $\eta u/p$. We also observed a decrease in the friction coefficient with a decrease in the amplitude of the microgeometry. We have studied the behavior of homothetic microgeometries (deduced one from the other by the same amplification in all directions). All homothetic microgeometries give exactly the same behavior if the classical Stribeck parameter $\eta u/p$ is transformed into a new parameter $\eta u/pS_c$ that includes the scale factor S_c of the microgeometry. From a quantitative perspective, this result is in very good agreement with the experimental results given in the literature.

From a practical perspective, this approach enables the prediction of the behavior of a family of new microgeometries from that obtained for a given single microgeometry. A necessary condition for the proposed extrapolation is to consider that the new geometry must be scaled from the initial one. A possible perspective of this work is to isolate the effects of an individual aspect of microgeometry (for example, only the vertical amplitude or periodicity).

Nomenclature

A_0:	Apparent area of contact
E_i:	Young's modulus of material "i"
E_q:	Equivalent Young's modulus
f_i $(i = b, f)$:	Coefficient of friction for boundary lubrication (b) and for fluid lubrication (f)
F_{regime}:	Normal load of individual asperity working in the corresponding regime (subscript regime can be EL, EP, PL, HL, and EHL)
$G = \alpha E_q$:	Material parameter
h:	Mean film thickness in the contact
h_0:	Local film thickness at an asperity
$H_0 = h_0/r$:	Dimensionless film thickness
$H_{0EHL\text{lim}}$:	Limiting film thickness in EHL regime
$H_{0PVR\text{lim}}$:	Limiting film thickness in PVR regime
Hp:	Constant used in EHL regime
m_z, s_z:	Mean value of the summit altitude and standard deviation
n_0:	Total number of asperities on the entire surface
N:	Normal load on the contact
N_{regime}:	Normal load on the contact for the considered regime (subscript regime can be EL, EP, PL, HL, and EHL)
p:	Apparent pressure in the contact
r:	Radius of asperity
Rp_e:	Yield stress of the softer material
Rq:	Standard deviation of the microgeometry
Rq_{asper}:	Standard deviation of the microrelief of asperities
S_c:	Scale factor of the microgeometry
T:	Friction force
u:	Sliding speed
$U = \eta u/2r$:	Dimensionless speed
$W_{mech} = F_{mech}/E_q r^2$:	Dimensionless normal load
z:	Altitude of summit
α:	Piezoviscosity coefficient of lubricant
β:	Density of asperities
δ:	Asperity indentation
$\Delta = \delta/r$:	Dimensionless deformation
$\Delta_{EL\text{lim}}$:	Elastic/elastoplastic limit for dimensionless deformation
$\Delta_{EP\text{lim}}$:	Elastoplastic/plastic limit for dimensionless deformation
η:	Viscosity of lubricant
ϕ:	Ratio of load transmitted by asperities working with local boundary regime
ν_i:	Poisson coefficient of surface "i".

Conflicts of Interest

The authors, Francois Robbe-Valloire, Robert Progri, and Tony Da Silva Botelho, declare that there are no conflicts of interest regarding the publication of this paper.

References

[1] C. J. Hooke, "The effect of roughness in EHL contacts," *Tribology and Interface Engineering Series*, vol. 48, pp. 31–46, 2005.

[2] A. Beheshti and M. M. Khonsari, "An engineering approach for the prediction of wear in mixed lubricated contacts," *Wear*, vol. 308, no. 1-2, pp. 121–131, 2013.

[3] M. B. Dobrica, M. Fillon, and P. Maspeyrot, "Mixed elastohydrodynamic lubrication in a partial journal bearing - Comparison between deterministic and stochastic models," *Journal of Tribology*, vol. 128, no. 4, pp. 778–788, 2006.

[4] A. Albers and S. Reichert, "On the influence of surface roughness on the wear behavior in the running-in phase in mixed-lubricated contacts with the finite element method," *Wear*, vol. 376-377, pp. 1185–1193, 2017.

[5] H. Zhang, M. Hua, G.-N. Dong, D.-Y. Zhang, and K.-S. Chin, "A mixed lubrication model for studying tribological behaviors of surface texturing," *Tribology International*, vol. 93, pp. 583–592, 2016.

[6] S. Fricke, C. Hager, S. Solovyev, M. Wangenheim, and J. Wallaschek, "Influence of surface form deviations on friction in mixed lubrication," *Tribology International*, vol. 118, pp. 491–499, 2018.

[7] W. C. Emmens, "The influence of surface roughness on friction," in *Proceedings of the 15th biennial Congress IDDRG at Dearborn*, pp. 63–70, Michigan USA, May 1988.

[8] E. Felder and V. Samper, "The generation by friction and plastic deformation on the restraining characteristics of draw beads in sheet metal forming," in *Proceedings of the 20th Leeds-Lyon Symposium on Tribology : Dissipative Processes in Tribology*, pp. 361–372, Lyon, France, September 1993.

[9] I. Nogueira, A. M. Dias, R. Gras, and R. Progri, "An experimental model for mixed friction during running-in," *Wear*, vol. 253, no. 5-6, pp. 541–549, 2002.

[10] A. O. Lebeck, "Parallel sliding load support in the mixed lubrication - Part 2: Evaluation of the mechanisms," *Journal of Tribology*, vol. 109, pp. 196–205, 1987.

[11] S. Xiong and R. F. Salant, "A dynamic model of contacting mechanical seal for down hole tools," *Journal of Tribology*, vol. 125, no. 2, pp. 391–402, 2003.

[12] J. Hol, V. T. Meinders, H. J. M. Geijselaers, and A. H. Van Den Boogaard, "Multi-scale friction modeling for sheet metal forming: The mixed lubrication regime," *Tribology International*, vol. 85, pp. 10–25, 2015.

[13] D. K. Karupannasamy, J. Hol, M. B. de Rooij, T. Meinders, and D. J. Schipper, "Modelling mixed lubrication for deep drawing processes," *Wear*, vol. 294-295, pp. 296–304, 2012.

[14] F. Robbe-Valloire and R. Progri, "Mechanisms developed at the asperity scale for mixed lubrication between parallel surfaces," *Journal of Engineering Tribology*, vol. 226, no. 12, pp. 1441–1453, 2012.

[15] F. Robbe-Valloire, B. Paffoni, and R. Progri, "Prediction of the amount of load transmitted by elastic, elastoplastic or

plastic deformed asperities at a rough interface," *Mechanics of Materials*, vol. 33, no. 11, pp. 617–633, 2001.

[16] K. L. Johnson, J. A. Greenwood, and S. Y. Poon, "A simple theory of asperity contact in elastohydro-dynamic lubrication," *Wear*, vol. 19, no. 1, pp. 91–108, 1972.

[17] K. L. Johnson, "Normal contact of inelastic solids," in *Contact Mechanics*, Cambridge University Press, 1985.

[18] W. R. Chang, I. Etsion, and D. B. Bogy, "An elastic-plastic model for the contact of rough surfaces," *Journal of Tribology*, vol. 109, no. 2, pp. 257–263, 1987.

[19] L. Kogut and I. Etsion, "Elastic-plastic contact analysis of a sphere and a rigid flat," *Journal of Applied Mechanics*, vol. 69, no. 5, pp. 657–662, 2002.

[20] R. Jackson, I. Chusoipin, and I. Green, "A finite element study of the residual stress and deformation in hemispherical contacts," *Journal of Tribology*, vol. 127, no. 3, pp. 484–493, 2005.

[21] D. E. Brewe and B. J. Hamrock, "Analysis of starvation on hydrodynamic lubrication in non-conform contact," *Journal of Lubrication Technology*, vol. 104, no. 3, pp. 410-411, 1982.

[22] B. J. Hamrock and D. Dowson, "Isothermal elastohydrodynamic lubrication of points contacts," *Journal of Lubrication Technology*, vol. 99, pp. 264–274, 1977.

[23] J. A. Greenwood and J. B. Williamson, "Contact of nominally flat surfaces," *Proceedings of the Royal Society A Mathematical, Physical and Engineering Sciences*, vol. 295, no. 1442, pp. 300–319, 1966.

[24] F. Robbe-Valloire, "Statistical analysis of asperities on a rough surface," *Wear*, vol. 249, no. 5-6, pp. 401–408, 2001.

14

Performance Evaluation of *Jatropha* and *Pongamia* Oil based Environmentally Friendly Cutting Fluids for Turning AA 6061

T. P. Jeevan[1,2] and S. R. Jayaram[1,2]

[1]*Department of Mechanical Engineering, Malnad College of Engineering, Hassan 573201, India*
[2]*Visvesvaraya Technological University, Belagavi, India*

Correspondence should be addressed to T. P. Jeevan; jeevantpmce@gmail.com

Academic Editor: Huseyin Çimenoğlu

Owing to the desirable properties of vegetable oils as cutting fluids, an attempt is made to explore the potentiality of plentifully available vegetable oils as a cutting fluid for turning AA 6061. Two nonedible vegetable oils, *Jatropha* and *Pongamia*, in their chemically modified (epoxidized) versions are used as straight cutting fluids. Cutting fluids are introduced to the machining zone with the aid of Minimal Quantity Lubrication (MQL) method. Taguchi's technique of orthogonal arrays is used to develop an effective design of experiments. The results obtained under epoxidized versions of *Jatropha* and *Pongamia* oils are compared with the results of mineral oil in terms of cutting forces and surface roughness. Experimental observations and statistical analysis show that, compared to mineral oil, the modified versions of vegetable oil-based cutting fluids are more effective in reducing the cutting forces and increasing surface finish. It is also observed that the modified *Pongamia* oil showed lesser flank wear compared to the other two tested oils.

1. Introduction

The application of cutting fluids in machining was first reported by F. Taylor in 1894 who observed that up to 33% increase in cutting speed can be achieved without affecting the life of cutting tool by applying large amounts of water as cutting fluid to the machining zone. Cutting fluids are extensively used in the machining process to carry away heat generated at machining zone, lubricate and take away chips, and prevent corrosion during machining operations [1]. Cutting fluids results in improved tool life, work quality, effective chip management, and reduced process variability. Thus, utilization of cutting fluids is increasing in the metal cutting operations [2]. Major share of cutting fluids being used across the globe are petroleum-based oils. Enormous use of petroleum-based oils has had a lot of negative environmental and health-related consequences like skin diseases [3]. Petroleum-based lubricants in 2016 have increased tremendously on high global consumption, showing at least 1% annual increments with 13,726 million tons of oil equivalent [4]. This has therefore, made another viewpoint of bad impact on the environmental pollution and the danger of large loss proportion (13–50%) of the lubricants in the aquatic and terrestrial ecosystems, including continuous depletion of global energy and natural resources, prevail [5]. The increasing consciousness for green manufacturing globally and consumer focus on environmentally friendly products has put increased pressure on industries to minimize the use of petroleum-based cutting fluids [6]. The demand for biodegradable and eco-friendly cutting fluids has opened an avenue for using vegetable oils as a potential alternative to petroleum-based cutting fluids [7]. In this regard, vegetable oils are emerging as potential replacements to conventional cutting fluids [8].

Vegetable oil-based cutting fluids are highly biodegradable, eco-friendly, renewable, less toxic, high flash point, low volatility, high viscosity index, wide production possibilities, and economical in the waste management [9–12]. Vegetable oils primarily consist of triglycerides; [13–15] the triglycerol structure of vegetable oil makes it a strong competitor as a base stock for lubricants and functional fluids [16]. Vegetable oils are triglycerides in which C18 carboxylic acids are dominant. Some of the fatty acids derived from these glycerides are unsaturated; those typically contain stearic,

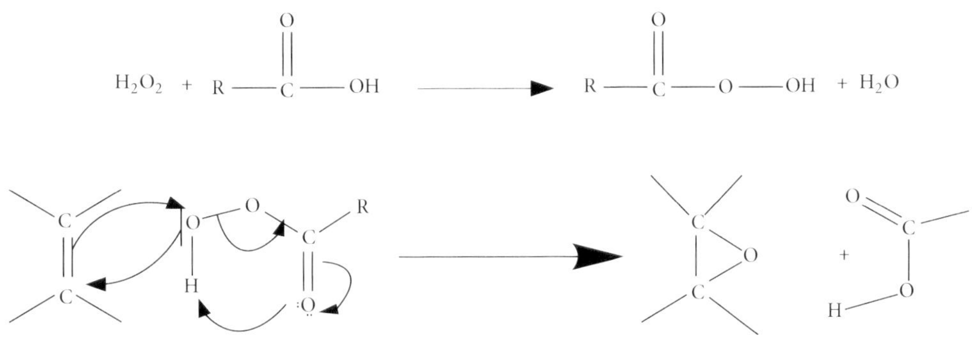

Figure 1: Mechanism of epoxidation reaction for vegetable oil [18].

oleic, linoleic, and linolenic acids in varying amounts. Three of these are unsaturated acids, namely, oleic (18 : 1), linoleic (18 : 2), and linolenic (18 : 3).

The use of edible vegetable oil to produce lubricants and cutting fluids is not feasible in view of the big gap in demand and supply of such oil. Hence, nonedible vegetable oils are finding importance due to their abundance and also, this would save large quantities of edible oils which are in great demand. The forecast for eco-friendly lubricants for next 10–15 years is a worldwide volume share of approximately 15% and in some regions up to 30% [17].

Many metal cutting processes are tested by researchers to employ the vegetable oil-based fluids as metal cutting oils and observed a better performance. The *Jatropha* oil proved to be the best one for being used in the machining AA7050-T7451 [19]. The coconut oil is impressive in reducing surface roughness at higher values of feed and depth of cut while turning hardened AISI 52100 [20] and also showed an improved tool life for turning of AISI 304 [21]. The significant reduction in tool wear, cutting force, better dimensional accuracy, and surface roughness by MQL and a favourable change in chip tool and work tool interaction are observed while turning AISI 1060 [22] and AISI 4340 [23] steel using vegetable oils. The reduction in coefficient of friction is observed while machining mild steel, brass, and aluminum under different cutting conditions using Palm kernel oil as cutting fluid [24]. A formulated cutting fluid of Canola with 8% extreme pressure additive showed lower turning forces, tool wear, and surface roughness values for turning AISI 304L [25]. Sunflower and Canola oil based cutting fluids [26] generate better surface finish and produce lower cutting and feed forces while turning AISI 304 L. A novel cutting fluid was developed from nonedible Neem [27] performed better with respect to temperature rise at tip, cutting forces, tool wear, and surface roughness. The modified *Jatropha* oil [11] showed the lowest values of cutting force, cutting temperature, and surface roughness, with a prolonged tool life and less tool wear. Coconut oil with 0.5% nanoparticle inclusions resulted in improved machining performance compared to Sesame oil and Canola oil [28]. The demand for a balance in meeting both the technological and environmental requirements of a new cutting fluid for machining process forms the basis of this research. In this regard, nonedible vegetable oil modified *Jatropha* (MJO) and modified *Pongamia* (MPO) are experimentally investigated and compared with mineral oil (MO) based on Taguchi's design of experiments, for their sustainability as metal cutting fluids during turning AA6061, using minimum quantity lubrication technique (MQL). MQL has proven to be a better alternative to a flood coolant system in which a mixture of air and cutting fluid is applied to the cutting zone. It has a better reach ability as a result of the high pressure [29].

2. Materials and Methods

2.1. Formulation of the Vegetable Oil-Based Cutting Fluids. Biofuels have already been accepted around the world for their advantages over conventional petroleum fuels, including the opportunity for energy independency. Now, similar growth is expected for biolubricants, which are derived from renewable vegetable oils for different applications [30]. *Jatropha* and *Pongamia* based cutting fluids have a huge potential owing to their abundant availability, renewability, and biodegradability. However, the challenges with these oils in meeting lubricant performance is their low oxidative stability [31]; they cannot be used in their raw form. Hence, they are chemically modified by the process of epoxidation. Epoxidation of fatty acids is a reaction of a carbon–carbon double bond with active oxygen, which results in the addition of an oxygen atom, converting the original double bond into a three membered epoxide (oxirane) ring (Figure 1). In epoxidation process a known volume of oil is blended with formic/acetic acid and hydrogen peroxide. The reaction takes place for 14 hours, with 7 hours of titration in the presence of formic acid and hydrogen peroxide and 7 hours of constant agitation in the temperature range of 50°C to 60°C, thus removing double bonds in the fatty acid chain of the oil. Epoxidized vegetable oil produces flexibility and elasticity due to the presence of the epoxy ring in the backbone chain [32]. The changes in the properties of vegetable oil after chemical modification can be identified in Table 1. The iodine value of oil confirms the process of epoxidation.

2.2. Experimental Design by Taguchi Technique and Machining Conditions. In this study, AA 6061 having a Vickers hardness of 107 was used as a workpiece material, which had a dimension of Ø 45 mm × 200 mm. The measured values of chemical composition of AA 6061 in percentage of weight

TABLE 1: Physicochemical properties of vegetable oils, their modified versions, and mineral oil.

Properties	MJO	MPO	MO
Kinematic viscosity @ 40°C (cSt)	11.22	20.15	33.00
Viscosity index	225.36	219.00	185.21
Saponification value (mg KOH g^{-1})	203.80	198.00	-
Total acid value (mg KOH g^{-1})	00.05	0.13	01.75
Flash point (0°C)	180.00	190.00	160.00
Pour point (0°C)	−03.00	−02.00	−7.00
Iodine value (mg I g^{-1})	22.00	21.41	6.50

TABLE 2: Compositions of AA6061.

Contents	Cu	Mg	Si	Fe	Mn	Ti	Zn	Cr	Al
% of weight	0.263	0.902	0.403	0.135	0.076	0.022	0.011	0.153	Remaining

TABLE 3: Machining conditions.

Machine	Engine lathe-LB-20
Work piece material	AA 6061
Process parameters	
Spindle speed (rpm)	800, 1270, and 1600
Feed (mm/rev)	0.100, 0.175, and 0.250
Depth of cut (mm)	0.5, 1.0, and 1.5
Cutting tool	
Material	Carbide cutting tool (uncoated)
Tool holder	High speed steel (HSS)
Tool nose radius (mm)	0.4
Cutting tool edge length (mm)	9
Cutting fluids	MJO, MPO, and MO
Cutting fluid supply	MQL method (3 to 4 bars)

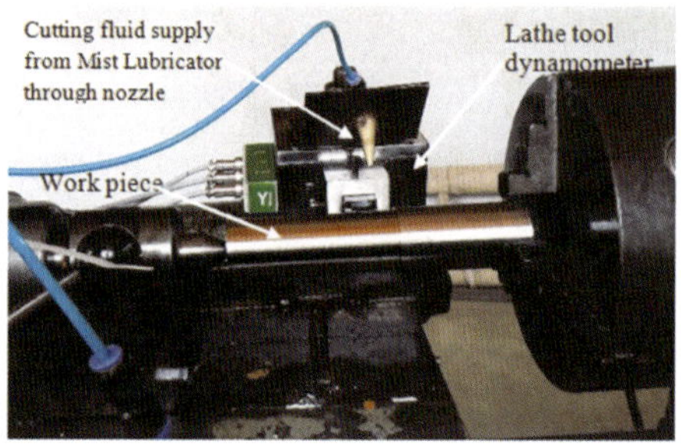

FIGURE 2: Experimental setup.

FIGURE 3: Measuring surface roughnesses with stylus on workpiece.

are given in Table 2. The machining parameters as per the recommendations from the tool manufacturer are selected. Spindle speed, feed, depth of cut, and type of cutting fluid are considered as machining parameters. The type of cutting fluid used is considered as one of the demanding input parameters while designing the experiments. Accordingly, four input parameters (the type of cutting fluid, spindle speed, feed rate, and depth of cut), and for each parameter three levels, were assumed (Table 3). For four factors, three-level experiments, Taguchi had specified L27 orthogonal array for experimentation. The data obtained from the trials conducted as per L27 array experimentation was recorded and further analyzed. Table 3 shows the parameters and their levels considered for the experiments. The experimental setup is shown in Figure 2.

2.3. Cutting Force and Surface Roughness Measurement. The cutting force influences the deformation of the machined work piece, its dimensional accuracy, chip formation, tool wear, surface roughness, and stability of the machining system. Machining parameters affecting cutting forces include depth of cut, cutting speed, and type of cutting fluid. Hence, the study of cutting force as a quality characteristic and the influence of cutting fluid on the cutting force is important. The cutting force generated during machining is measured with a strain gauge type lathe tool dynamometer at different speeds and feeds under modified vegetable oil-based cutting fluids. Surface roughness is a significant design specification that is known to have considerable influence on properties such as wear resistance and fatigue strength. Surface roughness (R_a) of the machined surface is measured with a SURFCOM FLEX-50. The surface roughness values are measured by moving the stylus along the axis of turned workpiece over the machined surface as shown in Figure 3.

TABLE 4: Cutting force and surface roughness for interfamily comparison.

Trial number	Cutting fluids	Depths of cut (mm)	Feed rates (mm/rev)	Spindle speeds (rpm)	Cutting force (N)	Surface roughness (μm)
1	MJO	0.5	0.100	0800	75	0.8
2	MJO	0.5	0.175	1270	105	0.9
3	MJO	0.5	0.250	1600	80	1.4
4	MJO	1.0	0.100	1270	145	0.7667
5	MJO	1.0	0.175	1600	215	0.8333
6	MJO	1.0	0.250	0800	345	1.3667
7	MJO	1.5	0.100	1600	200	0.7333
8	MJO	1.5	0.175	0800	350	1.2
9	MJO	1.5	0.250	1270	435	1
10	MPO	0.5	0.100	0800	70	0.8
11	MPO	0.5	0.175	1270	100	0.8
12	MPO	0.5	0.250	1600	145	1.6
13	MPO	1.0	0.100	1270	140	0.8
14	MPO	1.0	0.175	1600	235	1
15	MPO	1.0	0.250	0800	330	1.1
16	MPO	1.5	0.100	1600	215	0.8
17	MPO	1.5	0.175	0800	365	1
18	MPO	1.5	0.250	1270	420	1.4
19	MO	0.5	0.100	0800	75	1.033
20	MO	0.5	0.175	1270	110	1.1
21	MO	0.5	0.250	1600	145	1.6
22	MO	1.0	0.100	1270	155	0.7
23	MO	1.0	0.175	1600	230	1
24	MO	1.0	0.250	0800	340	1.5
25	MO	1.5	0.100	1600	210	1.1
26	MO	1.5	0.175	0800	345	1.2
27	MO	1.5	0.250	1270	315	1.4

2.4. Tool Wear Measurement. Tool wear was measured using educational optics transverse microscope which is modified into Tool Maker's Microscope having least count of 1 μm. Tool inserts were withdrawn after each continuous cut and were studied under Tool Maker's Microscope for the wear pattern and width of the flank wear. Flank wear most commonly results from abrasive wear of the cutting edge against the machined surface. Flank wear is measured in terms of average wear land size V_B.

3. Results and Discussion

Experimentally determined values of cutting force and surface roughness values (Table 4) are used to investigate the performance of vegetable based cutting fluids. Statistical analyses of experimental results obtained are carried out using MINITAB 17; statistical package established on Taguchi's principle is used for analysis.

3.1. Analysis of S/N and ANOVA. The statistical analysis is carried out to find the process parameters which extremely affect the quality characteristics (cutting force and surface roughness) and to identify the favourable level of each process parameter which results in the least cutting force and surface roughness.

The S/N ratio values for cutting force and surface roughness are obtained using the equation shown below and, for optimizing the process in this study, the smaller the S/N (dB) ratio, the better the characteristic.

$$\frac{S}{N} = -10 \log_{10} \left[\frac{1}{n} \sum_{i=1}^{n} y_i^2 \right], \quad (1)$$

where n is the number of measurements in a trial/row and yi is the ith measured value in a run. It includes the effect of noise as well.

From the S/N ratio analysis (Figure 4), optimal turning parameters for the cutting force are 0.5 mm of depth of cut, 0.1 mm/rev of feed rate, and 1600 rpm of spindle speed. MJO performed optimally followed by MPO and MO. The optimal turning parameters for surface roughness as obtained from main effect plot of the S/N ratio (Figure 5) are found to be 1.5 mm of depth of cut, 0.175 mm/rev of feed rate, and 800 rpm of spindle speed. MPO performed optimally in

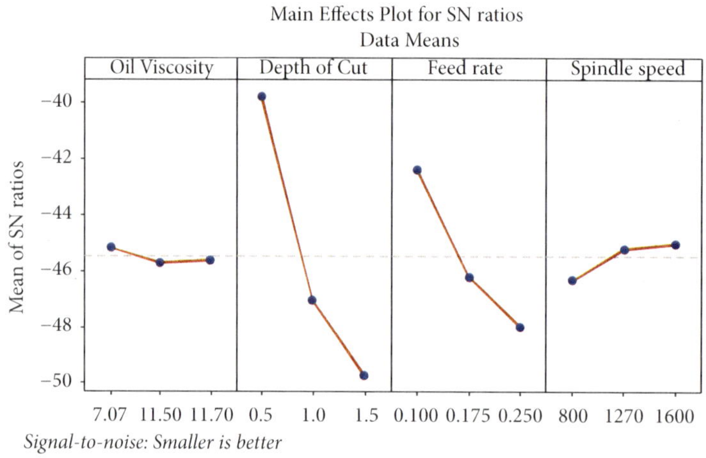

FIGURE 4: Main effects plot for cutting force.

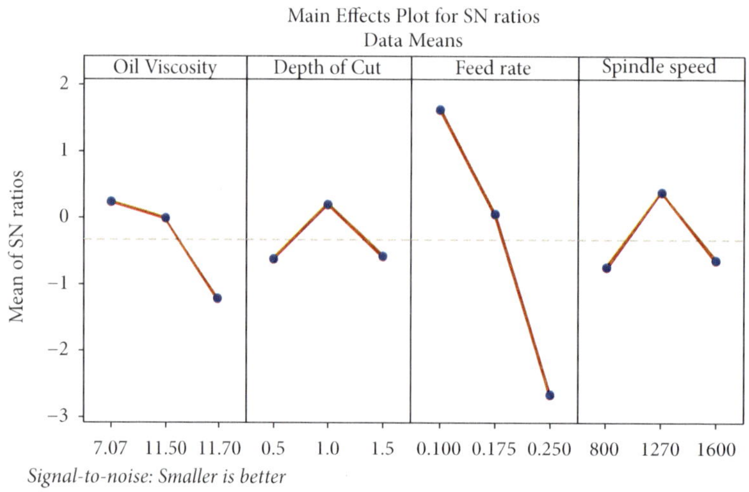

FIGURE 5: Main effects plot for surface roughness.

improving the machined surface quality followed by MJO and MO. Hence, both vegetable oil-based cutting fluids performed well compared to mineral oil with respect to cutting force and surface roughness. The lower cutting forces of vegetable oils can be attributed to better lubricity, higher viscosity index, and better thermal conductivity compared to mineral oils. This is because the modified version has more resistance to molecular breakdown, or a molecular rearrangement at a higher temperature, due to which the presence or absence of oxygen molecules was improved [7]. Previous studies also confirmed that, chemically modified vegetable oil exhibited better lubrication ability and stronger adsorption film onto metallic surface [33]. The better performance with respect to surface roughness is due to the fact that the longer carbon chains of vegetable oil corresponded to a stronger adsorption film which enhanced the surface quality [34].

The effect of viscosity, speed, and feed on response parameters, that is, cutting force and surface roughness, is measured using ANOVA. From Tables 5 and 6 it is observed that cutting fluid, speed, depth of cut, and feed rate influence the cutting force by 0.1%, 5%, 63%, and 28%, respectively. Effects of cutting fluid, speed, depth of cut, and feed rate on surface roughness is found to be 10.65%, 5.73%, 4.13%, and 61.68%, respectively.

3.2. Regression Analysis. Multiple regression analysis on the obtained data is done using Minitab 17. The independent variables are viscosity of cutting fluids, spindle speeds, feed rates, and depths of cut. The dependent variables are cutting force and surface roughness. Separate regression models are built for predicting cutting force and surface roughness. The regression equations obtained are

$$\begin{aligned} F_c &= -62.2905 + 0.474026 \text{ Viscosity} + 216.66 \text{ Depth} \\ &\quad + 940.741 \text{ Feed} - 0.08620 \text{ Speed} \\ R_a &= 0.12240 + 0.034549 \text{ Viscosity} - 0.1036 \text{ Depth} \\ &\quad + 3.531 \text{ Feed} + 5.844e - 005 \text{ Speed}. \end{aligned} \quad (2)$$

Table 5: ANOVA interfamily cutting force.

Source of variance	DOF	Sum of squares	% contribution
Viscosity	2	363	0.098066
Depth	2	235524	63.62814
Feed	2	104080	28.1178
Speed	2	21446	5.793758
Viscosity * Depth	4	837	02.999
Viscosity * Feed	4	131	0.22612
Viscosity * Speed	4	1781	0.03539
Error	6	5994	
Total	*26*	*370157*	

Table 6: ANOVA interfamily surface roughness.

Source of variance	DOF	Sum of squares	% contribution
Viscosity	2	0.23704	10.658
Depth	2	0.09188	04.131
Feed	2	1.37208	61.683
Speed	2	0.12748	05.732
Viscosity * Depth	4	0.12314	05.537
Viscosity * Feed	4	0.00981	00.441
Viscosity * Speed	4	0.04754	02.1375
Error	6	0.21509	
Total	*26*	*2.22405*	

R-Sq values for cutting force and surface roughness are 93% and 78%, respectively, where *R*-Sq is correlation coefficient and should be between 0.8 and 1 in multiple linear regression analyses [35]. It provides a correlation between the experimental and predicted results. The comparisons of actual and predicted values of the observed parameters, that is, cutting force and surface roughness, from the regression models are depicted in Figures 6 and 7. And it is seen that the cutting force and surface roughness observed experimentally and the predicted values by the regression models built are close.

The confirmation tests are conducted for that particular run. Using the regression model, the values of cutting force and surface roughness are predicted for randomly chosen trials. The results of confirmation experiments in Tables 7 and 8 show the difference between the predicted values and actual values. For reliable statistical analyses, error values must be smaller than 20%. The values of both the sets are compared and percentage errors are calculated. The maximum error of 11.39% is observed which is found within an acceptable range.

3.3. Tool Wear Behaviour. It is observed that modified *Pongamia* oil showed lesser tool wear (264.59 μm) for selected cutting parameters followed by mineral oil (558.15 μm) and modified *Jatropha* oil (577.51 μm). Figure 8 shows SEM images of tool flank wear for different oil samples. From Figures 8(a) and 8(b) it is clearly visible that wear of cutting tool is due to abrasion between tool and workpiece in the presence of vegetable oil as cutting fluids. In Figure 8(c), attrition or peeling away of the carbide insert is observed for mineral oil based cutting fluid. The chemical modification processes significantly increase the polar functionality of vegetable oils. The interactions between ester chains in both *Jatropha* and *Pongamia* molecules became stronger, hence increasing the adsorption potential on the metal surface. Thus, tool wear in modified versions of vegetable oils was reduced due to the polar structure of oil samples by forming a protective layer at the contact surfaces that reduced wear and friction [36].

4. Conclusion

The analysis of results by statistical approach based on Taguchi technique shows that the force developed during turning AA6061is said to be optimum for 0.1 mm/rev of feed rate, for 0.5 mm of depth of cut, and at 1600 rpm of speed in the presence of MJO as cutting fluid. A good surface finish is observed for MPO as cutting fluid at low speeds and MJO as cutting fluid at high speeds. Regression analysis showed that the percentage error for experimental and predicted values of cutting force and surface roughness are within an acceptable range. Thus, the modified versions of *Pongamia* and *Jatropha* oil as straight cutting fluids have a great potential for use as environmentally friendly and biodegradable metal cutting fluids.

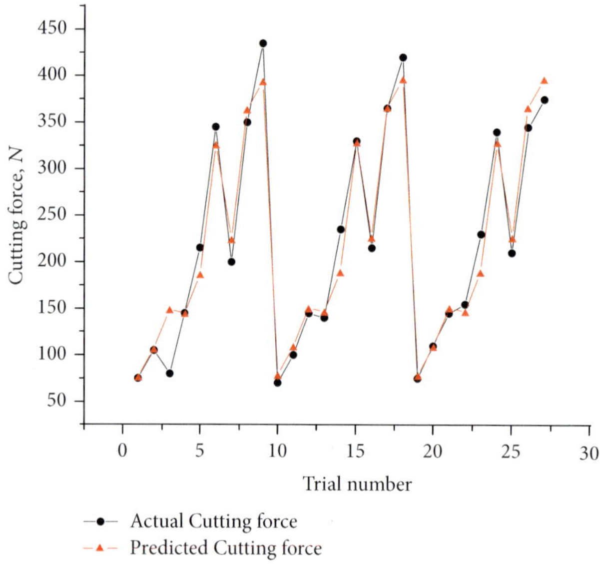

Figure 6: Actual and predicted values of cutting force.

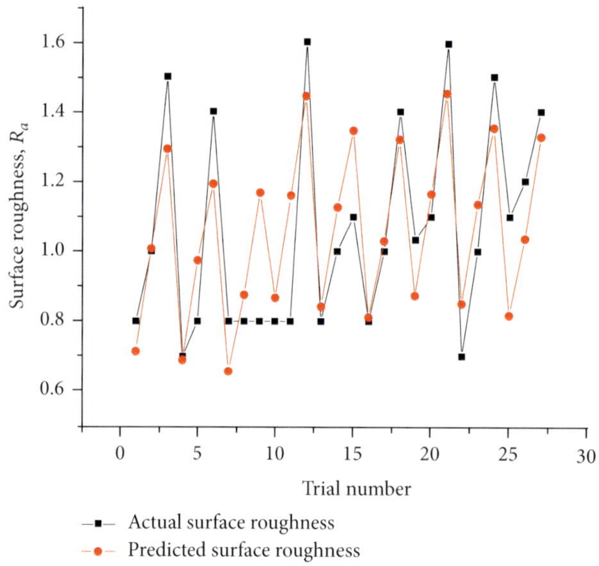

Figure 7: Actual and predicted values of surface roughness.

Table 7: Confirmation tests for cutting forces.

Cutting fluids	Depths of cut (mm)	Feed rates (mm/rev)	Speed (rpm)	Predicted force F_c (N)	Experimental force, F_c (N)	% error
MJO	1	0.175	800	253.39	235	7.821
MPO	0.5	0.175	1270	106.64	100	6.645
MO	1.5	0.25	800	434.47	465	6.563

Table 8: Confirmation tests for surface roughness.

Cutting fluids	Depths of cut (mm)	Feed rates (mm/rev)	Speed (rpm)	Predicted R_a (μm)	Experimental R_a (μm)	% error
MJO	0.5	0.1	1600	0.761	0.8	4.8186
MPO	1	0.25	800	1.345	1.3	3.5065
MO	1.5	0.175	1270	1.063296	1.2	11.392

(a) *Jatropha*

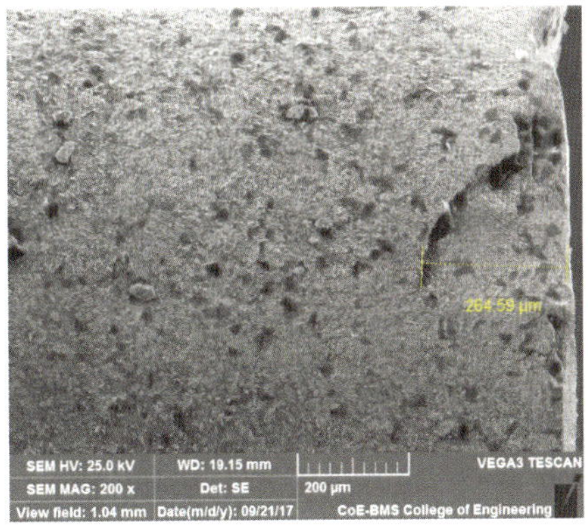

(b) *Pongamia*

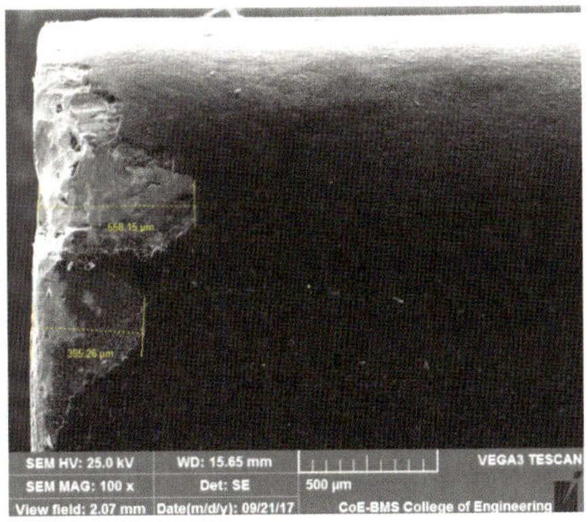

(c) Mineral oil

FIGURE 8: SEM images for flank wear.

Conflicts of Interest

The authors declare that they have no conflicts of interest.

References

[1] K. K. Gajrani and M. R. Sankar, "Past and current status of eco-friendly vegetable oil based metal cutting fluids," *Materials Today: Proceedings*, vol. 4, no. 2, pp. 3786–3795, 2017.

[2] G. Burton, C.-S. Goo, Y. Zhang, and M. B. G. Jun, "Use of vegetable oil in water emulsion achieved through ultrasonic atomization as cutting fluids in micro-milling," *Journal of Manufacturing Processes*, vol. 16, no. 3, pp. 405–413, 2014.

[3] R. Somashekaraiah, P. S. Suvin, D. P. Gnanadhas, S. V. Kailas, and D. Chakravortty, "Eco-friendly, non-toxic cutting fluid for sustainable manufacturing and machining processes," *Tribology Online*, vol. 11, no. 5, pp. 556–567, 2016.

[4] British Petroleum. BP statistical review of world energy june 2017. 2017. p. 48.

[5] S. Amiril, E. Rahim, Z. Embong, and S. Syahrullail, "Tribological investigations on the application of oil-miscible ionic liquids additives in modified Jatropha-based metalworking fluid," *Tribology International*, vol. 120, pp. 520–534, 2018.

[6] P. Rapeti, V. K. Pasam, K. M. Rao Gurram, and R. S. Revuru, "Performance evaluation of vegetable oil based nano cutting fluids in machining using grey relational analysis-A step towards sustainable manufacturing," *Journal of Cleaner Production*, vol. 172, pp. 2862–2875, 2016.

[7] Y. M. Shashidhara and S. R. Jayaram, "Experimental determination of cutting power for turning and material removal rate for drilling of AA 6061-T6 using vegetable oils as cutting fluid," *Advances in Tribology*, vol. 2013, Article ID 362931, 7 pages, 2013.

[8] Y. M. Shashidhara and S. R. Jayaram, "Vegetable oils as a potential cutting fluid-An evolution," *Tribology International*, vol. 43, no. 5-6, pp. 1073–1081, 2010.

[9] J. K. Mannekote, S. V. Kailas, K. Venkatesh, and N. Kathyayini, "Environmentally friendly functional fluids from renewable and sustainable sources-A review," *Renewable & Sustainable Energy Reviews*, vol. 81, pp. 1787–1801, 2018.

[10] Y. Singh, A. K. Singh, A. Singla, and A. K. Upadhyay, "Development and tribological characteristics of bio-based lubricant from Jatropha curcas oil," *Energy Sources, Part A: Recovery, Utilization, and Environmental Effects*, vol. 39, no. 16, pp. 1725–1732, 2017.

[11] N. Talib and E. A. Rahim, "Performance of modified jatropha oil in combination with hexagonal boron nitride particles as a bio-based lubricant for green machining," *Tribology International*, vol. 118, pp. 89–104, 2018.

[12] J. Sharma and B. S. Sidhu, "Investigation of effects of dry and near dry machining on AISI D2 steel using vegetable oil," *Journal of Cleaner Production*, vol. 66, pp. 619–623, 2014.

[13] B. Wilson, "Lubricants and functional fluids from renewable sources," *Industrial Lubrication and Tribology*, vol. 50, no. 1, pp. 6–15, 1998.

[14] N. J. Fox and G. W. Stachowiak, "Vegetable oil-based lubricants-A review of oxidation," *Tribology International*, vol. 40, no. 7, pp. 1035–1046, 2007.

[15] E. Kuram, B. Ozcelik, and E. Demirbas, "Environmentally Friendly Machining: Vegetable Based Cutting Fluids," in *Green Manufacturing Processes and Systems*, Materials Forming, Machining and Tribology, pp. 23–47, Springer Berlin Heidelberg, Berlin, Heidelberg, 2013.

[16] E. A. Rahim and H. Sasahara, "A study of the effect of palm oil as MQL lubricant on high speed drilling of titanium alloys," *Tribology International*, vol. 44, no. 3, pp. 309–317, 2011.

[17] P. Nagendramma and S. Kaul, "Development of ecofriendly/biodegradable lubricants: An overview," *Renewable & Sustainable Energy Reviews*, vol. 16, no. 1, pp. 764–774, 2012.

[18] Z. S. Petrović, A. Zlatanić, C. C. Lava, and S. Sinadinović-Fišer, "Epoxidation of soybean oil in toluene with peroxoacetic and peroxoformic acids— kinetics and side reactions," *European Journal of Lipid Science and Technology*, vol. 104, no. 5, pp. 293–299, 2002.

[19] C. A. S. Bork, J. F. D. S. Gonçalves, J. D. O. Gomes, and J. Gheller, "Performance of the jatropha vegetable-base soluble cutting oil as a renewable source in the aluminum alloy 7050-T7451 milling," *CIRP Journal of Manufacturing Science and Technology*, vol. 7, no. 3, pp. 210–221, 2014.

[20] S. Chinchanikar, A. Salve, P. Netake, A. More, S. Kendre, and R. Kumar, "Comparative Evaluations of Surface Roughness During Hard Turning under Dry and with Water-based and Vegetable Oil-based Cutting Fluids," *Procedia Materials Science*, vol. 5, pp. 1966–1975, 2014.

[21] M. A. Xavior and M. Adithan, "Determining the influence of cutting fluids on tool wear and surface roughness during turning of AISI 304 austenitic stainless steel," *Journal of Materials Processing Technology*, vol. 209, no. 2, pp. 900–909, 2009.

[22] M. M. A. Khan and N. R. Dhar, "Performance evaluation of minimum quantity lubrication by vegetable oil in terms of cutting force, cutting zone temperature, tool wear, job dimension and surface finish in turning AISI-1060 steel," *Journal of Zhejiang University SCIENCE A*, vol. 7, no. 11, pp. 1790–1799, 2006.

[23] S. A. Lawal, I. A. Choudhury, and Y. Nukman, "Evaluation of vegetable and mineral oil-in-water emulsion cutting fluids in turning AISI 4340 steel with coated carbide tools," *Journal of Cleaner Production*, vol. 66, pp. 610–618, 2014.

[24] S. J. Ojolo and O. S. Ohunakin, "Study of rake face action on cutting using palm-kernel oil as lubricant," *Journal of Emerging Trends in Engineering and Applied Sciences (JETEAS)*, vol. 2, no. 1, pp. 30–35, 2011.

[25] B. Ozcelik, E. Kuram, M. Huseyin Cetin, and E. Demirbas, "Experimental investigations of vegetable based cutting fluids with extreme pressure during turning of AISI 304L," *Tribology International*, vol. 44, no. 12, pp. 1864–1871, 2011.

[26] M. H. Cetin, B. Ozcelik, E. Kuram, and E. Demirbas, "Evaluation of vegetable based cutting fluids with extreme pressure and cutting parameters in turning of AISI 304L by Taguchi method," *Journal of Cleaner Production*, vol. 19, no. 17-18, pp. 2049–2056, 2011.

[27] R. Katna, K. Singh, N. Agrawal, and S. Jain, "Green manufacturing—performance of a biodegradable cutting fluid," *Materials and Manufacturing Processes*, vol. 32, no. 13, pp. 1522–1527, 2017.

[28] P. Rapeti, V. K. Pasam, K. M. Rao Gurram, and R. S. Revuru, "Performance evaluation of vegetable oil based nano cutting fluids in machining using grey relational analysis-A step towards sustainable manufacturing," *Journal of Cleaner Production*, vol. 172, pp. 2862–2875, 2018.

[29] V. S. Sharma, G. Singh, and K. Sorby, "A review on minimum quantity lubrication for machining processes," *Materials and Manufacturing Processes*, vol. 30, no. 8, pp. 935–953, 2015.

[30] E. K. Heikal, M. S. Elmelawy, S. A. Khalil, and N. M. Elbasuny, "Manufacturing of environment friendly biolubricants from vegetable oils," *Egyptian Journal of Petroleum*, vol. 26, no. 1, pp. 53–59, 2017.

[31] Y. M. Shashidhara and S. R. Jayaram, "Tribological studies on AISI 1040 with raw and modified versions of pongam and jatropha vegetable oils as lubricants," *Advances in Tribology*, Article ID 560175, 2012.

[32] T. Saurabh, M. Patnaik, S. L. Bhagt, and V. C. Renge, "Epoxidation of vegetable oils: a review," *International Journal of Advanced Engineering Technology*, vol. 2, no. 4, pp. 491–501, 2011.

[33] J. Salimon, B. M. Abdullah, R. M. Yusop, and N. Salih, "Synthesis, reactivity and application studies for different biolubricants," *Chemistry Central Journal*, vol. 8, no. 1, article no. 16, 2014.

[34] B. Li, C. Li, Y. Zhang, Y. Wang, D. Jia, and M. Yang, "Grinding temperature and energy ratio coefficient in MQL grinding of high-temperature nickel-base alloy by using different vegetable oils as base oil," *Chinese Journal of Aeronautics*, vol. 29, no. 4, pp. 1084–1095, 2016.

[35] D. C. Montgomery, *Design and Analysis of Experiments*, John Wiley & Sons, 5th edition, 2001.

[36] Y. Zhang, C. Li, D. Jia, D. Zhang, and X. Zhang, "Experimental evaluation of MoS2 nanoparticles in jet MQL grinding with different types of vegetable oil as base oil," *Journal of Cleaner Production*, vol. 87, no. 1, pp. 930–940, 2015.

15

A Novel Two-Disc Machine for High Precision Friction Assessment

Jakob Moder, Florian Grün, Michael Stoschka, and István Gódor

Chair of Mechanical Engineering, Montanuniversität Leoben, Franz-Josef-Straße 18, 8700 Leoben, Austria

Correspondence should be addressed to Jakob Moder; jakob.moder@unileoben.ac.at

Academic Editor: Michael M. Khonsari

The concept of two-disc model testing has proven to deliver valuable information for the applicability of new technologies, such as surface structuring, coatings, alternative fluids, or advanced materials, in actual machine elements. In this article an advanced two-disc machine with sophisticated control technology for dry and lubricated setup is presented. (i) All involved components are controlled via a powerful PLC unit leading to the possibility of realising extremely accurate SRRs down to 0.1%. (ii) High-speed data acquisition allows local insight into tribological phenomena by providing 72 data points along one shaft rotation. (iii) Several lubrication scenarios such as fluid, mixed, and starved lubrication, as well as dry contacts, can be considered. (iv) Raw-data of all sensors, including normal force, friction force, vibration speed, stiffness (wear), infrared temperature, contact potential, and motor speed, is presented. Some example results of dry polymer-steel tribosystems and lubricated experiments are shown to elucidate the capabilities of the novel test rig.

1. Introduction

The history of two-disc model testing goes back to 1935, when [1] introduced an apparatus for a simplified testing of worm gears. Ever since then this type of machines has become more popular among researchers, due to easy usage and comparability of test results to actual machine components. The simple cylindrical geometry enables a precise measurement of the friction force in the contact. Also, operating conditions, such as rolling velocity, slide-to-roll ratio, or contact pressure, in terms of normal force, can be adjusted easily. Furthermore, the machining of discs is more effective from an economic point of view in contrast to the complicated manufacturing process of gears. The tribological behaviour of machine elements, as for instance gears, is depending on operating conditions, lubrication (especially viscosity and additives of the oil), and surface structure [2]. The latter two factors can be depicted equivalently with a two-disc machine; however gears generally operate under transient conditions, due to the continuously changing normal force, sliding velocity, and contact radii, along the line of action [3]. A two-disc experiment typically depicts the operating states of a gear contact which are of greatest interest stationarily.

On the one hand such tests are carried out for a specific servicing point to determine the rolling contact fatigue (RCF) of certain tribological systems. Lifetime results of discs can be transferred to actual gears under certain conditions, as, for instance, explained in [4] or [5]. However, one has to keep in mind that transient phenomena seen in gears are impossible to reproduce with discs. Due to the high contact stresses, of at least 3 GPa, required for RCF testing, test-rigs are typically big and have to operate under severe loads, which requires high maintenance efforts and makes an accurate measurement of friction forces not feasible.

On the other hand, two-disc machines are employed for friction measurements of tribological contacts. It has been shown on numerous occasions [6, 7] that two-disc experiments can indicate how different aspects, such as fluid, surface finishing, surface texturing, or coatings, will affect the friction in gear systems.

With time machines have become more sophisticated, especially the advent of modern control technology opened the way to new testing concepts. While the older test-rigs were restricted to certain slide-to-roll ratios (SRR), typically determined by disc diameter or mechanical gear ratios, the

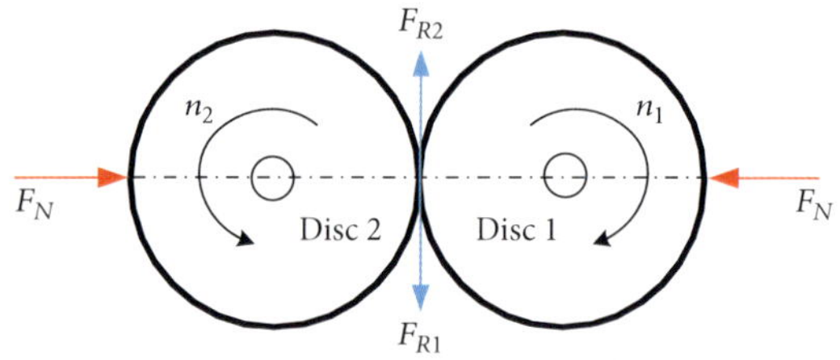

FIGURE 1: Basic principle of a two-disc machine.

usage of electrical drives allows a more precise setting of arbitrary SRRs. Nowadays numerous research facilities have engineered two-disc machines [8–12], which basically differ in type of frictional force measurement.

In contrast to other machines the presented test rig integrates several components in a programmable logic controller (PLC). This permits the evaluation and setting of parameters in real-time, enabling, for instance, a control loop between both motors leading to extremely accurate SRRs down to 0.1%. Furthermore, the option to perform fully automated tests and data evaluation procedures is implemented. Due to the high computational power of the PLC a high-speed data evaluation, allowing detailed insight into local phenomena of tribological systems, becomes feasible. Also, multiple sensors, for the analysis of vibrations, stiffness, wear, temperatures, and contact potential, are integrated in order to gain a profound understanding of the tested tribosystems. Raw-data of all sensors is presented, while other papers typically only publish postprocessed traction curves.

2. Machine Setup

The basic principle of a two-disc machine is depicted in Figure 1.

The discs are loaded with a normal force F_N and rotating with speeds n_1 and n_2. When the speeds of the discs are not equal, a friction force vector, with equal norm but different direction, results on the surfaces of both discs. This force ultimately determines the frictional performance of the tribological system and is therefore the desired quantity to be measured. In the case of cylindrical specimens, the direction of the friction force vector is time independent, contrary to gears, where the friction vector is changing with time along the line of action, which makes it difficult to diagnose.

Several methods exist in order to measure the friction force in two-disc machines. One way is the usage of torque transducers, which are applied to a shaft on which the machine element of interest is mounted. Papers [11, 12] are, for example, employing this technique for friction force determination in their test-rigs. However, one has to consider that parasitic influences, such as losses of roller bearings of the shaft, or sealing losses, are measured as well by this setup. Depending on the type of roller bearings and sealings in the machine, it can therefore be very difficult to determine the actual friction force accurately in the contact. Another possibility is the usage of a force transducer, which is positioned in line to measure the friction directly, as, for instance, described in [9]. This way, the measurement of parasitic forces can basically be avoided.

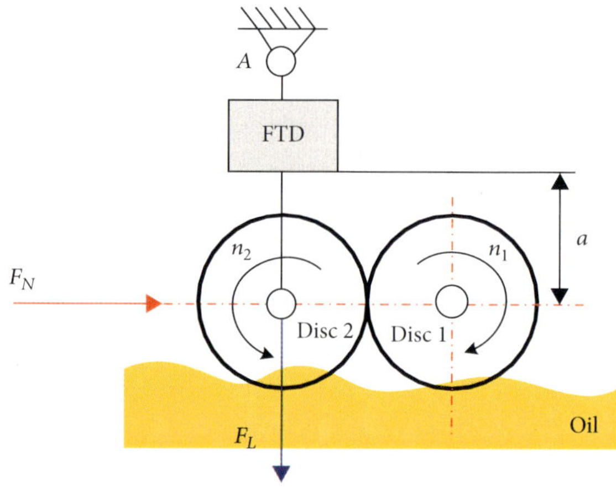

FIGURE 2: Measurement concept of the presented two-disc test rig.

The direct friction force measurement principle via force transducer is employed in the presented test rig, which is based on previous developments at the *Chair of Mechanical Engineering, Montanuniversität Leoben* [13, 14], as illustrated in Figure 2.

The shaft of disc 2 is mounted in a housing, which can rotate with respect to point A; therefore any friction force appearing on disc 2 is directed to the force transducer via shaft, bearings, and housing. The force measured is denoted as F_L, whereby it has to be taken into account that the weight force of the housing and its components is measured as well. This force is quantified by a highly precise force transducer, in our case a uniaxial, laterally compensated force transducer of type *HBM U10M*.

Figure 3 displays a cut through the fluid chamber, of the computer-aided-design (CAD) assembly, and a photograph of the actual test rig, which has been taken during an experiment with grease lubrication. Note that the view of the photograph is rotated ninety degrees clockwise with respect to the CAD image.

The disc specimens, which have a diameter of 50 mm, are mounted onto the respective shaft with a spring key and fixed axially by a screw. The normal force is induced into the system by a *Norgren* pneumatic cylinder via the rear of the housing, which also mounts shaft 1. On the left of Figure 3 the initiation of the normal force is illustrated by the vector F_N. The mechanism for the application of the normal force is presented in Figure 4. To make sure that the desired amount of force is applied, a force transducer is placed directly in the loading path next to the cylinder. A roller is used to transfer the load into the system in order to eliminate disturbances which can affect the friction measurement.

Flywheels can be mounted on the shafts optionally to increase the moment of inertia of the system in order to prevent motor control issues when high friction forces spontaneously (e.g., seizure) occur in the contact. For the results presented in this article no flywheels are used. Each shaft is connected to a motor by employing a coupling.

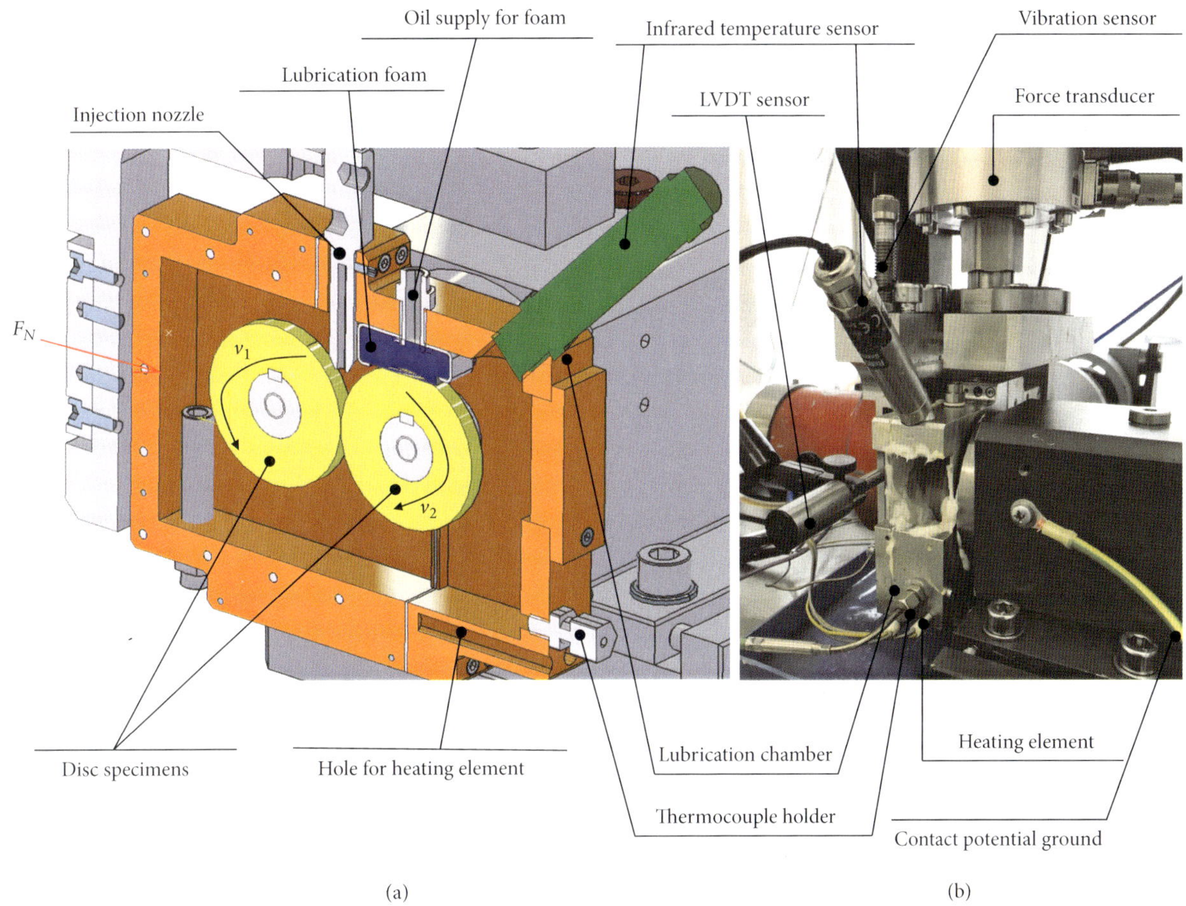

FIGURE 3: Cut through the lubrication chamber (a) and photograph of the machine (b).

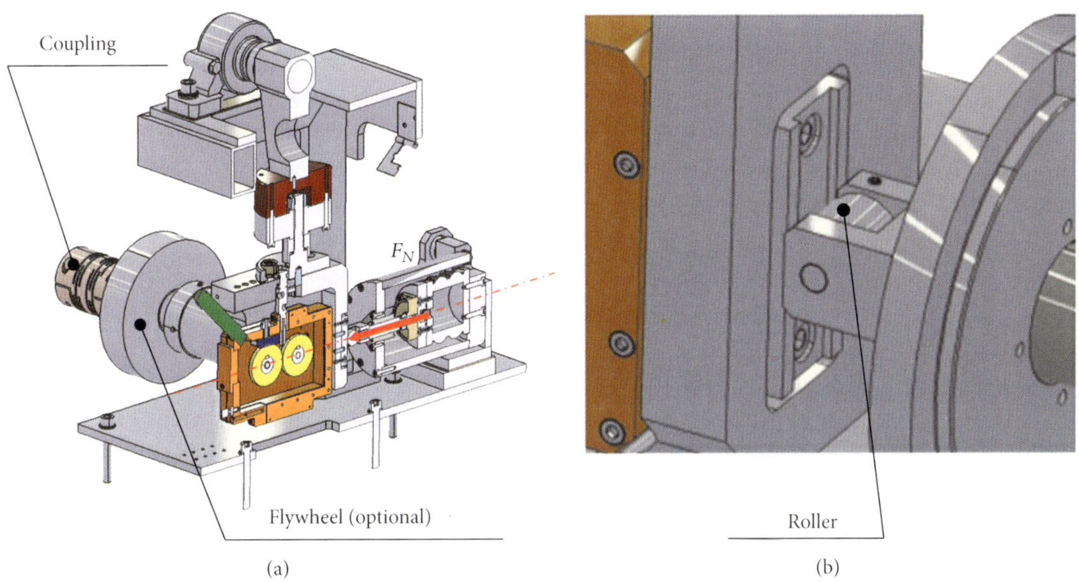

FIGURE 4: Application of the normal force by a pneumatic cylinder. Rear view (a) and force application via a roller (b).

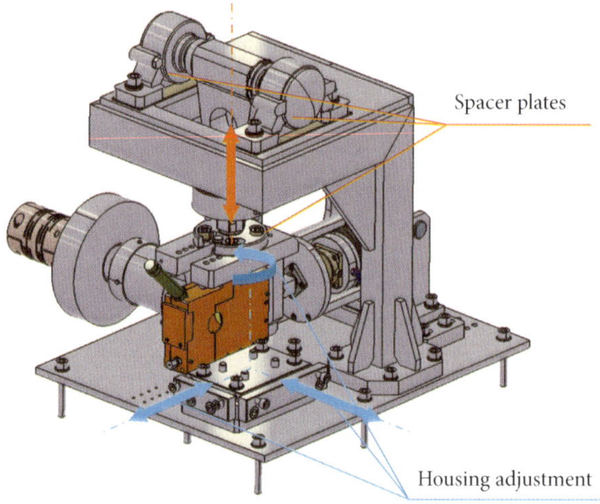

FIGURE 5: Options for contact alignment.

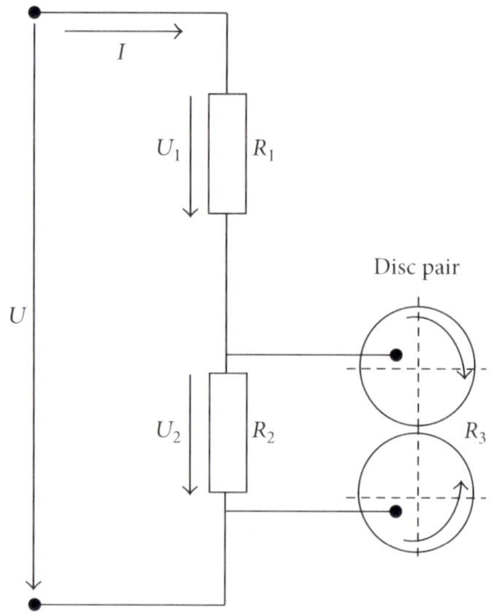

FIGURE 6: Electrical circuit for contact potential measurement.

In order to adjust the alignment of the disc specimens, which is paramount for performing high quality measurements, several options are available, which are highlighted in Figure 5.

The horizontal position of the housing, which mounts shaft 2, can be adjusted via screws and the vertical position of shaft 1 is precisely set by utilising spacer plates with a thickness of 0.1 mm. The alignment is considered successful when an application of the normal force is not altering the friction force (±1 N), whereby the motors are switched off. Furthermore, a pressure foil (*Fujifilm*), which locally determines the pressure by colouring, is utilised in order to guarantee an optimal contact pattern.

A piezoelectric vibration sensor is integrated into the machine for the detection of wear phenomena or specimen with intolerable eccentricity.

Several lubrication scenarios, from dry to fully flooded conditions, can be realised with this test rig. In the case of splash lubrication, the chamber is filled with a volume of about 250 ml fluid. This ensures that both discs are dipped 11 mm into the fluid. There is a small gap between the lubrication chamber and the shaft of disc 2, which can cause minor leakage when the machine is operating under high speed. However, additional fluid can be provided via the injection nozzle in this case. The primary field of use for the injection nozzle is the analysis of starved lubrication systems; therefore a *LECHLER* pneumatic nebuliser is installed in the test machine. The amount of injected fluid is controlled by a *FESTO LR DB* pneumatic control unit, which permits a fine dosage of fluid air mixture to be injected directly into the contact line of the discs. To investigate even severe starved lubrication conditions, a foam holder, which is pressing an oil soaked foam unit onto disc 2, is available as well.

Another important issue is the control of the lubrication chamber temperature. Hence, the machine is equipped with four heating elements with a power of 150 W each, which are mounted directly in the aluminium lubrication chamber for optimal heat transfer to the fluid. The agile control loop can heat the lubricant from room temperature to about 65°C in only forty seconds.

TABLE 1: Set variables for the two-disc machine.

Name (abbreviation)	Value range
Slide-to-roll ratio (s, SRR)	0–200%
Normal force (F_N)	100–2000 N
Speed (n_1, n_2)	0–4500 rpm
Temperature (T_0)	RT–150°C
Fluid injection ($\dot{V}_{in}$)	20 ml/min–300 ml/min

A linear variable differential transformer (LVDT) sensor is integrated into the machine for stiffness and wear measurements. This transducer can resolve length scales in range of microns and is therefore suitable for wear investigations. This sensor is positioned on the housing, which can rotate with respect to point A, as illustrated in Figure 2. Therefore, any wear or changes of stiffness, which can for instance occur temperature dependently, are directly translated into the housing and measured via the displacement sensor.

Three different temperature sensors are mounted around the chamber and a thermocouple holder is fixed in the chamber directly, allowing for Type-K thermoelements to measure the temperature of the oil.

Furthermore, an infrared sensor is measuring the temperature on the surface of disc 2 at an angle of 135° next to the contact line.

Housing and shaft of disc 1 are electrically isolated with respect to disc 2 which allows a contact resistance measurement of the tribological system and therefore binary judgement of the state of friction. A sketch explaining the contact potential measurement system is provided in Figure 6.

All set variables, as well as measured properties, are depicted in Tables 1 and 2, respectively.

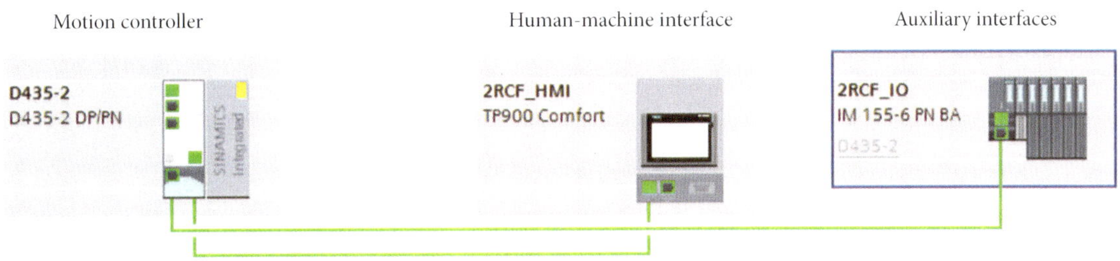

FIGURE 7: Sketch of the major electrical components.

TABLE 2: Measured variables.

Name (abbreviation)	Value range
Friction force (F_R)	0–1000 N
Infrared temperature (T_I)	0–160°C
Oil-bath temperature (T_O)	0–200°C
Lubrication chamber temperature (T_C)	0–200°C
Contact potential (c_{pot})	1–100 mV
LVDT distance (s_{LVDT})	1 μm–2 mm
Vibration sensor (v_{vbs})	0–10 mm/s

The highest possible *Hertzian* pressure is depending on the actual width of the discs. For the case of a disc thickness of 5 mm the contact pressure limit is 1.1 GPa. The corresponding velocity for a specimen diameter of 50 mm is in the range of 0–11 m/s.

2.1. Electrical Architecture. Figure 7 illustrates schematically the involved major electrical units of the presented two-disc machine. The system is operated by a *Siemens D435-2* digital control unit which permits real-time cycles of up to 1000 Hz. The controller not only acts as real-time controller for both drives, but also features motion chart controls for the sequential main flow as well as transient data logging and several force and temperature closed loop controls. The tribological test rig demands several analogue and digital in- and outputs for actuation and measurement. This peripheral IO is implemented as *Siemens ET200SP* station. In addition, the force signal is transmitted by a *HBM digiclip* device, which is connected to the main controller via Profibus. Finally, the digital control unit *Siemens D435-2* embeds also the motion controller axis featuring *SINAMICS Integrated* technology.

The human-machine-interface is realised as a *Siemens TP900* touch panel [15], which is connected via Profibus to the digital control unit. This panel is used for machine monitoring, specification of set values, and low speed data acquisition. Furthermore, fully automated tests can be performed via the recipe system [16]. In addition, the touch panel acts as VNC-server within the local area network and simplifies therefore user operating and supervision issues. Figure 8 displays the human-machine-interface covering the definition of set values based on the recipe system. Furthermore, the recipe based test programme can be looped to extend the test time. The test run is switched off if a certain number of steps or a certain value of test time, temperature, vibration speed, friction force range, or displacement is exceeded. The human-machine-interface supports manual as well as automated test programmes. The first are used for initial studies; in case of unknown material or lubricant behaviour, the latter are applied to maintain identical test settings.

Besides the reproducible definition of set values and shutdown criteria, the monitoring of the actual values and their data logging is important for scientific evaluations of the tribological system. The speed of the drive axis, slip, friction force, contact potential and displacement sensor values are of interest for the user. The measured data is displayed in Figure 9 based on a cyclic refresh rate of about five hundred milliseconds. Aside this quasi-stationary evaluation, a high-speed transient data logging, featuring one thousand measurement points per second, is also implemented in the digital control unit.

The *Sinamics S120* double motor module [17] is powering both *Siemens 1FT7* PESM machines [18], with a nominal power of 3.77 kW each. The synchronous motor has an integrated high-resolution incremental encoder which is connected via the *DRIVE CLiQ* technology to the motor module. This hardware design facilitates, in combination with the *Simotion* control technology, a highly accurate adjustment of SRRs down to 0.1%. A stable velocity control is especially demanding at low speed set points, since the local variations in friction forces may disturb the speed controlled axis. Hence, the speed control of the underlying *SINAMICS Integrated* drive technology is set up to handle such disturbance moments quickly and stably. The closed loop control of the drive unit is initially parametrised by automatic control routines available in *SIEMENS Scout/TIA Portal*. In addition, the bearing dependent friction losses for each axis are also compensated by this drive technology.

All previously mentioned auxiliary components, such as analogue and digital in- and outputs, amplifier units for force measurement, pneumatic control valves, temperature sensors, fuses and connector devices, are assembled in an electrical cabinet close to the two-disc test rig. The touch panel is also integrated in the cabinet to support an easy on-site user-operation.

2.2. Control Setup. For the configuration and programming, of both drive and control related functionalities, *SIEMENS Scout/TIA Portal* is used [16]. As mentioned before, the speed drives are based on a *SINAMICS Integrated* technology which

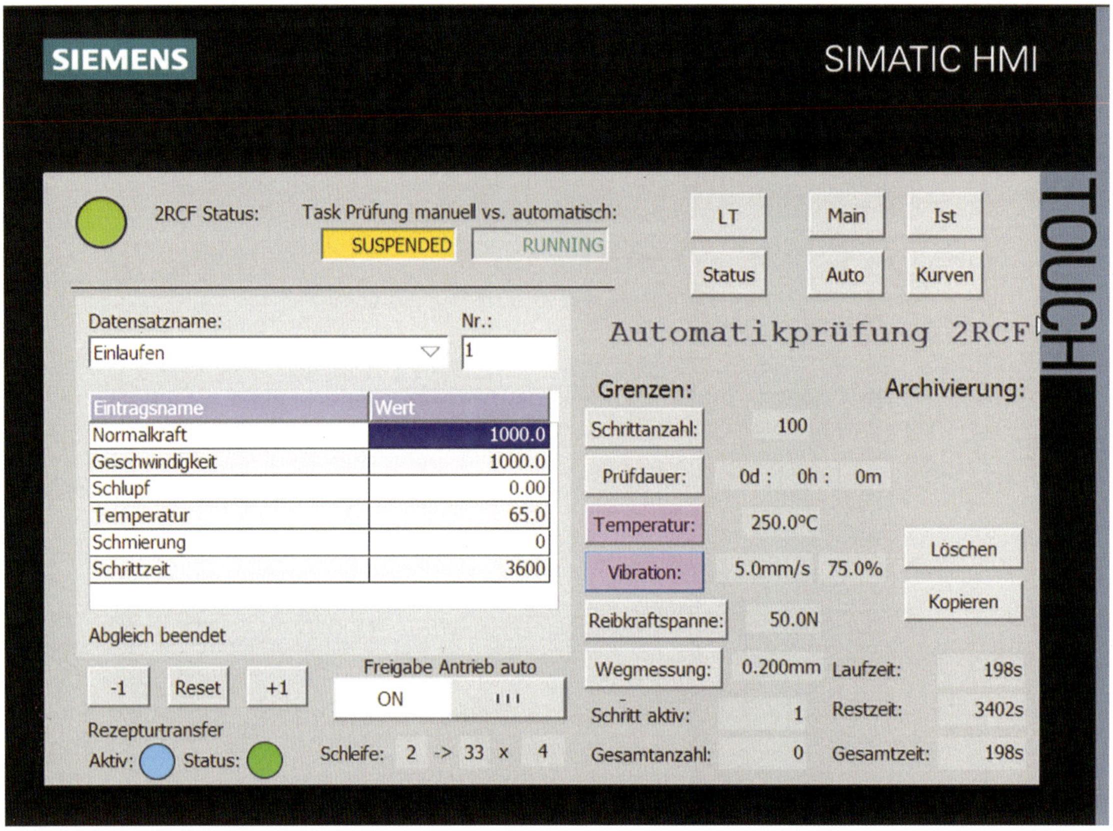

FIGURE 8: Recipe based set value interface including test limits settings.

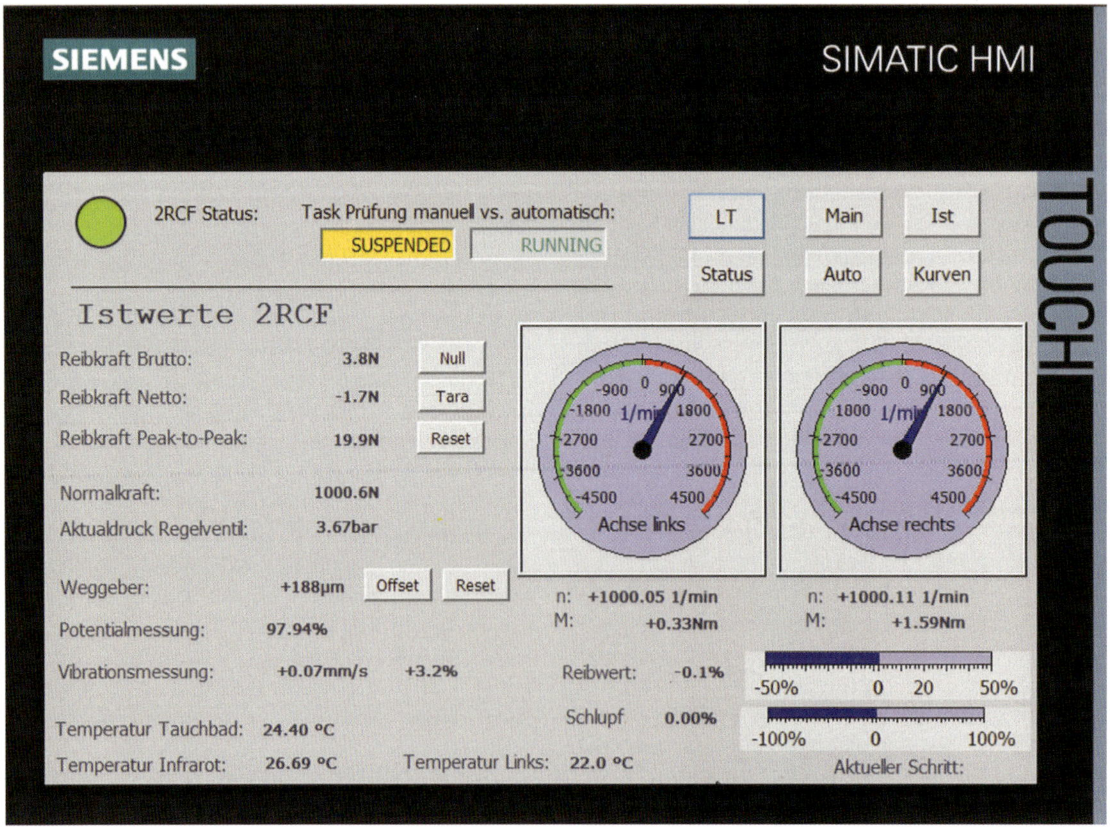

FIGURE 9: Actual value human-machine-interface.

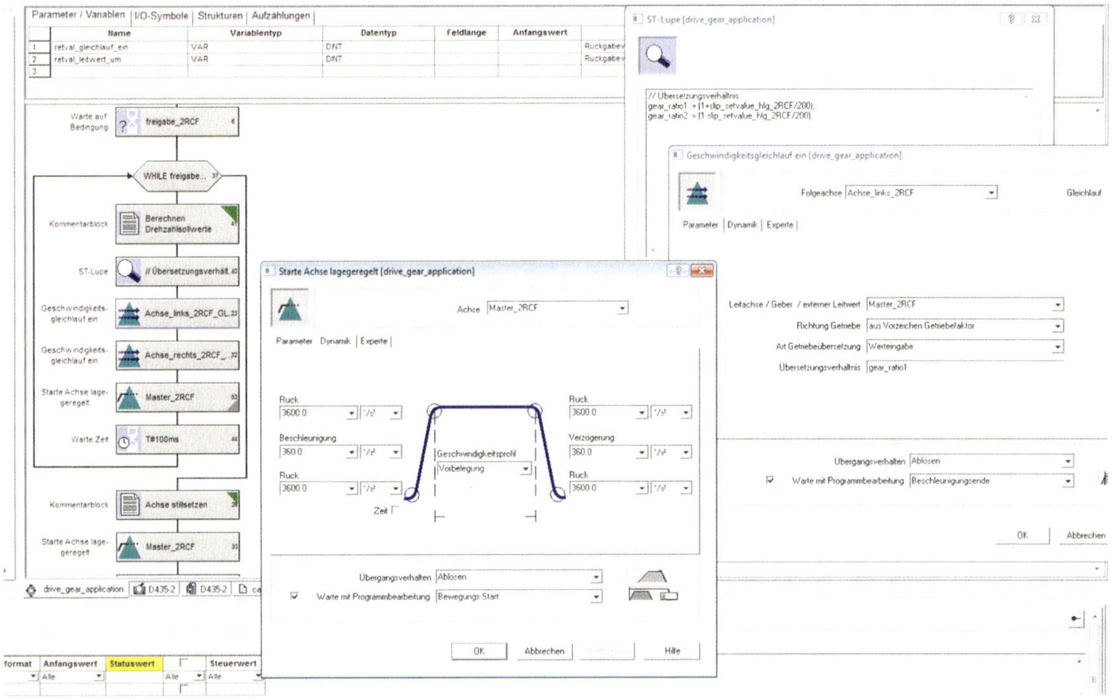

FIGURE 10: Motion control chart enabling speed synchronisation for varying slip set values by virtual master axis and incorporated gear transmission for each drive axis.

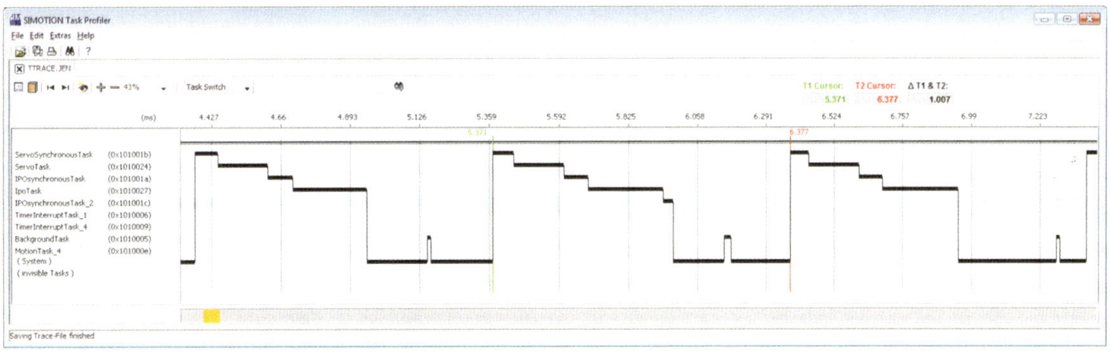

FIGURE 11: Task profiler of *D435-2* real-time control system of two-disc tester.

utilises an internal control clock frequency of eight thousand Hertz. The position control of the *D435-2* defines the motion of both axes by implementation of a virtual master axis. A common synchronism is guaranteed as the speed of this virtual master axis is modified by the given slip set value for each drive axis. The implemented programme section is shown in Figure 10. The speed set value for each axis is generated by the interpolator clock. To ensure a high dynamic response of the system, the interpolator clock is set equal to the servo clock pulse of one millisecond. To fasten the servo control further, the option *Dynamic Servo Control* is activated, which causes the superimposed position controller to be run in the clock pulse of the *Simotion* speed control unit. The speed set value is interpolated based on the data obtained from the virtual master axis. This guarantees a high bandwidth for the drive control of both test rig axes.

The real-time drive control system utilises therefore the servo task, the interpolator synchronous task, and finally the interpolator task with an identical clock pulse of one millisecond. The control programme is split up into motion tasks, which are processed on the basis of a certain motion requirement or started by other tasks; servo-synchronous tasks for the conversion of IO-devices related data: cyclic high-speed logging and evaluation of transient changes, for example, in angular position and friction force values, is taken into account by interpolator synchronous processing; timer-based interrupt tasks with decreased sample rates facilitate real-time processing of closed force or temperature control loops and logging issues; and finally, the background task handles the start and stop procedure of the overall test rig. Figure 11 exhibits the task profile of the *D435-2* control unit.

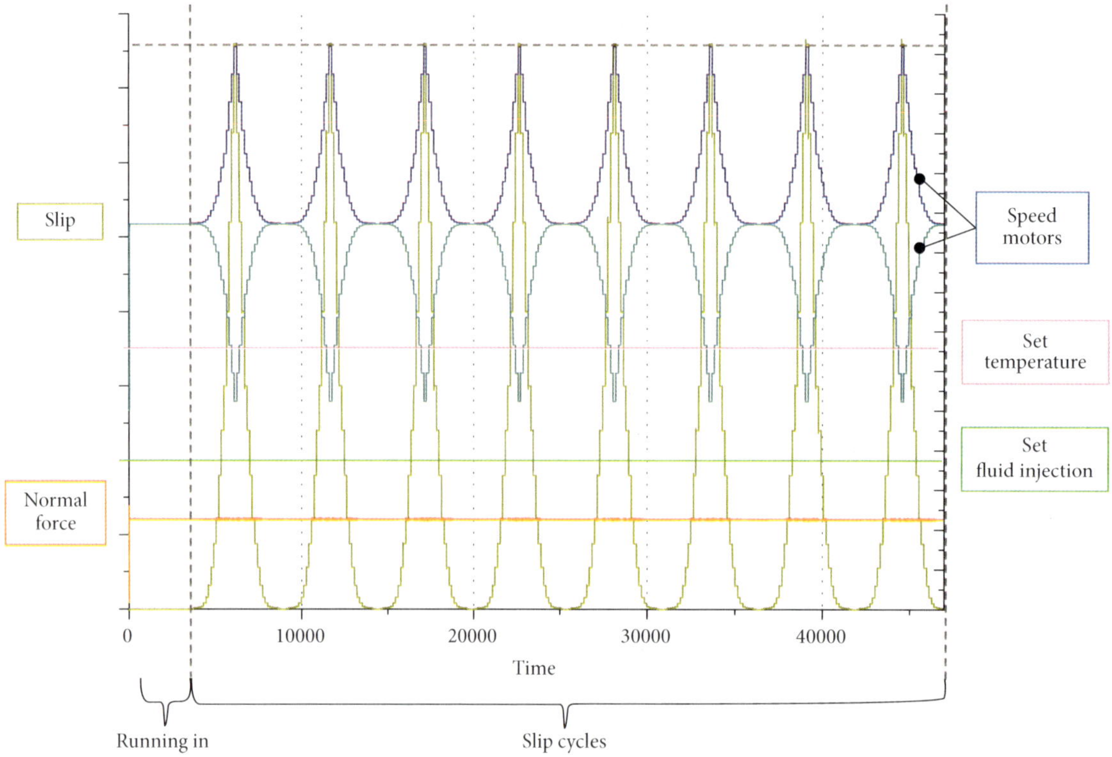

FIGURE 12: Test programme for presented experiments.

Utmost priority has the servo task, which handles the position control of each axis. The corresponding speed set value is continuously derived in interpolator task. The interpolator task covers several time-critical user-defined processes; hence it necessitates already about one-quarter of the overall available clock pulse for itself. Motion tasks are carried out on demand; the background task is processed dependent on the free time slot between each cycle. As transient data logging is implemented in the fast interpolator task, it is possible to gather transient values of friction force, vibration velocity, and others within a 4k-buffer. These buffer values are then periodically evaluated and mapped onto a result matrix covering steps of five degree angles. This real-time based, transient data evaluation enables an improved high-speed data measurement and gains additional insight into local changes of the tribological system.

3. Experimental Results and Discussion

In this section exemplary results are presented to emphasise functionalities of the developed two-disc machine.

3.1. Test Programme. The term test programme is used for the description of all available set values as a function of time. For the experiments presented in this article, a cyclic slip programme with running-in phase, as illustrated in Figure 12, is applied.

Firstly, typically one hour is spent for a running-in procedure of the system and also in order to reach a thermally stable state of the machine. Then, slip and therefore motor speed are increased (or decreased in case of the other axis) stepwise until the maximum slip is reached. After the maximum slip is maintained, an incremental decrease of the slip is performed in order to return to the initial state. Those slip cycles are repeated throughout the experiment, whereby total time for the test is typically in the range of four to twelve hours, depending on the number of slip cycles actually performed. The set parameters, including temperature, fluid injection rate, and normal force, are kept constant throughout the tests presented in this paper. The slide-to-roll ratio is defined according to

$$\mathrm{SRR} = s = \frac{u_1 - u_2}{u_m} \tag{1}$$

whereby the mean velocity is kept constant throughout each test:

$$u_m = \frac{u_1 + u_2}{2}. \tag{2}$$

3.2. Lubricated Steel-Steel Experiment. In this subsection experiments that are carried out with discs made of hardened 42CrMo4 and super-finished surface, with a roughness of R_a 0.03 μm, using a mineral oil under splash lubrication, are presented. The thickness of one disc has been reduced from 11 mm to 5 mm to improve contact alignment and enable the testing of increased contact pressures. The respective operating and material parameters are given in Table 3.

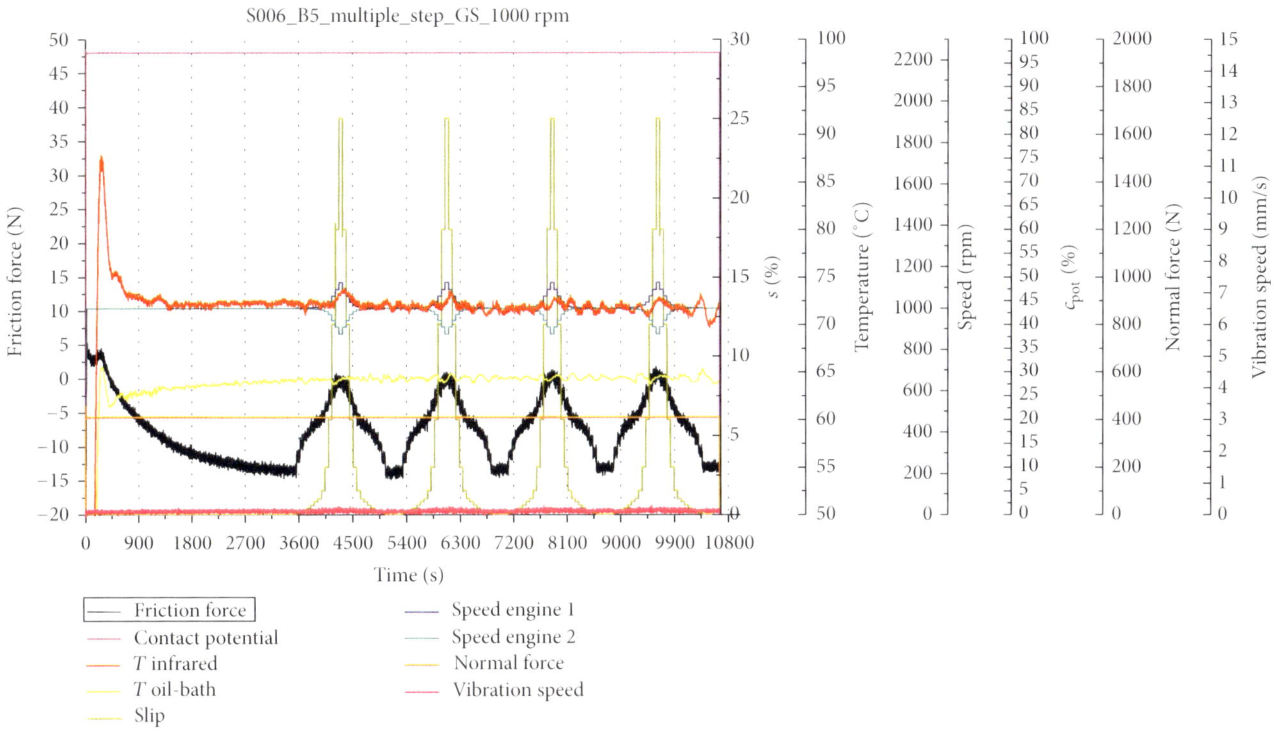

FIGURE 13: Results for a steel-steel experiment under splash lubrication.

TABLE 3: Operating and material parameters for lubricated steel-steel experiment.

$u_m = 2.6$ m/s	$s = 0$–25%
$r_1 = 25$ mm	$r_2 = 25$ mm
$E_1 = 210$ GPa	$E_2 = 210$ GPa
$\nu_1 = 0.3$	$\nu_2 = 0.3$
$b_1 = 5$ mm	$b_2 = 11$ mm
$R_a = 0.03\,\mu$m	$\eta(T = 40°\text{C}) = 95.52\,\text{mm}^2/\text{s}$
$T_0 = 64°$C	$p_{\max} = 0.5$ GPa

A graph of selected measured variables, depending on time, is shown in Figure 13. The test programme is started with a running-in phase, which lasts for one hour, followed by four slip increase-decrease cycles. It takes about 2,700 s for the temperature control loop to reach a steady state, during this time the whole machine, including the electric motors, is experiencing temperature changes, which lead to an elongation of components and therefore a continuous change of the measured friction force during this phase. However, afterwards a steady state of the measured friction force is reached, which indicates a thermally stable system.

During each cycle, the slip is increased stepwise to its maximum value and then decreased to zero again. The friction force is, at very low slip values, increasing linearly; however shear-thinning and thermal effects cause nonlinear effects at high slip rates [19]. The oil-bath temperature is constantly in the range of 64°C throughout the experiment; however the infrared temperature, which is actually measuring the fluid film on disc 2, is experiencing more pronounced changes, especially when slip is high. With increasing slip the frictional power of the system is raised; consequently higher temperatures can be expected. Small fluctuations, which occur independently of slip, are caused by the heating control loop. Also note that the contact potential is holding its maximum value throughout the experiment, indicating a continuous operation in the fluid friction regime.

The friction force measurements from Figure 13 are then postprocessed to derive traction curves, which show the coefficient of friction (COF) as a function of slip, as illustrated in Figure 14. Average values for each slip-step are calculated in order to generate these graphs. The vertical bars indicate the standard deviation occurring along one slip-step.

The excellent reproducibility of the traction curves indicates stable machine conditions and the absence of wear in the tribological system, which is also confirmed by the very low vibration speed values seen in Figure 13. The behaviour of the fluid is therefore *Newtonian* below a slip rate of 0.5%; thereafter thermal effects and viscosity dependence on shear rate significantly reduce the slope of the traction curve [19].

3.3. Validation. In this part of the article the influences of pressure, speed, and oil-bath temperature on traction curves are investigated and compared to state-of-the-art results as, for instance, presented in [6]. Assuming that all tests are operated under fluid lubrication, the following main findings have been concluded in [6].

(i) *Finding 1.* With increasing *pressure* the COF is shifting towards higher values for all slip values. This is due to the increasing viscosity at elevated pressure and the

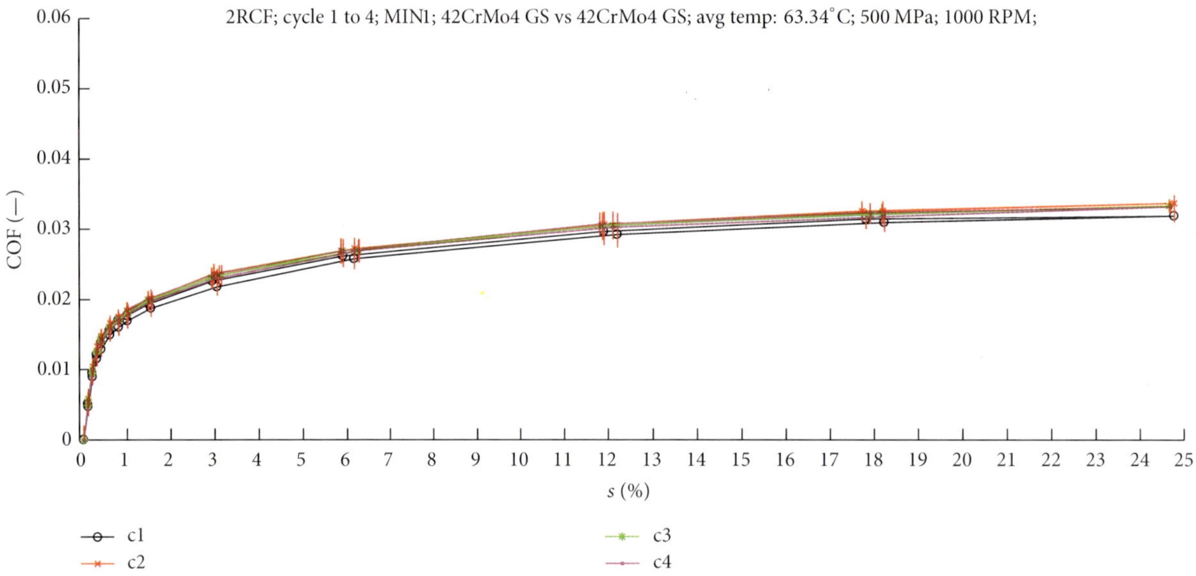

FIGURE 14: Traction curve for a steel-steel experiment under splash lubrication.

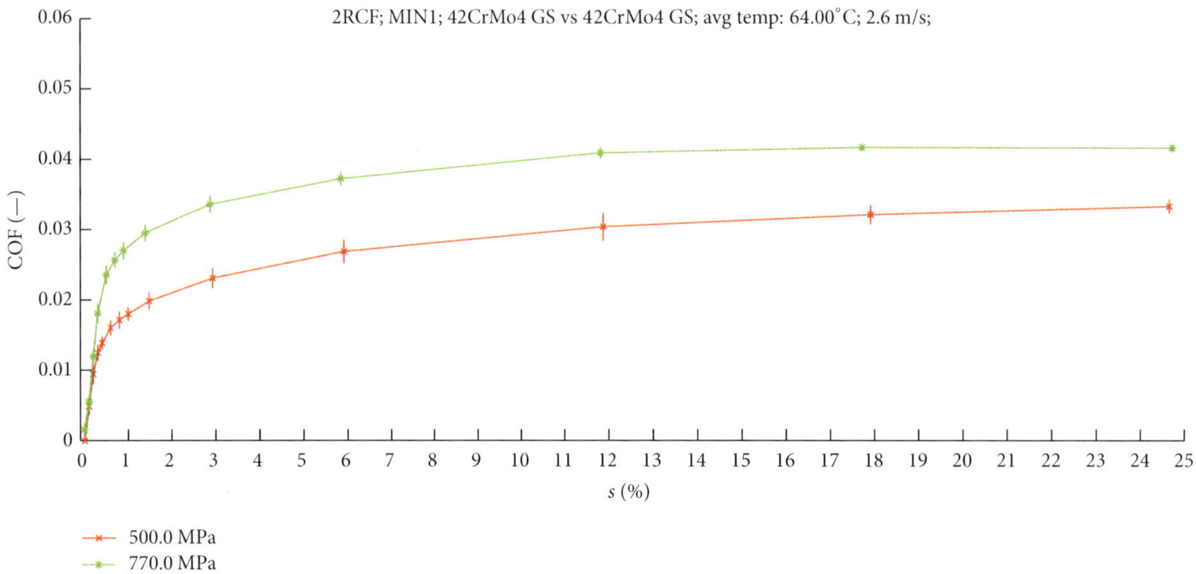

FIGURE 15: Influence of pressure on traction.

limiting shear stress of lubricants is known to linearly rise with pressure, which has been shown in [20].

(ii) *Finding 2*. A higher *velocity* (u_m) is linearly raising the frictional power at constant slip (SRR > 0). The film thickness is increasing as well when the entrainment speed is raised. Experiments have shown that a higher speed leads to a lower COF. This effect is growing with slip since the frictional power increases with slip.

(iii) *Finding 3*. The oil-bath *temperature* is a highly influential parameter since it strongly affects the lubricant's viscosity. Therefore an elevated temperature decreases the COF unless the mixed lubrication regime is reached.

Results of experiments, carried out with parameters as listed in Table 3, with exception of the currently investigated influence, are presented and compared with the findings made by other authors.

In comparison to Figure 14 only the rising traction curve of the third cycle of the respective experiment is illustrated in the graphs in order to improve readability.

3.3.1. Pressure Influence. Figure 15 investigates the influence of pressure on traction. Values of 500 MPa and 770 MPa are compared under otherwise same conditions.

Conclusions drawn in *Finding 1* can be confirmed by this test. Also note that the 770 MPa traction curve has a

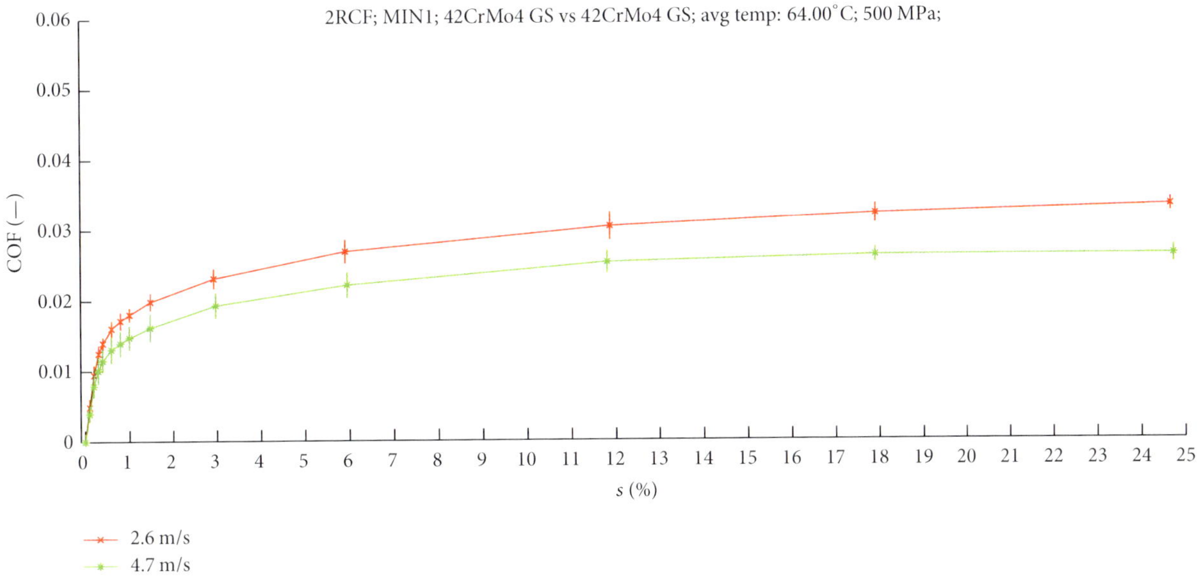

FIGURE 16: Influence of speed on traction.

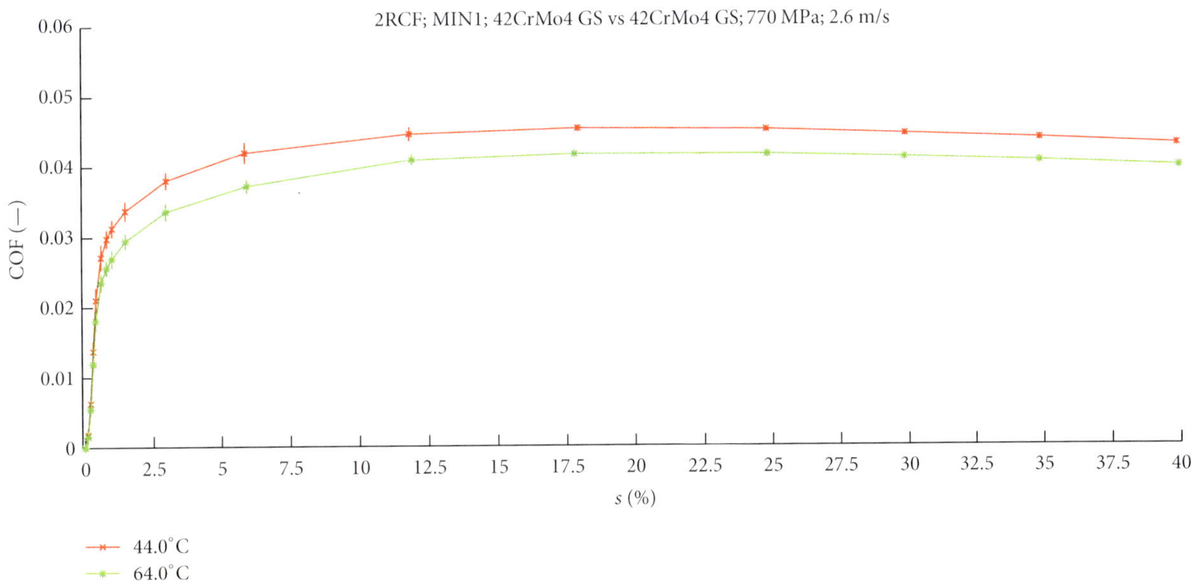

FIGURE 17: Influence of temperature on traction.

maximum at SRR = 18% due to the increased viscosity, which leads to a higher frictional power, while the 500 MPa curve is steadily increasing throughout the investigated slip range.

3.3.2. Speed Influence. Two different velocities (u_m), 2.6 m/s and 4.7 m/s, are examined in Figure 16.

The curve representing a speed of 4.7 m/s yields a lower COF, whereby this effect is more pronounced at higher SRRs. This is corresponding well to *Finding 2*.

3.3.3. Temperature Influence. Finally, experiments with two different oil-bath temperatures are compared in Figure 17. Note that these experiments are carried out with a *Hertzian* pressure of 0.77 GPa and an increased slip range up to SRR = 40%, in order to raise the frictional power of the system.

One can note that in both cases a COF maximum is reached, indicating a transition into the thermal area, where the COF is lowered predominantly by thermal effects. As expected an oil-bath temperature of 44°C increases the COF which confirms the statements made in *Finding 3*.

Since a mineral oil with higher viscosity (ISO VG-150) has been used in [6] a direct comparison of COF values is not possible. However, qualitative trends obtained are very

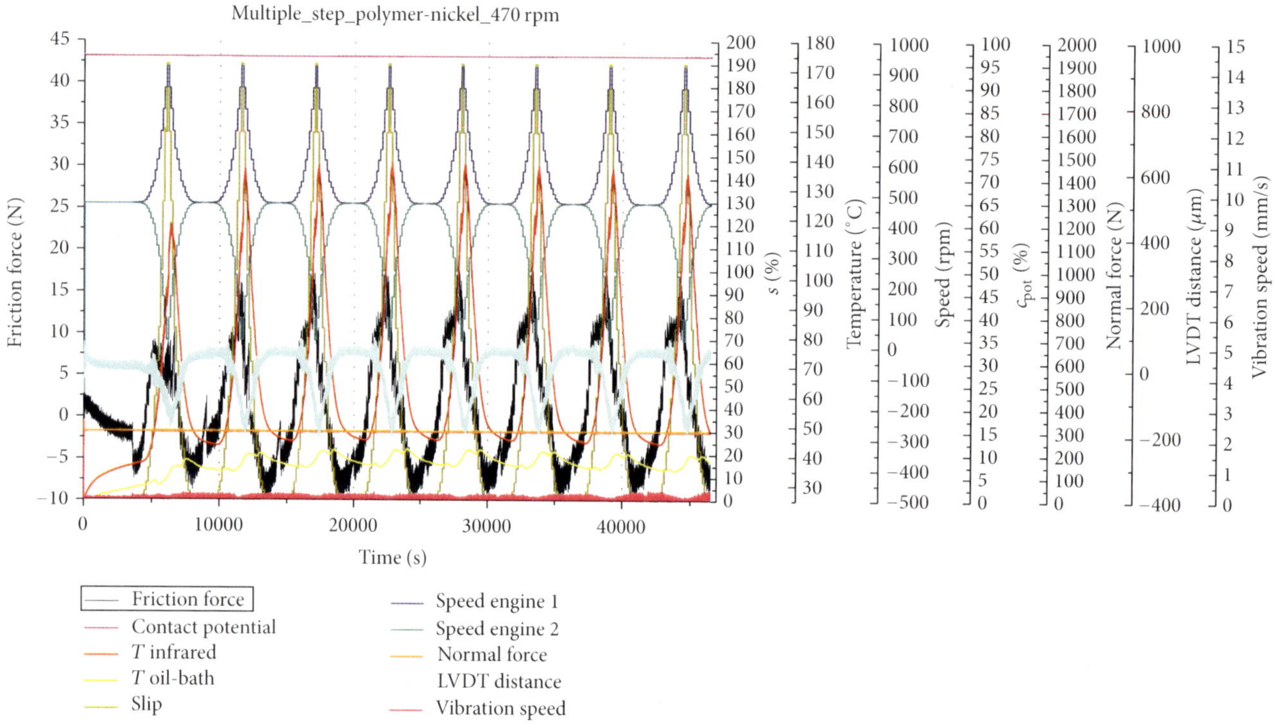

FIGURE 18: Results for a polymer-nickel coated experiment under dry conditions.

TABLE 4: Operating and material parameters for dry polymer-steel experiment.

$u_m = 1.23$ m/s	$s = 0$–190%
$r_1 = 25$ mm	$r_2 = 25$ mm
$E_1 = 210$ GPa	$E_2 = 3$ GPa
$\nu_1 = 0.3$	$\nu_2 = 0.3$
$b_1 = 11$ mm	$b_2 = 11$ mm
$R_{a1} = 1.3\,\mu$m	$R_{a2} = 1.8\,\mu$m
$T_0 = $ RT	$p_{max} = 30$ MPa

similar. Also, the same trends are, for instance, observed in [21] even though a ball-on-disc tribotester was used.

3.4. Dry Polymer-Steel Experiments. High performance polymers, in combination with a metal counterpart, nowadays represent promising solutions for moderately loaded contacts under high sliding speed, whereby fully dry systems can be realised under certain conditions. Here, we present results of a polymer with polytetrafluoroethylene (PTFE) modification, rubbing against a 42CrMo4 disc, which has been coated with a nickel-phosphor layer, performed under room temperature (RT). The corresponding material and operating parameters are listed in Table 4.

After an one-hour-lasting running-in phase eight slip cycles are carried out, which results in a total test time of twelve hours. Main findings are shown in Figure 18. Compared to the lubricated experiment high dependency of the infrared temperature, which reaches values of up to 135°C, on the sliding speed can be examined. Due to the absence of oil, all heat has to be transferred through the discs. Furthermore, the signal of the LVDT distance sensor is correlating directly to the temperature signal; since the polymer suffers under a stiffness reduction at higher temperatures, deformations of up to 200 μm can be measured.

Since polymers are electrically resistant, the contact potential is at its maximum value throughout the test.

The last three cycles of this experiment are depicted in Figure 19. One can note that at a temperature of 120°C the infrared temperature signal shows discontinuities at each cycle. After this event the friction is slightly decreasing, even though slip is raised further. The examined behaviour could be linked to phase changes of the polymer, most likely partial melting of PTFE. Also, due to thermal inertia, a time gap between the maximum of slip, which corresponds to maximum frictional power, and infrared temperature is detected in the graph.

Again, friction force and slip are postprocessed to a traction curve, which is depicted in Figure 20. Cycles five to eight are shown in this diagram.

Compared to the results shown in Figure 14 a hysteresis between the slip increase and decrease steps can be detected.

During the slip increase period local phase changes do occur, as indicated in Figure 19, leading to a self-lubricated system which reduces the COF during the slip decrease steps. Overall excellent reproducibility can be found again by comparison of the traction curves.

For a dry system, the measured friction is low, which can be attributed to the PTFE filling and also, the applied electroless nickel-phosphor coatings are known to have beneficial tribological effects when used with polymers [22, 23].

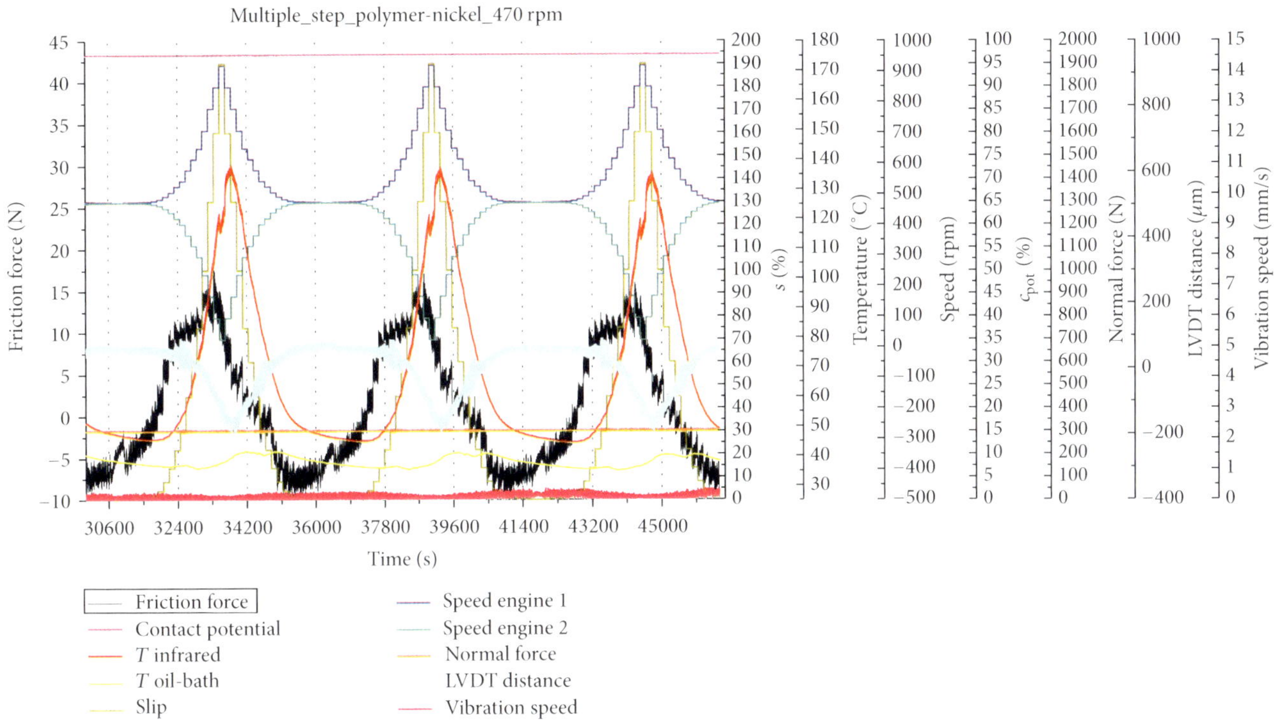

Figure 19: Results for a polymer-nickel coated experiment under dry conditions. Last three cycles.

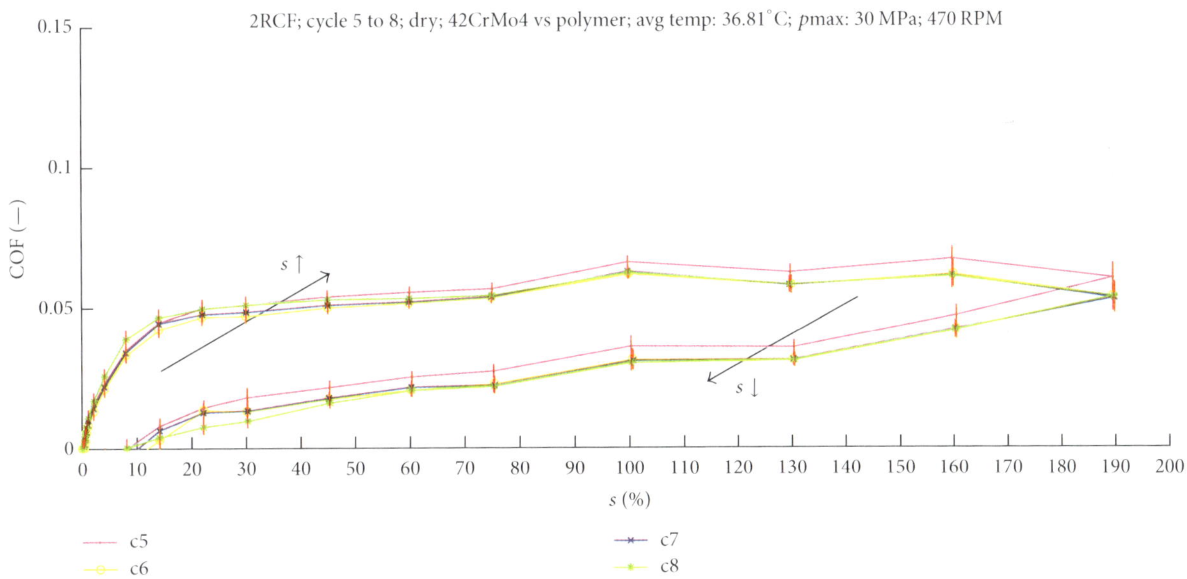

Figure 20: Traction curve for a polymer-nickel coated experiment under dry conditions.

3.5. High-Speed Data Measurements. To conclude this demonstration section, polar plots of measured signals are presented as a function of rotation angle. These measurements are transiently performed and logged periodically throughout the test at certain time steps. During one rotation of shaft 1, positions and velocities of motors, friction force, contact potential, and vibration velocity are evaluated continuously in an array with 72 entries each, which means that data is available in steps of five degrees independent of the current speed and slip values. Figure 21 presents polar plots of friction force and vibration velocity for a polymer-steel experiment with conditions reported in Table 4 at a slip rate of seven percent.

An evenly distributed friction force along rotation angle can be concluded from this high-speed measurement, which indicates stationary friction conditions and therefore a good

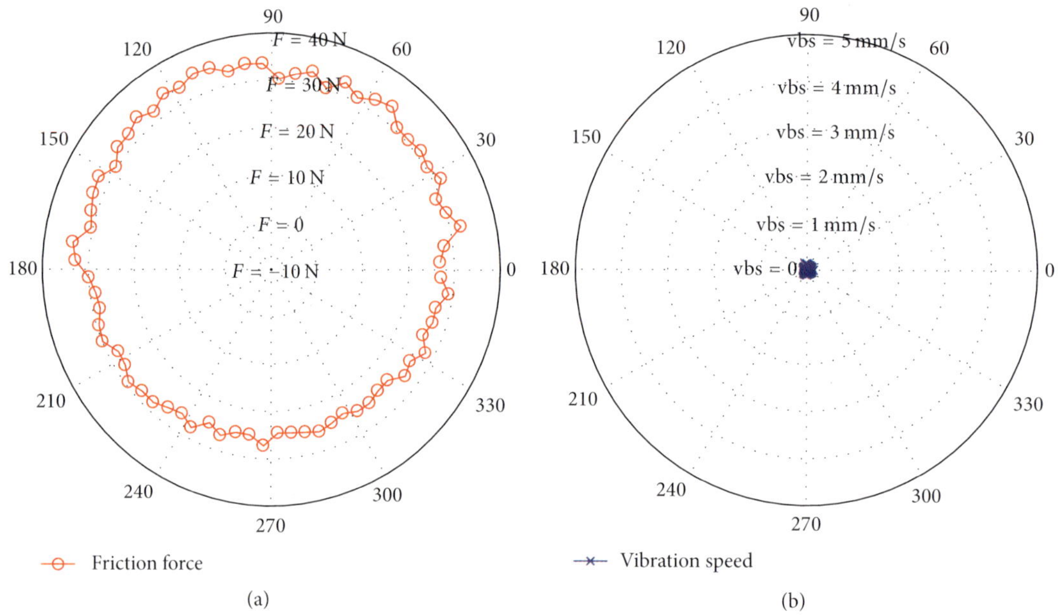

Figure 21: Friction force (a) and vibration velocity (b) along one rotation of shaft 1 for $s = 7\%$.

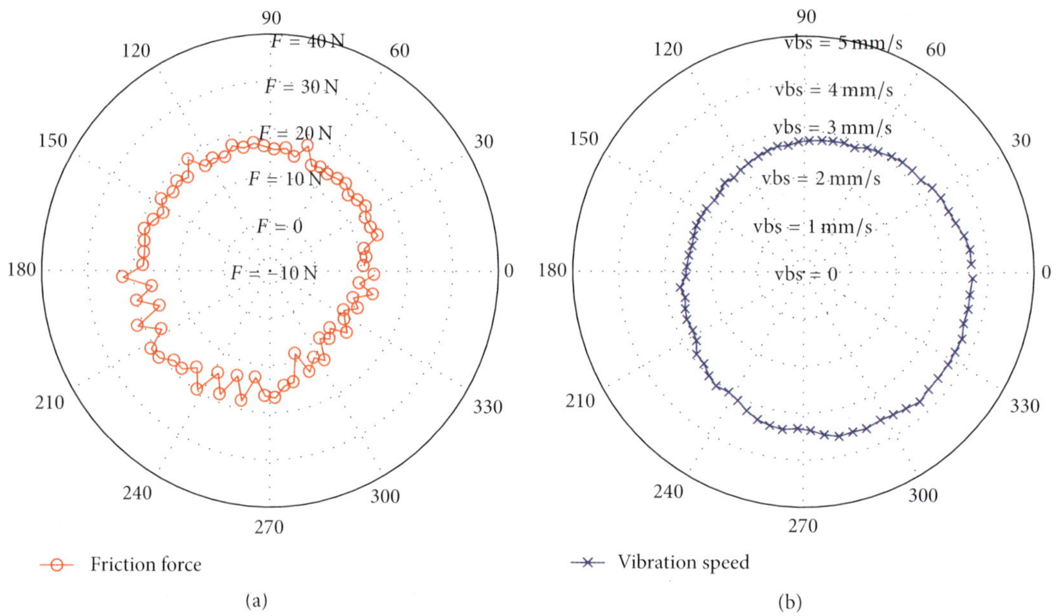

Figure 22. Friction force (a) and vibration velocity (b) along one rotation of shaft 1 for $s = 0\%$.

manufacturing quality of the specimens. The vibration velocity is close to zero, which indicates smooth test conditions as well.

Figure 22 depicts results of a polymer-steel high-speed measurement, whereby the polymer disc has experienced significant heat loads through high friction.

The damaged surface triggers significant vibrations of up to 3 mm/s and the data is less homogeneously distributed along the rotating angle compared to Figure 21. Note that this damage is hardly detectable by just considering the friction force, which emphasises the importance of vibration sensors and subsequent transient evaluation in two-disc machines.

Figure 23 shows images, obtained by an *Olympus SZX12* light microscope, of the undamaged and damaged polymer specimens, used in the experiments shown in Figures 21 and 22. While in the undamaged case only very mild wear, which occurs evenly distributed along the surface, can be detected, the damaged specimen shows localised loss of material. Note that the area above the white dashed line is filled with investment material. In this case the inhomogeneous material

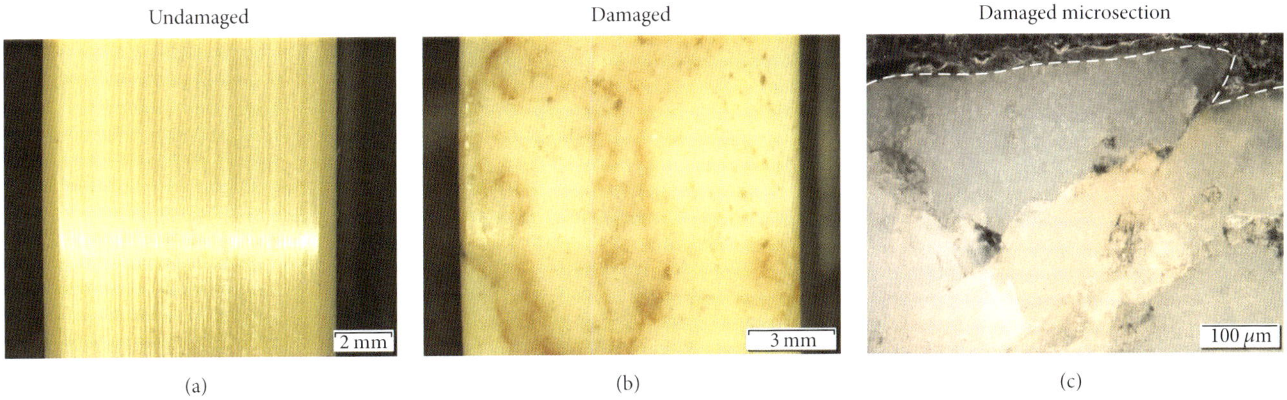

FIGURE 23: Microscopy images of undamaged and damaged polymer specimens. Overviews (a and b) and damaged microsection (c).

TABLE 5: Weight measurements of polymer specimens.

Specimen	Initial weight	Final weight	Difference
Undamaged	22.61250 g	22.59388 g	0.01862 g
Damaged	23.96540 g	23.92409 g	0.04131 g

distribution triggers stress concentrations which lead to a rupture of small particles, resulting in vibrations as measured in Figure 22.

The weight of the specimens before and after testing is measured with a 5-digit digital scale; results are presented in Table 5.

As expected the wear loss of the damaged specimen is higher. Due to the mild wear and the strong temperature dependence of polymers on the elastic modulus, no link between the weight loss and LVDT sensor measurements can be established. However, as already illustrated in Figure 22, local rupture of material can be detected by utilising the developed high-speed data acquisition method.

4. Conclusions

A new two-disc machine, which is capable of performing experiments under dry and lubricated conditions, is presented. Highly precise control of set values can be realised due to the usage of modern soft- and hardware. Experiments under lubricated and dry conditions are conducted, to emphasise the functionalities of the test rig. High-speed measurements are performed in order to gain additional insight into the local tribological changes during the test phase. The machine will be employed for further research in the area of friction reduction in highly loaded contacts, by investigating polymer materials, coatings, and alternative lubrication scenarios.

Nomenclature

$\dot{V}_{in}$: Fluid flow rate injected by the nozzle in ml/min
η: Kinematic viscosity in mm^2/s
ν_1, ν_2: Poison's ratios of the discs
b_1, b_2: Thickness of the discs mm
c_{pot}: Contact potential in %
COF: Coefficient of friction, $COF = F_R/F_N$
E_1, E_2: Young's modulus of the discs GPa
F_L: Force measured by the friction force transducer in N
F_N: Normal force in N
F_R: Friction force in N
n_1, n_2: Rotational speed of bodies in 1/min
p_{max}: Maximum *Hertzian* pressure in GPa
r_1, r_2: Radii of the investigated discs mm
s, SRR: Slide-to-roll ratio, $s = (u_1 - u_2)/u_m$
s_{LVDT}: Distance measured by the LVDT distance sensor in μm
T_0: Set temperature °C
T_I: Temperature measured by the infrared sensor in °C
u_1, u_2: Velocity of the discs in m/s
u_m: Entrainment velocity $u_m = (u_1 + u_2)/2$
v_{vbs}: Speed measured by the vibration sensor in m/s.

Conflicts of Interest

The authors declare that there are no conflicts of interest regarding the publication of this paper.

Acknowledgments

Financial support by the Austrian Federal Government (in particular from Bundesministerium für Verkehr, Innovation und Technologie and Bundesministerium für Wissenschaft, Forschung und Wirtschaft) represented by Österreichische Forschungsförderungsgesellschaft mbH and the Styrian and the Tyrolean Provincial Government, represented by Steirische Wirtschaftsförderungsgesellschaft mbH and Standortagentur Tirol, within the framework of the COMET Funding Programme, is gratefully acknowledged.

References

[1] H. E. Merritt, "Worm gear performance," *Proceedings of the Institution of Mechanical Engineers*, vol. 129, no. 1, pp. 127–194, 1935.

[2] Deutsches Institut für Normung, Din 3990-1, 1987, https://shop.austrian-standards.at/action/en/public/details/111662/DIN_3990-1_1987_12.

[3] H. Wittel, D. Muhs, D. Jannasch, and J. Voßiek, *Roloff/Matek Maschinenelemente, Normung, Berechnung, Gestaltung; 21. vollständig überbearbeitete Auflage*, Springer, 2013.

[4] A. Trausmuth, I. Godor, M. Stoschka, A. Dietrich, and W. Eichlseder, "Load-carrying capacity prediction of different plasma nitrided, case hardened and quenched specimens under rolling contact loading," in *Proceedings of the European Conference on Tribology*, 2011.

[5] A. Trausmuth, *Rolling contact fatigue of nitrided, case-hardened and quenched specimens [Ph.D. thesis]*, Montanuniversität Leoben, 2012.

[6] A. Doleschel, *Wirkungsgradberechnung von Zahnradgetrieben in Abhängigkeit vom Schmierstoff [Ph.D. thesis]*, Technische Universität München, 2003.

[7] K. Bobzin, T. Brögelmann, K. Stahl, K. Michaelis, J. Mayer, and M. Hinterstoißer, "Friction reduction of highly-loaded rolling-sliding contacts by surface modifications under elasto-hydrodynamic lubrication," *Wear*, vol. 328-329, pp. 217–228, 2015.

[8] G. Meneghetti, A. Terrin, and S. Giacometti, "A twin disc test rig for contact fatigue characterization of gear materials," *Procedia Structural Integrity*, vol. 2, pp. 3185–3193, 2016.

[9] O. Kreil, *Einfluss der Oberflächenstruktur auf Druckverteilung und Schmierfilmdicke im EHD-Kontakt [Ph.D. thesis]*, Technische Universität München, 2008.

[10] S. Akbarzadeh and M. M. Khonsari, "Experimental and theoretical investigation of running-in," *Tribology International*, vol. 44, no. 2, pp. 92–100, 2011.

[11] M. Masjedi and M. M. Khonsari, "Theoretical and experimental investigation of traction coefficient in line-contact EHL of rough surfaces," *Tribology International*, vol. 70, pp. 179–189, 2014.

[12] J. Kleemola and A. Lehtovaara, "Experimental evaluation of friction between contacting discs for the simulation of gear contact," *Lubrication Science*, vol. 13, no. 1, pp. 13–20, 2007.

[13] H. Krampl, F. Grün, and I. Godor, "Numerical and experimental investigation of non-conformal lubricated contacts," in *Proceedings of the Nordic Symposium on Tribology*, 2012.

[14] J. Moder, H. Krampl, and F. Grün, "Tribological analysis of lubricated non-conforming contacts," in *Proceedings of the ÖTG Symposium*, 2014.

[15] SIEMENS, "SIMATIC HMI Devices, Operating Instructions," 2015.

[16] SIEMENS, "Simotion Scout TIA Configuration Manual," 2014.

[17] SIEMENS, "S120 Drive Functions Manual," 2015.

[18] SIEMENS, "1FT7 Synchronous Motors Operating Instructions," 2015.

[19] R. Gohar, *Elastohydrodynamics*, World Scientific, 2001.

[20] W. Habchi, P. Vergne, S. Bair, O. Andersson, D. Eyheramendy, and G. E. Morales-Espejel, "Influence of pressure and temperature dependence of thermal properties of a lubricant on the behaviour of circular TEHD contacts," *Tribology International*, vol. 43, no. 10, pp. 1842–1850, 2010.

[21] M. Björling, *Friction in elastohydrodynamic lubrication [Ph.D. thesis]*, Luleå Tekniska Universitet, 2014.

[22] R. Parkinson, "Properties and applications of electroless nickel," 2001.

[23] P. Sahoo and S. K. Das, "Tribology of electroless nickel coatings—a review," *Materials and Design*, vol. 32, no. 4, pp. 1760–1775, 2011.

16

The Application of Response Surface Methodology in the Investigation of the Tribological Behavior of Palm Cooking Oil Blended in Engine Oil

M. H. Sakinah,[1] A. K. Amirruddin,[1] K. Kadirgama,[1,2] D. Ramasamy,[1] M. M. Rahman,[1,2] and M. M. Noor[1,2]

[1] *Faculty of Mechanical Engineering, University Malaysia Pahang, Pekan, Pahang, Malaysia*
[2] *Automotive Engineering Research Group (AERG), University Malaysia Pahang, Pekan, Pahang, Malaysia*

Correspondence should be addressed to K. Kadirgama; kumaran@ump.edu.my and M. M. Noor; muhamad@ump.edu.my

Academic Editor: Shyam Bahadur

The purpose of this study was to determine the optimal design parameters and to indicate which of the design parameters are statistically significant for obtaining a low coefficient of friction (COF) and low wear rate with waste palm oil blended with SAE 40. The tribology performance was evaluated using a piston-ring-liner contact tester. The design of experiment (DOE) was constructed by using response surface methodology (RSM) to minimize the number of experimental conditions and to develop a mathematical model between the key process parameters such as rotational speeds (200 rpm to 300 rpm), volume concentration (0% to 10% waste oil), and applied loads (2 kg to 9 kg). Analysis of variance (ANOVA) test was also carried out to check the adequacy of the empirical models developed. Scanning electron microscopy (SEM) was used to examine the damage features at the worn surface under lubricant contact conditions.

1. Introduction

Tribological Studies of Waste Oil Biolubricant. In the late 1800s, petroleum had been discovered and that led to the replacement of animal fats, vegetable oils, and mineral oils with synthetic oils. Petroleum oil gradually started to be the main lubricant base stocks and that was because of their low cost and superior performance. Lubricants are being used widely in all fields of manufacturing and industrial applications. Studies showed that more than thirty-eight million metric tons of oils was used for lubrication techniques in 2005 for different industrial applications in the United States (USA). Lubricants are commonly used to reduce overheating and friction in various engines, machinery, turbines, and gear.

The excessive usage of petroleum-based oils has significantly contributed to the environmental pollution and triggered awareness from the environmental sectors [1]. Besides that, the demands for fossil fuel and oil products are increasing in numerous areas. Based on the reported works, alternative oil should increase to cover about 36 billion gallons in 2022 [2]. In other words, there is a great demand for oil in the coming few years and high attention should be paid to find alternative resources. To overcome such issue, researchers start developing an alternative fuel and/or oil products from natural resources aiming to replace the fossil products which become the main goal of many researchers, environmental and government bodies especially in the developed countries such as Australia, US, and Europe.

In the current decade, there are a few attempts aiming to study the potential of using bio-oil such as sunflower oil, castor oil, soybean oil, and pollock oil as biofuel for diesel engines. Most of the works showed good and promising results. However, there is a tribological issue raised by most of the researchers in which biofuels deteriorate the engine components. On the other hand, currently there is an effort to try to use pure bio-oil as a lubricant. From 2010 until recently, several biolubricants have been investigated in different countries [3], for example, soybean oils (the USA and South America), rapeseed oil (Europe), and palm oil (Asia) [4–6]. Those studies are still at the initial stage and there are many issues and limitations that need to be addressed before using such oil [7]. Moreover, the literature highly recommends deep investigation on the performance and the potential of using biolubricants. Developing a friendly low cost biolubricant attracts the attention of researchers to use waste cooking oil as the main resources of lubricant.

Waste cooking oil can be considered the most promising bio-oil feedstock despite its drawbacks, that is, high free fatty acid (FFA) and water contents [8]. As reported by many researchers, biofuels produced from waste cooking oils have numerous advantages such as low pollution (CO_2, CO, and NO_x), low cost, and acceptable brake specific fuel consumption. An interest can be drawn to use the waste cooking oil as a lubricant. Kalam et al. [9] experimentally investigated the friction and wear characteristics of normal lubricants, which is an additive-added lubricant and waste vegetable oil-(WVO-) contaminated lubricants. The WVO-contaminated lubricants with amine phosphate as antiwear additive reduced the wear and friction coefficient and increased the viscosity; thus palm oil waste with a normal lubricant and amine phosphate additive could be used as a substitution for lubricant (maximum 4%). Based on the four-ball tribo testing result, the WVO-contaminated lubricant with the presence of antiwear additives showed promising results due to better thermal and oxidative properties of waste vegetable oils which consist of long chain saturated fatty acids [10]. Masjuki and Maleque [11] have experimented the effect of palm oil diesel (POD) fuel contaminated lubricant on sliding wear of cast iron against mild steel and investigated the sliding contact using the pin-on-disc type of friction and wear apparatus. Based on the results, the use of pure commercial (0% POD contamination) lubricant resulted in a moderate wear rate while pure POD 100% lubricant produced the highest wear rate compared to other contaminated lubricants.

2. Experimental Setup

2.1. Lubrication and Material Preparation. Base oil used in this experiment was SAE 40. Palm oil was chosen because it is commonly used in Malaysia. Volume concentrations of 0% and 10% waste oil were blended with base oil using magnetic stirrer and ultrasonic bath. For the preparation of waste cooking oil as biolubricant, the waste cooking oil underwent three types of processes: coarse filtering, dewatering, and fine filtering. Wear and friction performance for biolubricant were evaluated using a piston ring-liner contact tester and the material use was aluminium 6061 which is the common material for a piston ring. After the lubrication preparation had been completed, the details of lubricant compositions which are viscosity, density, and moisture content for all lubrication were determined.

2.2. Evaluation of Tribological Properties. Wear test involves making linear movements similar to the pair of cylinder-piston ring operating under real conditions. Figures 1–3 show the picture of the wear tester and setup. The type of material for specimen used in this experiment was aluminium 6061 which are the material commonly used for a piston ring.

TABLE 1: Tribology test condition.

Test specifications	Values
Load, kg	2.0–9.0
Engine speed, rpm	200–300
Temperature, °C	Room temperature
Operating time, min	10 minutes per specimen

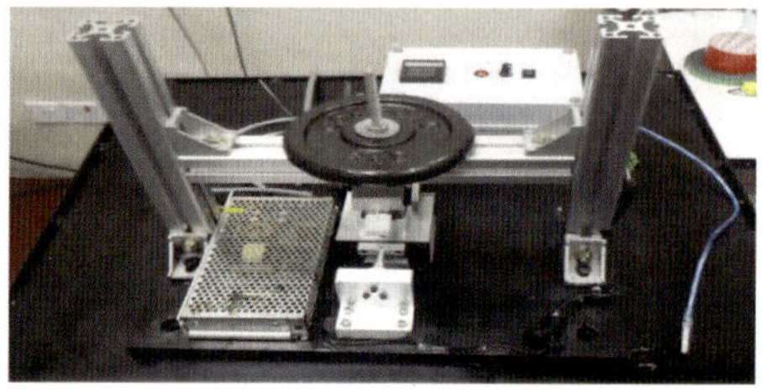

FIGURE 1: Piston ring reciprocating liner test machine, contact geometry, and test specimen.

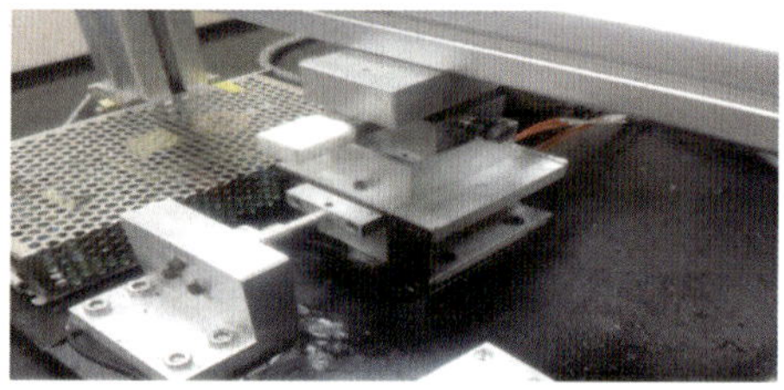

FIGURE 2: Lubricant bath for specimen facilitates the linear movement.

Normal loads were applied to the device by hanging weights on the bearing lever where the piston ring sample is attached in order to produce the desired loads. The load chosen was between 2.0 kg and 9 kg. Low engine-speed intervals (200 rpm and 300 rpm) were selected during testing because such conditions generate the greatest friction in engines, particularly during the first movement and at the top dead centre (TDC) [12]. The temperature used was the same as room temperature and the operating time was 10 minutes per specimen. The coefficient of friction (COF) was measured using ARDUINO Software and wear rate was determined via weight difference using weight scale with sensitivity 0.1 mg. Calculation of soefficient is shown in Figure 4. The test conditions are presented in Table 1.

2.3. Calculation of Coefficient of Friction and Specific Wear Rate. Consider

$$\mu_k = \frac{F_k}{N}, \tag{1}$$

where μ_k is coefficient of kinetic friction, F_k is applied force, and N is the load.

For specific wear rate evaluation,

$$\Delta w = (w_1 - w_2), \tag{2}$$

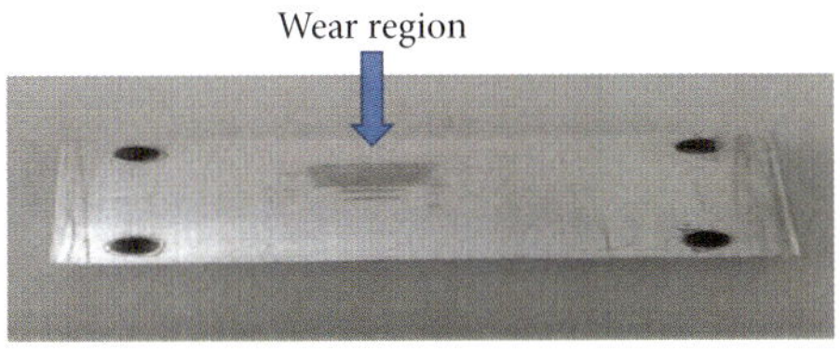

FIGURE 3: The wear region of a specimen.

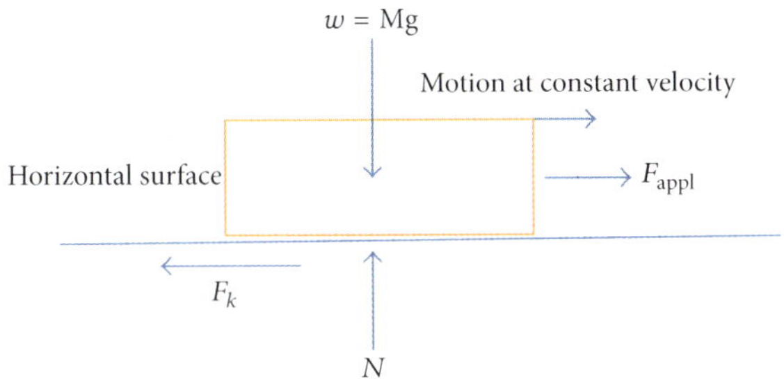

FIGURE 4: Coefficient of friction evaluation.

where Δw is weight loss of the specimen, w_1 is weight of the specimen before test, and w_2 is weight of the specimen after test.

Volume loss (ΔV) of the specimen is computed as per below:

$$\Delta V = \frac{1}{\rho}(w_1 - w_2), \quad (3)$$

where ρ is experimental density of the specimen.

The specific wear rate (w_s) of the specimen was calculated using the following equation:

$$w_s = \frac{\Delta V}{F_n \times s_s}, \quad (4)$$

where s_s is sliding distance and (m), F_n is normal load (N).

2.4. Design of Experiment (DOE)

2.4.1. Response Surface Methodology. The design of experiment (DOE) for this study was constructed using response surface methodology (RSM) to obtain the optimization for different parameters in the tribological behavior using Minitab software. RSM is the procedure to determine various relationships between process parameters and tribological criteria and explore the effect of these process parameters on the coupled responses Montgomery [13]. RSM techniques are based on the use of factorial design in which the main effect of the factor is defined as the variation in response caused by a change in the level of the factor considered, while the other ones are kept constant [14]. In order to study the effects of the tribological parameters, the two most important tribological criteria which are wear rate (WR) and coefficient of friction (COF) act as the response. Table 2 shows the suitable levels of the factors used to design the parameters for a tribological experiment while Table 3 shows the design values obtained from the Minitab.

TABLE 2: Process parameter and its level.

Parameter	−1	0	+1
Volume concentration (%)	0	5	10
Speed (rev/min)	200	250	300
Load (kg)	2	5.5	9

TABLE 3: Design values obtained from a Minitab.

Experiment	Speed (rev/min)	Load (kg)	Volume concentration (%)
1	300	5.5	10
2	300	2	5
3	200	9	5
4	250	2	10
5	250	5.5	5
6	200	5.5	10
7	250	9	0
8	250	2	0
9	300	5.5	0
10	300	9	5
11	250	5.5	5
12	250	9	10
13	200	2	5
14	200	5.5	0
15	250	5.5	5

2.4.2. Mathematical Modelling Based on RSM. Response surface regression was used to construct a complete quadratic mathematical equation for wear rate and average COF. A second-order polynomial response surface empirical model can be developed as follows to evaluate the parametric effects on the various tribological criteria:

$$f(x) = \beta_0 + \sum_{i=1}^{n} \beta_i x_i + \sum_{i=1}^{n} \beta_{ii} x_i^2 + \sum_{i=1}^{n} \beta_{ij} x_i x_j + \varepsilon, \quad (5)$$

where $f(x)$ is the response which is wear rate (WR) and coefficient of friction (COF). It is created by various process variables of tribological parameters. β_0, β_i, β_{ii}, and β_{ij} are the regression coefficients for intercept, linear, quadratic, and interaction terms, respectively. x_i and x_j are the independent variables. Contour plots were obtained using the fitted model by keeping the least effective independent variable at a constant value while changing the other two variables [15]. A Box-Behnken design with three levels of variables was used for the current study. 15 tribology tests were conducted according to the design as shown in Table 3. To ensure that the quadratic mathematical models for the wear rate and average COF of the analysis were adhered to, all of the experimental data were checked through the residual plot to verify that the mathematical models displayed standard normal distribution.

TABLE 4: Properties of baseline oil.

Properties	SAE 40	Palm oil	Waste oil
Viscosity (mpa·s)	179.2	72.7	180.6
Density (g/cm^3)	0.8609	0.9033	0.9049
Moisture content (%)	0.19	0.24	0.28

TABLE 5: Properties of blended volume concentration for waste oil.

Properties	5% waste oil + SAE 40	10% waste oil + SAE 40
Viscosity (mpa·s)	192.1	169.6
Density (g/cm^3)	0.8632	0.8639
Moisture content (%)	0.14	0.19

3. Results and Discussion

3.1. Physicochemical Properties of Waste Cooking Oil in Base Lubricant. The data were used to evaluate the differences between base lubricant stock (SAE 40) and blended lubricant of palm oil and waste oil. Table 4 shows the properties of base oil (SAE 40), palm oil, and waste oil.

A good lubricant should have a high boiling point, adequate viscosity, low freezing point, high oxidation resistant, noncorrosive properties, and good thermal stability. The most important property of oil is viscosity. It indicates the resistance to flow and is directly related to temperature, pressure, and film formation. High viscosity indicates low resistance of flow [16]. Lubricants are generally less dense than water. If the density of an object is less than that of water, then that object will float. This is why if there is a moisture problem in the lube system, the water settles at the bottom of the sump and is drained out first whenever the plug is pulled or the valve is opened. The density of a lubricant fluid can provide indication of its composition and nature [17]. The presence of water does not only have a direct harmful effect on the machine components but it can also trigger the progress of oxidation up to tenfold increase and thus resulted in premature aging of oil [18]. Less moisture content in lubricating oil indicates rust and corrosion prevention. Table 5 shows the results of blended lubricant composition for 5% and 10% waste palm oil.

3.2. Analysis of the Developed Empirical Models and Regression Analysis. Table 6 represents the results of experiments conducted to investigate the tribological properties of waste cooking oil blended with SAE 40 engine oil for different factor variables.

The analysis of variance (ANOVA) and the F-ratio test were performed to justify the goodness of fit of the empirical models. The calculated values of F-ratio for lack of fit were compared to the standard values of F-ratio corresponding to their degrees of freedom to find the adequacy of different empirical models. The F-ratio was calculated as a ratio of mean sum of the experimental error [19].

Tables 7 and 8 represent the estimated regression coefficient and analysis of variance (ANOVA) for average COF for blended waste cooking oil with SAE oil. The fit summary recommends that the empirical model is statistically significant for the analysis of COF. The value of R^2 was more than 99.10% which means that the empirical model provides an excellent explanation of the relationship between the independent variables (factors) and the response (COF). Based on Table 7, the associated P value for the model was lower than 0.05 (95% confidence interval). This indicates that the model was considered statistically significant. Meanwhile, the lack of fit of P values for the average COF models was also significant as they were less than 0.05. Figure 5 shows the residual plot for coefficient of friction.

Tables 9 and 10 represent the estimated regression coefficient and analysis of variance (ANOVA) for the specific wear rate for blended waste cooking oil with SAE oil. The fit summary recommends that the empirical model was statistically significant for the analysis of COF. The value of R^2 was over 95.99% which means the empirical model provides an excellent explanation of the relationship between the independent variables (factors) and the response (WR). Based on Table 6, the associated P value for the model was lower than 0.05 (95% confidence interval). This indicates that the model was considered to be statistically significant. Meanwhile, the lack of fit of P values for the average COF models were not significant as they were more than 0.05. Figure 6 shows the residual plot for coefficient of friction.

The t values and P values in the estimated regression coefficient of wear rate in Table 10 denote the significant influence of each input variable in the models. The smaller numerical values of "P" and larger values of "t" signify that the related regression coefficient is highly significant [13]. Equations (6) are the empirical equation for the average COF and the wear rate for the lubricant as the functions of independent variables of speed (S), load (L), and volume concentration (VC) in coded units:

$$\begin{aligned} f(x), \mathrm{COF} = {} & 0.337598 - 0.000174S - 0.0064568L \\ & + 0.005517\mathrm{VC} + 0.000002S^2 \\ & + 0.004496L^2 + 0.000039\mathrm{VC}^2 \\ & - 0.0005SL - 0.000033S\mathrm{VC} \\ & + 0.000236L\mathrm{VC}, \\ f(x), \mathrm{WR} = {} & -0.783168 + 0.008006S - 0.117524L \\ & - 0.012697\mathrm{VC} - 0.000009S^2 \\ & + 0.000042L^2 + 0.002760\mathrm{VC}^2 \\ & - 0.000392SL - 0.000068S\mathrm{VC} \\ & + 0.001647L\mathrm{VC}. \end{aligned} \tag{6}$$

According to the COF model, the highest significant level was quadratic load, followed by linear load and lastly the interaction of speed and applied load, while, for specific average wear rate, quadratic volume concentration showed

TABLE 6: Experimental design and results (uncoded factors).

Experiment	Speed (rev/min)	Load (kg)	Volume composition (%)	Coefficient of friction (μ)	Specific wear rate (mm^3/Nm)
1	300	5.5	10	0.09103	0.89984
2	300	2	5	0.26564	0.07589
3	200	9	5	0.06056	0.85635
4	250	2	10	0.23418	0.33661
5	250	5.5	5	0.09542	0.46867
6	200	5.5	10	0.09566	0.79715
7	250	9	0	0.06079	0.86458
8	250	2	0	0.25055	0.77947
9	300	5.5	0	0.09118	0.72354
10	300	9	5	0.06097	0.83164
11	250	5.5	5	0.09542	0.88928
12	250	9	10	0.06089	0.97986
13	200	2	5	0.26146	0.74241
14	200	5.5	0	0.09293	0.68133
15	250	5.5	5	0.09563	0.88928

TABLE 7: Analysis of variance for coefficient of friction (COF).

Source	DF	Seq SS	Adj SS	Adj MS	F value	P value
Regression	9	0.069229	0.069229	0.007692	462.37	0.000
Linear	3	0.058443	0.002365	0.000788	47.39	0.005
Square	3	0.010525	0.007339	0.002446	147.04	0.001
Interaction	3	0.000261	0.000261	0.000087	5.23	0.104
Residual error	5	0.000050	0.000050	0.000017		
Lack of fit	3	0.000050	0.000050	0.000050	3741.13	0.000
Pure error	2	0.000000	0.000000	0.000000		
Total	14	0.069279				

$S = 0.00407874$, PRESS = ∗, $R^2 = 99.93\%$, R^2 (pred) = ∗%, and R^2 (adj) = 99.71%
DF: degrees of freedom; Seq SS: sequential sum of squares; Adj SS: adjusted sum of squares; Adj MS: adjusted mean squares.

TABLE 8: Estimated regression coefficients for COF.

Term	Coef	SE Coef	t-ratio	P value
Constant	0.337598	0.077232	4.371	0.022
Speed	−0.000174	0.000520	−0.335	0.760
Load	−0.064568	0.006435	−10.033	0.002
Volume composition	0.005517	0.002745	2.010	0.138
Speed ∗ speed	0.000002	0.000001	1.571	0.214
Load ∗ load	0.004496	0.000240	18.712	0.000
Volume composition ∗ volume composition	0.000039	0.000118	−2.790	0.761
Speed ∗ load	−0.000050	0.000018	−2.790	0.068
Speed ∗ volume composition	−0.000033	0.000012	−2.667	0.076
Load ∗ volume composition	0.000236	0.000117	2.023	0.136

the highest significance level, followed by interaction of speed and applied load and finally the linear load.

3.2.1. Effect of Control Parameters on Coefficient of Friction (COF) and Specific Wear Rate (WR). Figures 7(a), 7(b), and 7(c) represent the three-dimensional response surface plots and the contour plots of COF regarding speed, load, and volume composition. Based on Figure 7(a), as the load and speed increase, the value of COF increases. Figure 7(b) shows the relationship of volume composition and load by which the COF is gradually decreasing as the volume composition decreases. For Figure 7(c), concerning the volume composition, the coefficient of friction reduces to the lowest at a certain speed and then increases as the speed increases even

TABLE 9: Analysis of variance for wear rate.

Source	DF	Seq SS	Adj SS	Adj MS	F value	P value
Regression	9	0.456675	0.456675	0.009888	173.19	0.000
Linear	3	0.198500	0.198500	0.024624	431.29	0.000
Square	3	0.236568	0.236568	0.005015	87.84	0.000
Interaction	3	0.021607	0.000073	0.000024	0.43	0.742
Residual error	5	0.054272	0.000285	0.000057		
Lack of fit	3	0.051876	0.000285	0.000095	6486.86	0.000
Pure error	2	0.002396	0.000000	0.000000		
Total	14	0.510947				

$S = 0.0330377$, PRESS = $*$, $R^2 = 95.99\%$, R^2 (pred) = $*\%$, and R^2 (adj) = 83.96%.

TABLE 10: Estimated regression coefficients for wear rate.

Term	Coef	SE Coef	t-ratio	P value
Constant	−0.783168	0.625581	−1.252	0.299
Speed	0.008006	0.004209	1.902	0.153
Load	0.117524	0.052127	2.255	0.109
Volume composition	−0.012697	0.022237	−0.571	0.608
Speed * speed	−0.000009	0.000008	−1.085	0.357
Load * load	0.000042	0.001946	0.022	0.984
Volume composition * volume composition	0.002760	0.000954	2.894	0.063
Speed * load	−0.000392	0.000144	−2.718	0.073
Speed * volume composition	−0.000068	0.000101	−0.673	0.549
Load * volume composition	0.001647	0.000944	1.745	0.179

more under low applied load conditions. Thus, volume composition had more significant effect than speed and load.

Figures 8(a), 8(b), and 8(c) represent the three-dimensional response surface plots and the contour plots of wear rate regarding speed, load, and volume composition. Based on Figure 8(a), as the load and speed increase, the value of specific wear rate increases. Meanwhile Figure 8(b) shows the volume composition and load by which the specific wear rate gradually increases as the volume composition increases. For Figure 8(c), concerning the volume composition and load, the specific wear rate gradually increases as the volume composition increases. Thus, volume composition had more significant effect than speed and load.

3.3. Multiobjective Optimization Using Response Surface Methodology. The main advantage of using response surface methodology (RSM) is that the response can be optimized by controlling the input parameters [20]. The performances of wear and coefficient of friction depend not only on the lubricant properties but also on the sliding conditions of material under lubricant contact condition. In this study, the optimization was carried out in order to determine the minimum wear and friction of the blended waste oil with SAE 40 contact with aluminium 6061. Optimization of the process parameters was carried out using RSM optimization technique. Desirability for the whole process of optimization was calculated to show the feasibility of optimization to examine whether all parameters are within the working range or not. The goal was to minimize COF and WR. Table 11 shows the target value and the upper value for both response, average COF, and wear rate.

TABLE 11: Target value and upper value of average COF and wear rate.

Response	Target value	Upper value
Average COF	0.0605	0.2656
Wear rate ($\times 10^{-3}$ mm^3/Nm)	0.68133	0.97986

Figure 9 exhibits the optimization plot for both COF and WR responses. The optimum value shown in the plot is 0.0717 μ for COF and 0.7380 for WR. The relevant parameters such as speed, load, and volume composition are 200 rev/min, 6.3712 kg, and 0.2020% of volume composition respectively. The composite shown in the plot is 0.87250.

3.4. Surface Texture Analysis. There are various types of wear in mechanical systems such as abrasive wear, adhesive wear, fatigue wear, and corrosive wear. Since the lubricant regime occurred in this experiment was boundary lubrication, thereby, abrasive wear, adhesive wear, fatigue wear, and corrosive wear were observed in the wear regions [21]. All these wear mechanisms were found in this experiment but most of the wear phenomena were abrasive and adhesive wear. The

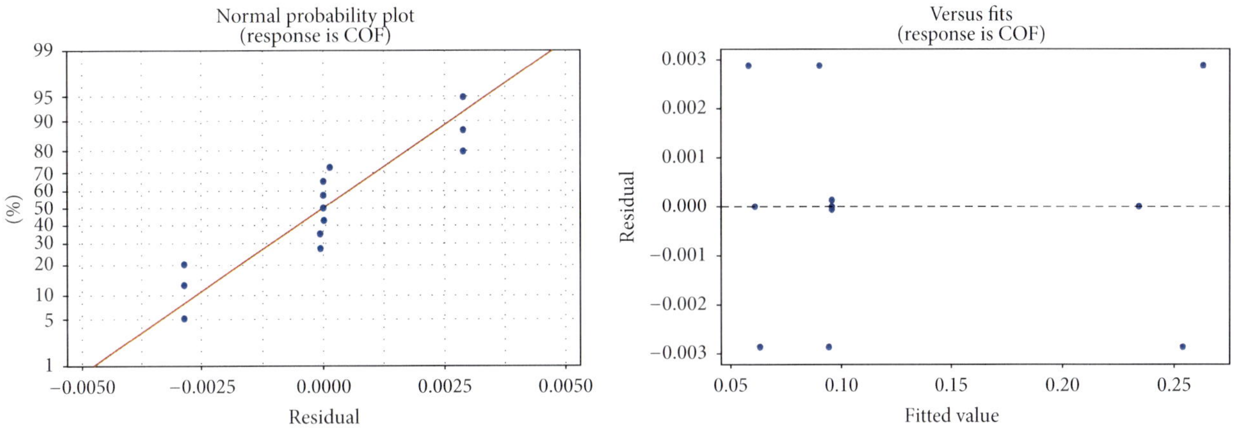

FIGURE 5: Normal probability plot and versus fits for average COF.

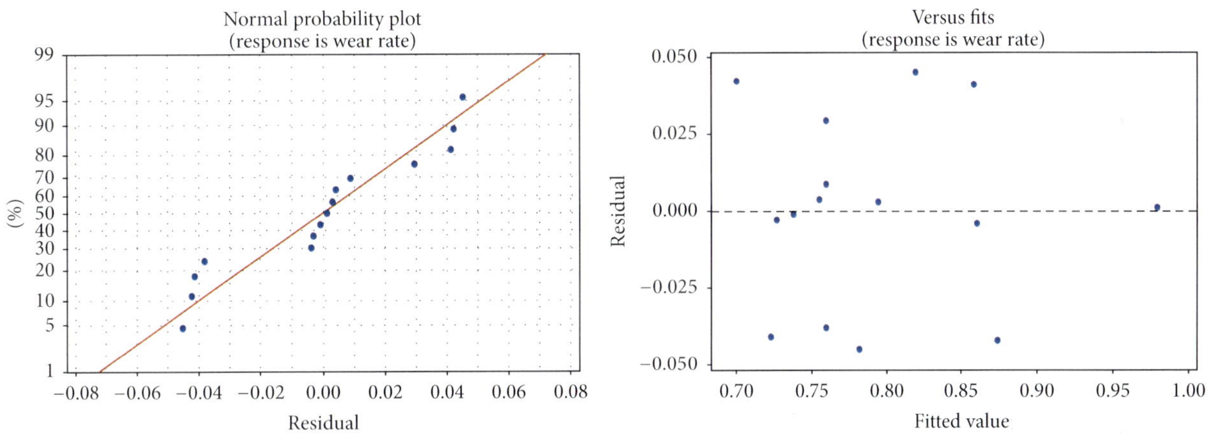

FIGURE 6: Normal probability plot, versus fits, histogram, and versus order for wear rate.

major wear mechanisms that can be found in the specimen were observed to be wear grooves that resulted from abrasive wear because the asperities on the hard surface of the liner samples touched the soft surface of the ring samples and had a close relationship with the thickness of lubricant film. The SEM images of the aluminium plate shown used various types of volume concentration of waste cooking oil blended with engine oil. Referring to Figure 10, it was found that the wear decreases at 5% concentration compared to SAE 40 and wear started to increase when 10% waste oil concentration was used. This is due to the 5% concentration of waste oil that showed the highest viscosity results compared to 10% concentration because high viscosity (thick) engine oil helps to maintain a barrier between moving part and also drag the movements between two contact surfaces.

4. Conclusion

As conclusions, the study examined the effects of various control parameters, namely, speed, load, and volume composition on the responses of coefficient of friction and wear rate. The following conclusion can be derived based on the results obtained:

(i) The correlations between the control parameters (speed, load, and volume composition) and responses (specific wear rate and coefficient of friction) of waste oil added with standard lubricant were successfully developed using RSM. The model showed that the speed, load, and volume composition have a significant effect on coefficient of friction (COF) and specific wear rate (WR). According to the COF model, the highest significance level was quadratic load, followed by linear load and lastly the interaction of speed and applied load, while for specific average wear rate, quadratic volume concentration showed the highest significance level, followed by interaction of speed and applied load, and finally the linear load.

(ii) The predicted optimized volume composition for the input variables to produce the lowest response of specific wear rate and average COF in the range tested for waste oil blended with SAE 40 was speed (200 rev/min), load (6.3712 kg), and volume composition (0.2020%).

(iii) According to SEM analysis on the worn surfaces, the maximum wear occurred at 10% concentration of

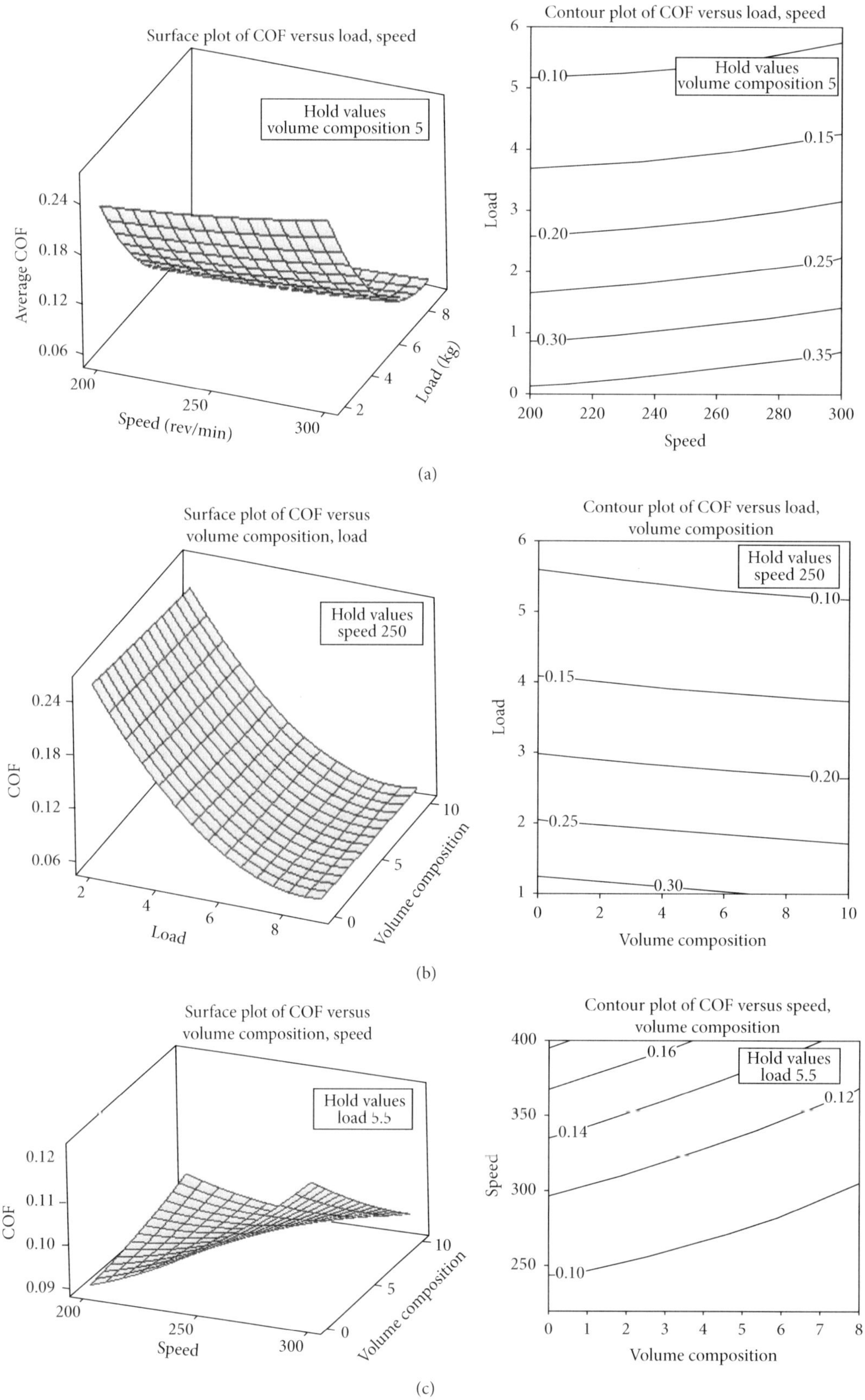

FIGURE 7: (a) Influence of load and speed on COF. (b) Influence of volume composition and load on COF. (c) Influence of volume composition and speed on COF.

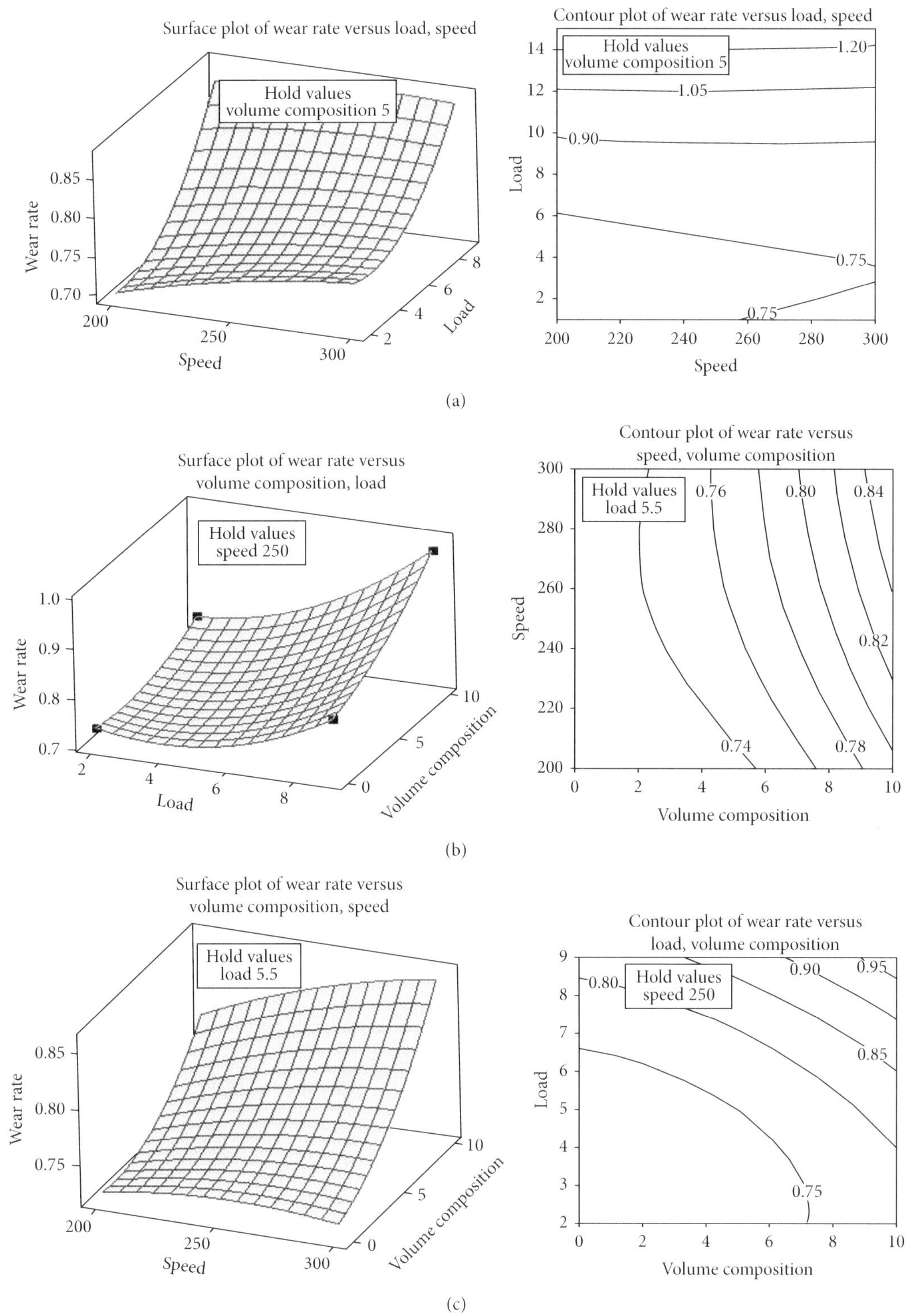

FIGURE 8: (a) Influence of load and speed on specific wear rate. (b) Influence of volume concentration and load on specific wear rate. (c) Influence of volume composition and speed on specific wear rate.

waste oil while minimum wear occurred at 5% concentration of waste oil. This shows that waste cooking oil has antiwear characteristics in the small amount of waste cooking oil and also means that using waste cooking oil as an additive to the engine oil will not have any severe wear causing premature failure.

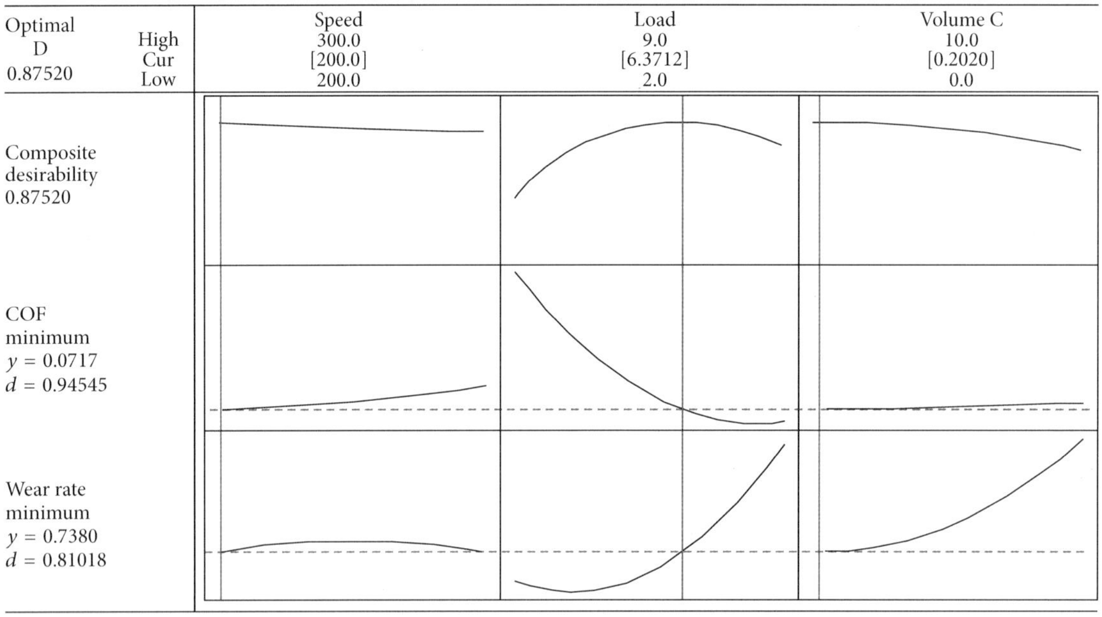

FIGURE 9: Optimal conditions for control variables on the wear and friction responses of blended waste oil with SAE.

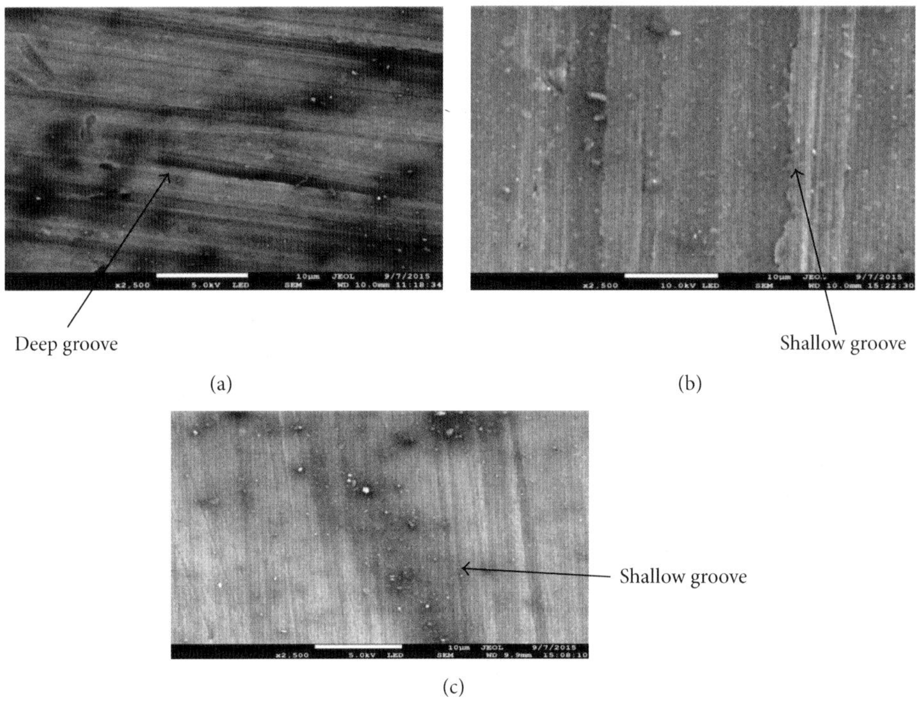

FIGURE 10: SEM images of the surface of aluminum plate for different volume concentrations of biolubricant; (a) 10% waste oil; (b) SAE 40; and (c) 5% waste oil.

Conflict of Interests

The authors declare that there is no conflict of interests regarding the publication of this paper.

Acknowledgments

The authors would like to acknowledge University Malaysia Pahang (UMP) and Ministry of Higher Education (MOHE) of Malaysia for RDU Grants RDU140125 and RDU150351 which made this study possible.

References

[1] L. Pop, C. Puşcaş, G. Bandur, G. Vlase, and R. Nuţiu, "Basestock oils for lubricants from mixtures of corn oil and synthetic diesters," *Journal of the American Oil Chemists' Society*, vol. 85, no. 1, pp. 71–76, 2008.

[2] S. Sadaka and A. A. Boateng, *Pyrolysis and Bio-Oil, Cooperative Extension Service*, US Department of Agriculture and County Governments Cooperating, University of Arkansas, Fayetteville, Ark, USA, 2009.

[3] C.-C. Ting and C.-C. Chen, "Viscosity and working efficiency analysis of soybean oil based bio-lubricants," *Measurement*, vol. 44, no. 8, pp. 1337–1341, 2011.

[4] K. Cheenkachorn and B. Fungtammasan, "Development of engine oil using palm oil as a base stock for four-stroke engines," *Energy*, vol. 35, no. 6, pp. 2552–2556, 2010.

[5] M. H. Jayed, H. H. Masjuki, R. Saidur, M. A. Kalam, and M. I. Jahirul, "Environmental aspects and challenges of oilseed produced biodiesel in Southeast Asia," *Renewable and Sustainable Energy Reviews*, vol. 13, no. 9, pp. 2452–2462, 2009.

[6] S. Syahrullail, B. M. Zubil, C. S. N. Azwadi, and M. J. M. Ridzuan, "Experimental evaluation of palm oil as lubricant in cold forward extrusion process," *International Journal of Mechanical Sciences*, vol. 53, no. 7, pp. 549–555, 2011.

[7] R. Kreivaitis, M. Gumbyte, K. Kazancev, J. Padgurskas, and V. Makarevičiene, "A comparison of pure and natural antioxidant modified rapeseed oil storage properties," *Industrial Crops and Products*, vol. 43, pp. 511–516, 2013.

[8] B. Balasubramaniam, A. Sudalaiyadum Perumal, J. Jayaraman, J. Mani, and P. Ramanujam, "Comparative analysis for the production of fatty acid alkyl esterase using whole cell biocatalyst and purified enzyme from *Rhizopus oryzae* on waste cooking oil (sunflower oil)," *Waste Management*, vol. 32, no. 8, pp. 1539–1547, 2012.

[9] M. A. Kalam, H. H. Masjuki, M. Varman, and A. M. Liaquat, "Friction and wear characteristics of waste vegetable oil contaminated lubricants," *International Journal of Mechanical and Materials Engineering*, vol. 6, no. 3, pp. 431–436, 2011.

[10] A. Zeman, A. Sprengel, D. Niedermeier, and M. Späth, "Biodegradable lubricants—studies on thermo-oxidation of metal-working and hydraulic fluids by differential scanning calorimetry (DSC)," *Thermochimica Acta*, vol. 268, pp. 9–15, 1995.

[11] H. H. Masjuki and M. A. Maleque, "The effect of palm oil diesel fuel contaminated lubricant on sliding wear of cast irons against mild steel," *Wear*, vol. 198, no. 1-2, pp. 293–299, 1996.

[12] İ. Cesur, V. Ayhan, A. Parlak, O. Sava, and Z. Aydin, "The effects of different fuels on wear between piston ring and cylinder," *Advances in Mechanical Engineering*, vol. 6, Article ID 503212, pp. 1–8, 2014.

[13] D. C. Montgomery, *Design and Analysis of Experiments*, Wiley, New Delhi, India, 2007.

[14] C. P. Koshy, P. K. Rajendrakumar, and M. V. Thottackkad, "Evaluation of the tribological and thermo-physical properties of coconut oil added with MoS_2 nanoparticles at elevated temperatures," *Wear*, vol. 330-331, pp. 288–308, 2015.

[15] M. M. Noor, A. P. Wandel, and T. Yusaf, "A review of mild combustion and open furnace design consideration," *International Journal of Automotive and Mechanical Engineering*, vol. 6, no. 1, pp. 730–754, 2012.

[16] H. M. Mobarak, E. Niza Mohamad, H. H. Masjuki et al., "The prospects of biolubricants as alternatives in automotive applications," *Renewable and Sustainable Energy Reviews*, vol. 33, pp. 34–43, 2014.

[17] N. W. M. Zulkifli, M. A. Kalam, H. H. Masjuki, M. Shahabuddin, and R. Yunus, "Wear prevention characteristics of a palm oil-based TMP (trimethylolpropane) ester as an engine lubricant," *Energy*, vol. 54, pp. 167–173, 2013.

[18] Water in Oil Contamination, http://www.machinerylubrication.com/Read/192/water-contaminant-oil.

[19] P. Rajasekhar, G. Ganesan, and C. Senthilkumar, "Studies on tribological behavior of polyamide filled jute fiber-nano-ZnO hybrid composites," *Procedia Engineering*, vol. 97, pp. 2099–2109, 2014.

[20] B. P. Chang, H. M. Akil, M. G. Affendy, A. Khan, and R. B. M. Nasir, "Comparative study of wear performance of particulate and fiber-reinforced nano-ZnO/ultra-high molecular weight polyethylene hybrid composites using response surface methodology," *Materials & Design*, vol. 63, pp. 805–819, 2014.

[21] M. Shahabuddin, H. H. Masjuki, M. A. Kalam, M. M. K. Bhuiya, and H. Mehat, "Comparative tribological investigation of bio-lubricant formulated from a non-edible oil source (Jatropha oil)," *Industrial Crops and Products*, vol. 47, pp. 323–330, 2013.

17

Recirculation Flow and Pressure Distributions in a Rayleigh Step Bearing

Feng Shen, Cheng-Jin Yan, Jian-Feng Dai, and Zhao-Miao Liu

College of Mechanical Engineering and Applied Electronics Technology & Institute for Advanced Mechanics in Engineering, Beijing University of Technology, Beijing 100124, China

Correspondence should be addressed to Zhao-Miao Liu; lzm@bjut.edu.cn

Academic Editor: Enrico Ciulli

Flow characteristics in the Rayleigh step slider bearing with infinite width have been studied using both analytical and numerical methods. The conservation equations of mass and momentum were solved utilizing a finite volume approach and the whole flow field was simulated. More detailed information about the flow patterns and pressure distributions neglected by the Reynolds lubrication equation has been obtained, such as jumping phenomenon around a Rayleigh step, vortex structure, and shear stress distribution. The pressure distribution of the Rayleigh step bearing with optimum geometry has been numerically simulated and the results obtained agreed with the analytical solution of the classical Reynolds lubrication equation. The simulation results show that the maximum pressure of the flow field is at the step tip not on the lower surface and the increment of the strain rate from Navier-Stokes equation is approximately 49 percent greater than that from Reynolds theory at the step tip. It is also shown that the position of the maximum pressure of the lower surface is a little less than the length of the first region. These results neglected by the Reynolds lubrication equation are important for designing a bearing.

1. Introduction

Rayleigh step bearing has been widely used in industry due to its highest load capacity among all other possible bearing geometries. Many researches on improving its load capacity were carried out using an analytical method by solving the classical Reynolds lubrication equation, assuming that the bearing length should be at least 100 times of the film thickness. In 1918, the theory of a step bearing was firstly discussed by Lord Rayleigh [1], determining the optimum geometry with maximum load capacity per unit width for a given film thickness and bearing length. This configuration is now referred to as Rayleigh step bearing.

Because of its high load capacity and cheap manufacturing, the Rayleigh step bearing has been widely used in industry, such as thrust and pad bearings [2–5]. In addition, a series of Rayleigh steps are used in journal bearings to form a grooved bearing with higher performance. Since then, researches into optimum design and fluid dynamic characteristics of this bearing have attracted much attention [6, 7]. Rahmani et al. [8] comprehensively studied the Rayleigh step slider bearing including the effect of variations of pressure at the boundaries on the optimum parameters. The bearing is also optimized considering the lubricant flow rate, friction force, and friction coefficient. Auloge et al. [9] studied the optimum design of Rayleigh step bearing and determined the relationships between step location and height along with non-Newtonian lubricants. Artiles et al. [10] described the analysis and design of a 50mm diameter floating-ring seals with the Rayleigh step lift pads, considering influences of the surface speed, gas pressure, and inertia. Hong et al. [11] investigated the flow of a Newtonian fluid and Bingham fluid in a Rayleigh step bearing.

Through the numerical solutions of the conservation equations of mass and momentum, they found some special jumping phenomena around a Rayleigh step, which are important for designing a bearing and the study of wear characteristics. Zhu [12] investigated both numerically and experimentally the response of a rotor supported on Rayleigh step gas bearing using Galerkin finite element method. Constantinescu et al. [13] analyzed the pressure variation due to fluid inertia effects in Rayleigh step bearings. Faria and San Andrés [14] calculated the bearing load capacity, static stiffness coefficients, and frequency-dependent force coefficients for the gas-lubricated plane and Rayleigh step slider bearings using both finite element and finite difference methods. Lee and Kim [15] calculated the air film temperature of Rayleigh step air foil thrust bearing by solving the Reynolds equation and 3D energy equation with thermohydrodynamic boundary conditions at the top foil, thrust disc, and cooling air plenum. By using the non-Newtonian fluid as a lubricant, the pressure distribution and load capacity for Rayleigh step bearing have also been investigated [11, 16–19].

In previous literature, most researchers used an analytical method by solving the Reynolds lubrication equation, which is derived from fundamental equations governing fluid flows and based on several simplifying assumptions. Thus, the solutions of Reynolds lubrication equation for the pressure distribution, load capacity, and frictional coefficient were obtained through a relatively simple calculation and some detailed flow information on Rayleigh step bearing has been neglected [20]. Only a few investigations used the calculation of the Navier-Stokes equations with finite volume method [11] or infinite element method [12, 14]. In recent years, the full Navier-Stokes equations have been increasingly applied to solve some lubrication problems [21–23]. The trend is driven by a growing interest in describing lubricant flow behavior within a whole lubricated system. The Reynolds equation is confined to just the contact region, but the Navier-Stokes equations can be applied throughout the flow field [24, 25]. In order to optimize Rayleigh step bearing design scientifically, it is beneficial to obtain detailed flow information of the whole lubricant flow field [26].

In this study, the flow characteristics in the infinitely wide Rayleigh step bearing have been investigated by solving Navier-Stokes equations and comparing the numerical results with those results obtained by analytical solutions deduced from the Reynolds equation. From the comparison, more detailed flow characteristics and pressure and shear stress distributions have been obtained, and there are some new differences between the results obtained by two different methods. Moreover, the effects of step geometry and velocity of the lower surface on the recirculation flow and pressure and shear stress distributions in the whole flow field are investigated for better control of these bearings.

2. Analytical Solution

The geometry of a two-dimensional infinite Rayleigh step slider bearing is shown in Figure 1. There are two opposed surfaces: the upper surface separates the bearing into two regions and the lower surface has a pure tangential sliding motion relative to the upper surface. The one-dimensional Reynolds equation for a steady flow of an incompressible fluid of constant viscosity can be expressed in the following form [27]:

$$\frac{\partial}{\partial x}\left(h^3(x)\frac{\partial p(x)}{\partial x}\right) = 6\mu U_1\frac{\partial h(x)}{\partial x} \quad (1)$$

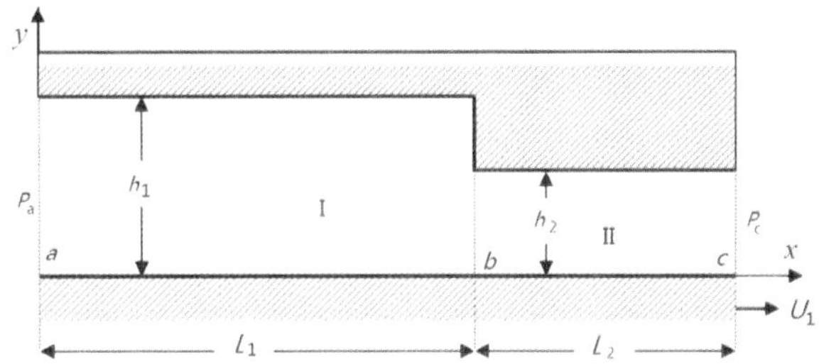

FIGURE 1: The Rayleigh step bearing.

Here p is the oil film pressure, h is the oil film thickness, μ is the fluid dynamic viscosity, U_1 is the velocity of the lower surface, and x is the Cartesian position variable.

With reference to Figure 1, the film thickness is given by

$$\begin{aligned} h(x) &= h_i \quad (i = 1,\ 0 \le x \le L_1) \\ h(x) &= h_i \quad (i = 2,\ L_1 < x \le L) \end{aligned} \quad (2)$$

Here L is the characteristic length of the bearing, given by

$$L = L_1 + L_2 \quad (3)$$

The dimensionless scales of the Rayleigh step bearing can be defined as

$$\varepsilon = \frac{L_1}{L}; \quad (0 < \varepsilon < 1) \quad (4)$$

and

$$\xi = \frac{h_1}{h_2}; \quad (\xi > 1) \quad (5)$$

The boundary conditions for the velocity and the pressure are

$$\begin{aligned} &u = U_1, \\ &v = 0 \\ &\qquad at\ y = 0 \\ &u = 0, \\ &v = 0 \\ &\qquad at\ y = h_i\ (i = 1, 2) \end{aligned} \quad (6)$$

and

$$\begin{aligned} p\big|_{x=c} &= p_0 \\ p\big|_{x=a} &= p_a - p_c \end{aligned} \quad (7)$$

Here p_0 is the ambient pressure. Double integrating of (1) with respect to x, the pressure distribution along the bearing surface can be expressed as

$$p(x) = 6\mu U_1 \int_a^c \frac{1}{h(x)^2}dx + c_1 \int_a^c \frac{1}{h(x)^3}dx + c_2 \quad (8)$$

Here a and c are the coordinates of the bearing leading and trailing ends (see Figure 1); c_1 and c_2 are integration constants. Substituting boundary conditions (6) and (7) into (8), the integration constants were obtained as

$$c_1 = \frac{(p_a - p_c) - 6\mu U_1 \int_a^c \left(1/h(x)^2\right) dx}{\int_a^c \left(1/h(x)^3\right) dx} \tag{9}$$

$$c_2 = 0$$

Employing film thickness (2), the pressure distribution for the two regions of the bearing can be expressed in the following forms, respectively:

$$p(x)\big|_{x\in[a,b]} = 6\mu U_1 \frac{x-a}{h_1^2} + c_1 \frac{x-a}{h_1^3},$$

$$p(x)\big|_{x\in[b,c]} = 6\mu U_1 \left[\frac{L_1}{h_1^2} + \frac{x-b}{h_2^2}\right] + c_1 \left[\frac{L_1}{h_1^3} + \frac{x-b}{h_2^3}\right] \tag{10}$$

In addition, substituting (2) into (9), the integration constant c_1 was obtained as follows:

$$c_1 = \frac{(p_a - p_c) - 6\mu U_1 \left(L_1/h_1^2 + L_2/h_2^2\right)}{L_1/h_1^3 + L_2/h_2^3} \tag{11}$$

Using pressure distribution (10), the load capacity of the bearing was calculated as

$$w = \int_a^c p(x)\, dx = 3\mu U_1 \left[\frac{L_1(L_1 + 2L_2)}{h_1^2} + \frac{L_2^2}{h_2^2}\right] + \frac{c_1}{2}\left[\frac{L_1(L_1 + 2L_2)}{h_1^3} + \frac{L_2^2}{h_2^3}\right] \tag{12}$$

When there is no pressure difference ($p_a = p_c = p_0$) between two ends of the bearing, the load capacity of the bearing is entirely supported by the hydrodynamic pressure distribution inside the bearing and the hydrostatic pressure support no longer exists. Equations (10), (11), and (12) will be reduced to a simpler form.

$$c_1' = -6\mu U_1 h_1 \frac{L_1 + L_2\left(h_1^2/h_2^2\right)}{L_1 + L_2\left(h_1^3/h_2^3\right)} \tag{13}$$

The derivation of (10) is constant:

$$\left.\frac{dp(x)}{dx}\right|_{x\in[a,b]} = \frac{6\mu U_1}{h_1^2} + \frac{c_1'}{h_1^3} > 0,$$

$$\left.\frac{dp(x)}{dx}\right|_{x\in[b,c]} = \frac{6\mu U_1}{h_2^2} + \frac{c_1'}{h_2^3} < 0 \tag{14}$$

So the maximum pressure p_m of the lower surface is at $x = b$; substituting (4), (5), and (13) into (10), p_m was obtained as

$$p_m = \frac{6\mu U_1 L (a-1)\varepsilon(1-\varepsilon)a^2}{h_1^2\left[\varepsilon + (1-\varepsilon)a^3\right]} \tag{15}$$

where the total load on unit width is

$$w = \frac{1}{2}Lp_m = \frac{3\mu U_1 L^2 (a-1)\varepsilon(1-\varepsilon)a^2}{h_1^2\left[\varepsilon + (1-\varepsilon)a^3\right]} \tag{16}$$

3. Governing Equation and Numerical Methods

For steady state incompressible flow, the momentum and continuity equations of film region are

$$u\frac{\partial u}{\partial x} + v\frac{\partial u}{\partial y} = -\frac{1}{\rho}\frac{\partial p}{\partial x} + \frac{\mu}{\rho}\left(\frac{\partial^2 u}{\partial x^2} + \frac{\partial^2 u}{\partial y^2}\right) \tag{17}$$

$$u\frac{\partial v}{\partial x} + v\frac{\partial v}{\partial y} = -\frac{1}{\rho}\frac{\partial p}{\partial y} + \frac{\mu}{\rho}\left(\frac{\partial^2 v}{\partial x^2} + \frac{\partial^2 v}{\partial y^2}\right) \tag{18}$$

$$\frac{\partial u}{\partial x} + \frac{\partial v}{\partial y} = 0 \tag{19}$$

Here ρ is the fluid density, u and v are the velocity, μ is the fluid viscosity, and p is pressure. The pressure at the inlet and outlet boundaries of the Rayleigh step bearing was set to be atmospheric (10^5Pa) and zero velocity gradient in the direction normal to sliding was assumed. This defines a fully developed flow approximation through these boundaries and it is important to set the boundaries relatively far from the region of interest so that they do not influence the numerical solution. The entire domain was fully flooded. At the solid walls, the "no-slip" boundary condition was assumed for the momentum equations. The film region was considered to be a fully developed thin film flow. The bearing working condition is specified by the velocity of the moving wall (U_1), varied from 1m/s to 8m/s. The fluid is considered as being incompressible and isothermal with $\mu = 0.188 P_a \cdot s$ and ρ=850kg/m^3. This approach was adopted because the present work is the first part of a study aimed at enlightening the flow characteristics in the Rayleigh step bearing. Meshes with 210295, 334793, and 5604066 cells were tested. The final refined mesh of the Rayleigh step bearing was built using 334793 cells and the solutions are independent of the refined mesh.

To solve the above equations, a CFD software package CFD-ACE+ (v2010, ESI Corporation) used for multiphysics computational analysis was employed. A finite volume method is utilized to turn the governing partial differential equations (PDE) into a system of algebraic equations and numerically integrated over each of the computational cells using a collocated cell-centered variable arrangement [28, 29]. A second-order upwind scheme is used for the momentum equations. For pressure calculations, SIMPLEC scheme [30] was used. The equation for pressure correction

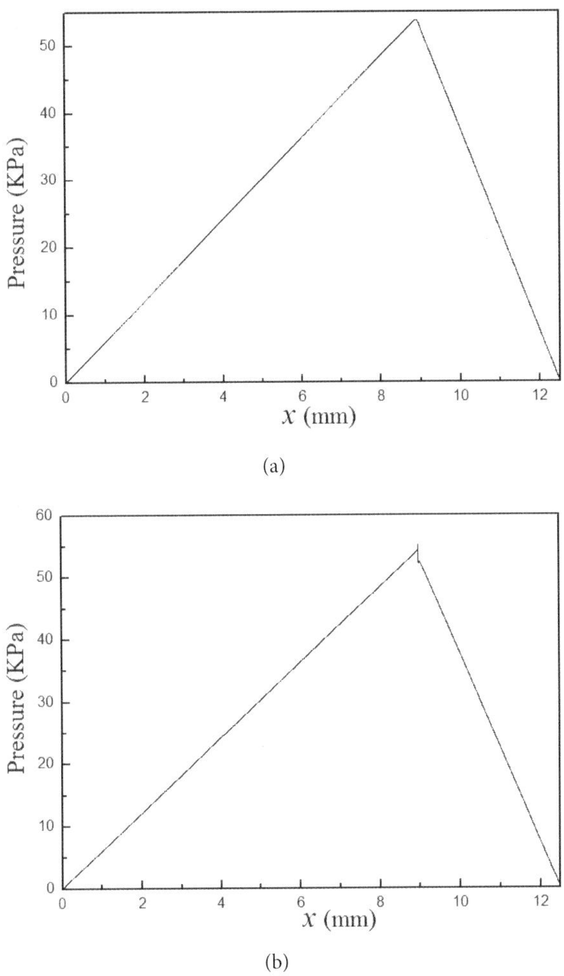

Figure 2: Pressure distributions on lower (a) and upper surfaces (b).

is obtained from the continuity equation, and the scheme of velocity and pressure calculations is fundamentally iterative in nature. The residuals of continuity and momentum equations are required to be below 10^{-5}.

4. Results and Discussion

The dynamics of flow through a Rayleigh step was investigated. The simulations visualized both the velocity and pressure profiles, while varying system parameters including the depth and length of the step, inlet conditions (pressure and velocity), and shear speed.

4.1. Effects of Length Ratios on Pressure. The geometry and operation conditions are h_1 =250μm, L=12.5mm, h_2 =133.976μm, l_1=8.975mm, U_1=1m/s, and $\mu = 0.188P_as$, which are the same as those reported in a previous publication [11]. From the analytic solution of Reynolds lubrication theory, the optimum values of ε and ξ according to maximum load for a Newtonian fluid are $\varepsilon = 0.718$ and $\xi = 1.866$ [1], respectively. The maximum pressure p_m of 54KPa can be obtained with the analytic expression of (14). Figure 2 shows the pressure curves on upper and lower surfaces. The maximum pressure

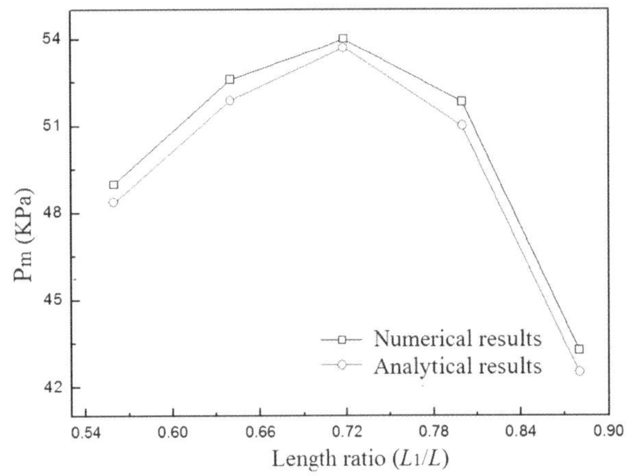

Figure 3: p_m values with different length ratios L_1/L.

(p_m) in Figure 2(a) is 53.7KPa close to the analytical value of 54KPa, which is the same as what is reported in a previous publication [11]. Figure 2(b) shows that pressure distribution on the upper surface changes approximately 5.5 percent at the point of $x = b$. The reason is that there is a vortex that appeared near the step and the pressure distribution of flow field is changed significantly, which will be discussed in next section. The results were compared with the simulations of other CFD software (ANSYS-FLUENT, V15.0) in the Supplementary Materials (available here).

The p_m values with different length ratios (ε=0.56, 0.64, 0.718, 0.8, and 0.88) are plotted in Figure 3. The maximum pressure p_m can be obtained for ε=0.718 by both analytical and numerical methods. Figure 3 shows that the analytical and numerical results are similar and the validity of the numerical results is further verified.

4.2. Effects of Height Ratios on Pressure. The effects of height ratios on the maximum pressure were simulated. The Rayleigh step parameters are h_1=250μm, L=12.5mm, ε=0.718, U_1=1m/s, and $\mu = 0.188P_as$. The p_m values with different height ratios (ξ=1.56, 1.786, 1.866, 2.273, 2.778, 3.57, 5, and 9.33) are plotted in Figure 4. Figure 4 shows that the maximum pressure p_m increases evidently as the value of ξ increases, and both the analytic and numerical results are similar.

Pressure contours around the steps with different height ratios (ξ=1.866 and ξ=9.33) are plotted in Figure 5. Figure 5 shows that the pressure values increase evidently as the value of ξ increases. It is found that the maximum pressure of the flow field is at the step tip instead of being on the lower surface. The maximum pressures of Figures 5(a) and 5(b) are 57KPa and 154.6KPa, which are higher than the p_m values of the lower surface: 53.7KPa and 152KPa. These details of pressure distribution are not predicted if the Reynolds lubrication equation is used.

4.3. Vortex and Flow Details. The flow characteristics of Rayleigh step bearing with different geometric structures were studied. Some new phenomena like the vortex and

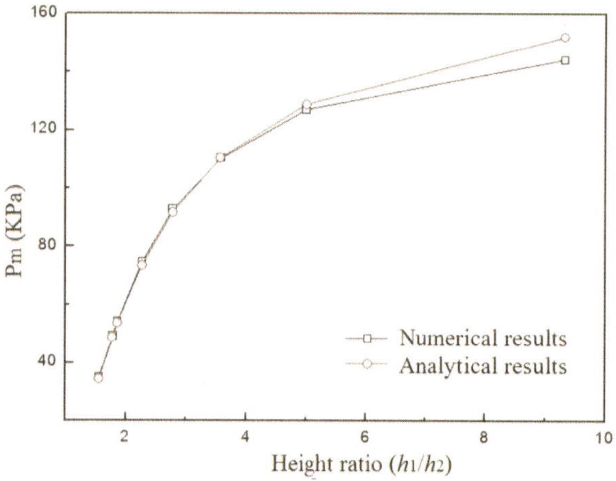

FIGURE 4: p_m values with different height ratios h_1/h_2.

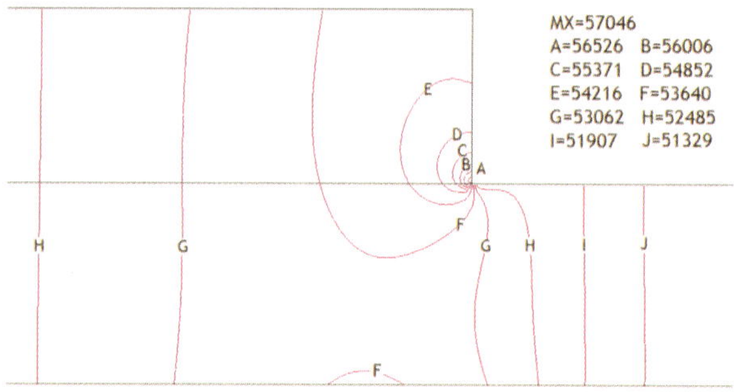

(a)

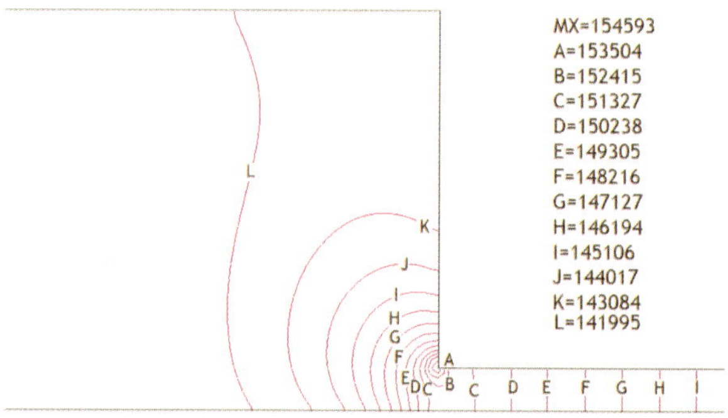

(b)

FIGURE 5: Pressure contours (Pa) of ξ=1.866 (a) and ξ=9.33 (b).

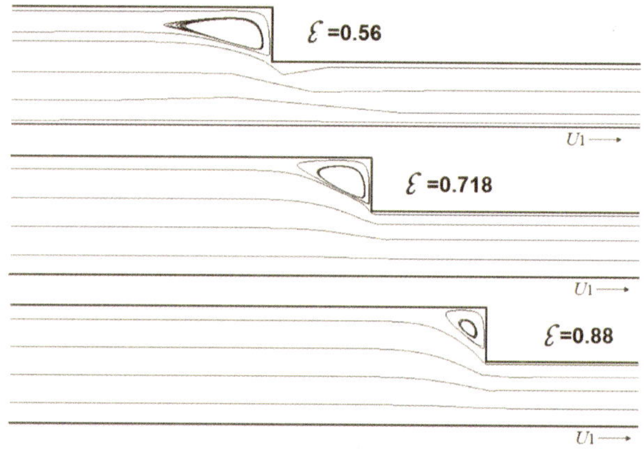

(a)

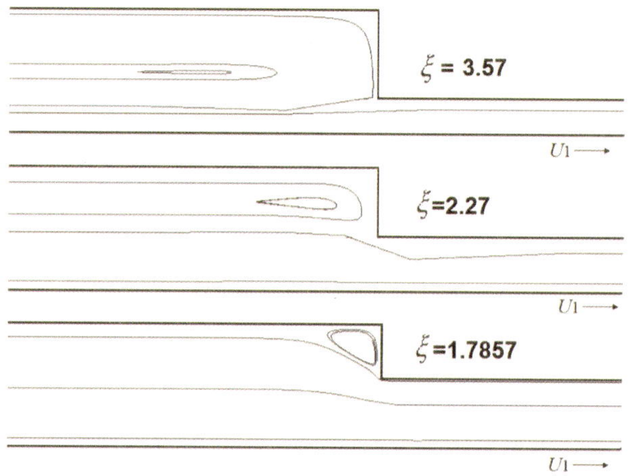

(b)

FIGURE 6: Streamlines of different geometric structures.

reversed flow zone which are not predicted by the Reynolds lubrication equation were investigated. Figure 6 shows the streamlines of different geometric structures in parameters of U_1=1m/s and μ = $0.188P_a s$. In Figure 6(a), a reversed flow zone appears near the step with ε=0.56. When ε=0.718 and 0.88, the reversed flow zone disappeared and vortices are formed, and the vortices become smaller with the increase of length ratios (ε). In Figure 6(b), two reversed flow zones appear with ξ=3.57 and 2.27, and the center of the reserved flow zone moves rightward with the decrease of the height ratios (ξ). When ξ=1.7857, a vortex appears near the step. In the reversed flow zone, the velocity direction near the upper surface is opposite to the velocity of the lower surface U_1.

The simulation results show that the pressure and shear stress distributions of the Rayleigh step bearing are significantly influenced by the vortex and velocity field. In Figure 7, the strain rates of fluid in Rayleigh step bearing (ε = 0.718 and ξ = 1.866) are changed obviously near the surfaces. Near the upper surface, the strain rate of fluid in region I is almost zero while that in region II is very high, as shown in Figure 7(a). It is found that the highest value of strain rate appears near the step tip. The strain rate from Navier-Stokes equation is approximately 49 percent greater than that from Reynolds theory at step tip and too strong to be omitted. This jumping phenomenon is not predicted by the Reynolds lubrication equation. Near the lower surface, the strain rate of fluid in region I is higher than that in region II. These results can supply more accurate boundary condition for a better solid analysis of the step tip.

4.4. Effects of U_1 on Pressure. The velocity of the lower surface U_1 has important effects on the pressure distribution. Figure 8 shows the lower surface pressures with different U_1 values in a Rayleigh step bearing with optimum geometry. The pressure

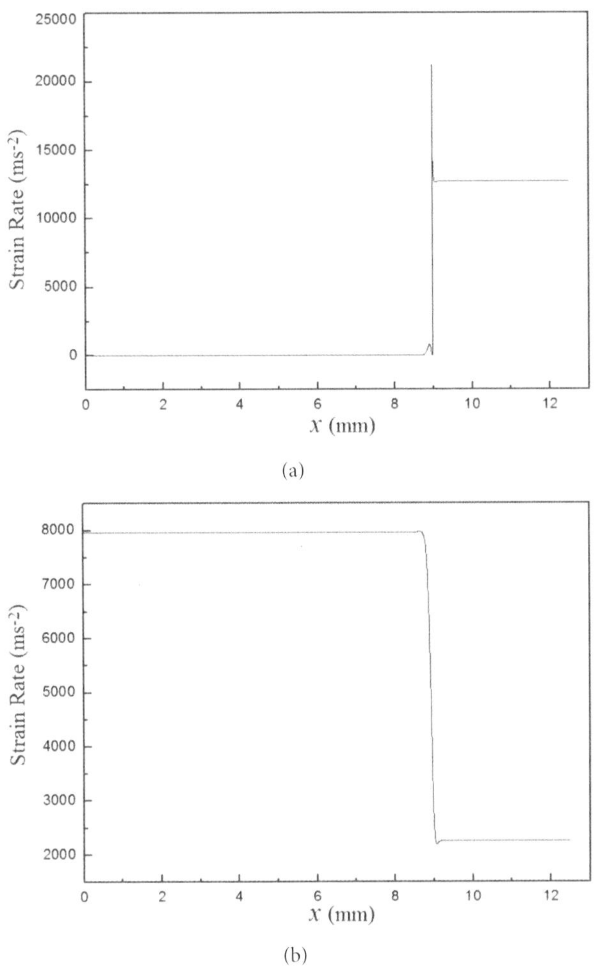

FIGURE 7: Strain rates of fluid near the upper (a) and lower (b) surfaces.

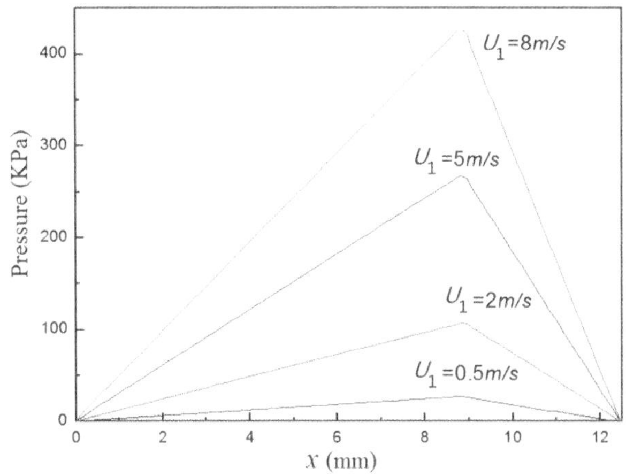

FIGURE 8: Lower surface pressures of different U_1 values.

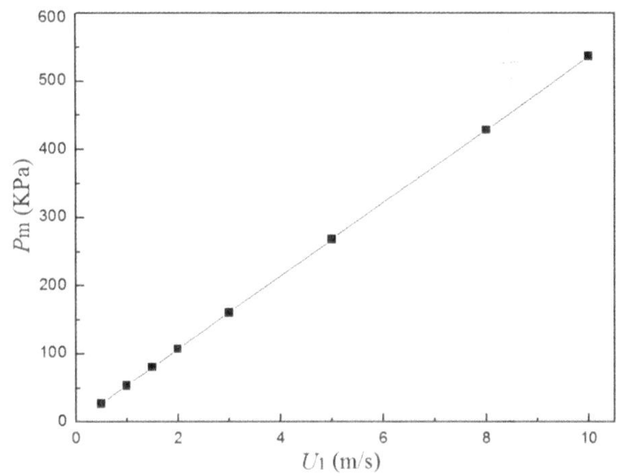

FIGURE 9: The p_m values of the lower surface of different U_1 values.

values increase evidently as the value of U_1 increases. The maximum pressure (p_m) of the lower surface was at x=8.8269 mm a little less than $L_1 = b$, which can be explained in Figure 5. A linear increase of p_m values of the lower surface was noted with increasing U_1 values, shown in Figure 9.

5. Conclusions

The current work presents a parametric study of the Rayleigh step bearing with infinite width using two different approaches: (a) analytical solution of the Reynolds equation; (b) the numerical solution of the Navier-Stokes equations by CFD method. The numerical results agree well with the analytical results with parameters of $\varepsilon = 0.718$ and $\xi = 1.866$. It was shown that the optimum length rate is $\varepsilon = 0.718$ with no pressure difference ($p_a = p_c = 0$) between two ends of the bearing, and the maximum pressures (p_m) of the lower surface increase evidently as the value of height ratio increases. More detailed information on flow characteristics neglected by Reynolds lubrication equation, such as vortices, reversed flow, shear stress, and pressure distributions, has been obtained. It is found that the maximum pressure of the flow field is at the step tip instead of being on the lower surface, and the highest value of strain rate appears near the step tip. The velocity of the lower surface U_1 has important effects on the pressure distribution. It is also found that the position of the maximum pressure (p_m) of the lower surface is a little less than $x = b$. These results are helpful to supply more accurate flow details in the Rayleigh step bearing when designing an optimum Rayleigh step bearing geometry.

Nomenclature

a: Coordinate of bearing leading end (mm)
a_{nb}: Link coefficients
b: Coordinate of region I leading end (mm)
c: Coordinate of bearing trailing end (mm)
c_1: Integration constant
c_2: Integration constant
h_1: Height of region I (μm)
h_2: Height of region II (μm)
L: Length of the bearing (mm)
L_1: Length of region I (mm)

L_2: Length of region II (mm)
p: Pressure of fluid (Pa)
p_a: Pressure at bearing leading end (Pa)
p_c: Pressure at bearing trailing end (Pa)
p_m: The maximum pressure of the lower surface at $x = b$ (Pa)
$p(x)$: Pressure distribution along the bearing lower surface (Pa)
U_1: Velocity of the lower surface (m/s)
u: Velocity in the x direction (m/s)
v: Velocity in the y direction (m/s)
w: Load capacity of the bearing (N)
μ: Viscosity (Pa.s)
ρ: Fluid density (Kg/m^3)
ε: Dimensionless length (L_1/L)
ξ: Dimensionless height (h_1/h_2).

Conflicts of Interest

The authors declare that they have no conflicts of interest regarding the publication of the paper.

Acknowledgments

This work was supported by the National Natural Science Foundation of China (11572013), Beijing Municipal Natural Science Foundation (7152012), and Development Project of Beijing Municipal Education Commission (KM201610005002).

Supplementary Materials

Description of each file of Supplementary Materials: Verification of the results compared with the ANSYS-FLUENT simulations. The results of pressure distributions in Figures 2 and 8 and the results of streamline patterns in Figure 6(a) are simulated and compared with the FLUENT simulations, finding the same results. Figure 1: FLUENT results of pressure distributions on lower surfaces under the same conditions as that in Figure 2(a), finding the deviation of the maximum pressure is 2.4%. Figure 2: FLUENT results of pressure distributions on upper surfaces corresponding to the result of Figure 2(b), finding similar pressure distributions. Figure 3: FLUENT results of lower surface pressures of different U1 values corresponding to the result of Figure 8, finding the same pressure distributions. Figure 4: streamlines in a Rayleigh step bearing with optimum geometry at different U1 values corresponding to Figure 6(a), finding similar flow phenomena. *(Supplementary Materials)*

References

[1] L. Rayleigh, "Notes on the theory of lubrication," *Philosophical Magazine*, vol. 35, pp. 1–12, 1918.

[2] Y. Sato and J. D. Knight, "Performance characteristics of shrouded rayleigh-step and spiral groove viscous pumps," *Journal of Tribology*, vol. 114, no. 3, pp. 499–503, 1992.

[3] Y. Y. Boldyrev, S. V. Lupulyak, and Y. K. Shinder, "Numerical solution of the variational Rayleigh problem of gas lubrication theory," *Journal of Fluid Dynamics*, vol. 30, no. 6, pp. 826–832, 1995.

[4] D. G. Farmer and J. J. Shepherd, "Slip flow in the gas-lubricated Rayleigh step-slider bearing," *Internal Journal of Applied Mechanics and Engineering*, vol. 11, pp. 593–608, 2006.

[5] M. B. Dobrica and M. Fillon, "About the validity of Reynolds equation and inertia effects in textured sliders of infinite width," *Proceedings of the Institution of Mechanical Engineers, Part J: Journal of Engineering Tribology*, vol. 223, no. 1, pp. 69–78, 2009.

[6] M. Oike, M. Nosaka, Y. Watanabe, M. Kikuchi, and K. Kamijo, "Experimental study on high-pressure gas seals for a liquid oxygen turbopump," *Tribology Transactions*, vol. 31, no. 1, pp. 91–97, 1988.

[7] A. Lipschitz, P. Basu, and R. P. Johnson, "A bi-directional gas thrust bearing," *Tribology Transactions*, vol. 34, no. 1, pp. 9–16, 1991.

[8] R. Rahmani, A. Shirvani, and H. Shirvani, "Analytical analysis and optimisation of the Rayleigh step slider bearing," *Tribology International*, vol. 42, no. 5, pp. 666–674, 2009.

[9] J. Y. Auloge, P. Bourgin, and B. Gay, "The optimum design of one-dimensional bearings with non-Newtonian lubricants," *Journal of Lubrication Technology*, vol. 105, pp. 391–395, 1983.

[10] A. Artiles, W. Shapiro, and H. F. Jones, "Design analysis of Rayleigh-step floating-ring seals," *Tribology Transactions*, vol. 27, pp. 321–331, 1984.

[11] Y. Hong, D. Chen, J. Wang, and X. Li, "Numerical modeling of a Newtonian and a Bingham fluid in a Rayleigh step bearing," *Journal of Tribology*, vol. 125, no. 1, pp. 206–210, 2003.

[12] X. Zhu, *Experimental response of a rotor supported on Rayleigh step gas bearings [M. S. thesis]*, Texas A&M University, College Station, TX, USA, 2004.

[13] V. N. Constantinescu and S. Galetuse, "Pressure Drop Due to Inertia Forces in Step Bearings," *Journal of Lubrication Technology*, vol. 98, no. 1, p. 167, 1976.

[14] M. T. C. Faria and L. San Andrés, "On the numerical modeling of high-speed hydrodynamic gas bearings," *Journal of Tribology*, vol. 122, no. 1, pp. 124–130, 2000.

[15] D. Lee and D. Kim, "Three-dimensional thermohydrodynamic analyses of rayleigh step air foil thrust bearing with radially arranged bump foils," *Tribology Transactions*, vol. 54, no. 3, pp. 432–448, 2011.

[16] K. Yagi and J. Sugimura, "Elastohydrodynamic simulation of Rayleigh step bearings in thin film hydrodynamic lubrication," *Tribology International*, vol. 64, pp. 204–214, 2013.

[17] A. F. Elkouh and D.-F. Yang, "Flow of a power-law fluid in a rayleigh step," *Journal of Tribology*, vol. 113, no. 3, pp. 428–433, 1991.

[18] T. H. Leek, S. Lingard, R. J. Atkini, and W. A. Bullough, "An experimental investigation of the flow of an electro-rheological fluid in a rayleigh step bearing," *Journal of Physics D: Applied Physics*, vol. 26, no. 10, pp. 1592–1600, 1993.

[19] N. B. Naduvinamani, "Non-Newtonian effects of second-order fluids on double-layered porous Rayleigh-step bearings," *Japan Society of Fluid Mechanics: Fluid Dynamics Research*, vol. 21, no. 6, pp. 495–507, 1997.

[20] M. Dobrica and M. Fillon, "Reynolds' model suitability in simulating rayleigh step bearing thermohydrodynamic problems," *Tribology Transactions*, vol. 48, no. 4, pp. 522–530, 2005.

[21] M. J. Braun and M. B. Dzodzo, "Three-Dimensional Flow and Pressure Patterns in a Hydrostatic Journal Bearing Pocket," *Journal of Tribology*, vol. 119, no. 4, pp. 711–719, 1997.

[22] M. Helene, M. Arghir, and J. Frene, "Numerical three-dimensional pressure patterns in a recess of a turbulent and compressible hybrid journal bearing," *Journal of Tribology*, vol. 125, no. 2, pp. 301–308, 2003.

[23] J.-S. Kim and K.-W. Kim, "Effects of distance between pads on the inlet pressure build-up on pad bearings," *Journal of Tribology*, vol. 124, no. 3, pp. 506–514, 2002.

[24] P. Brajdic-Mitidieri, A. D. Gosman, E. Ioannides, and H. A. Spikes, "CFD analysis of a low friction pocketed pad bearing," *Journal of Tribology*, vol. 127, no. 4, pp. 803–812, 2005.

[25] A. V. Radulescu, "Theoretical Calculus Model for the Finite Length Rayleigh Step Bearings Lubricated with Greases," in *Proceedings of the BALTTRIB 2015*, Aleksandras Stulginskis University, Akademija, Lithuania, November 2015.

[26] R. Kumar, M. S. Azam, S. K. Ghosh, and H. Khan, "Effect of surface roughness and deformation on Rayleigh step bearing under thin film lubrication," *Industrial Lubrication and Tribology*, vol. 69, no. 6, pp. 1016–1032, 2017.

[27] Y. Hori, *Hydrodynamic Lubrication*, Springer-Verlag, Tokyo, Japan, 2006.

[28] D. F. Deng, *A numerical and experimental investigation of Taylor flow instabilities in narrow gaps and their relationship to turbulent flow in bearings [Ph.D. thesis]*, University of Akron, Akron, Ohio, USA, 2007.

[29] F. E. Horvat, *Experimental and numerical analysis of flow and pressure fields inside a variable depth single pocket hydrostatic bearing [M. S. thesis]*, University of Akron, Akron, Ohio, USA, 2008.

[30] J. P. van Doormaal and G. D. Raithby, "Enhancements of the SIMPLE Method for predicting incompressible fluid flows," *Numerical Heat Transfer, Part B: Fundamentals*, vol. 7, no. 2, pp. 147–163, 1984.

18

The Contact Mechanics of Novikov's Surface-Hardened Gearing during Running-in Process

Alexey Beskopylny,[1] **Nikolay Onishkov**,[1] **and Viktor Korotkin**[2]

[1]*Department of Transport, Construction and Road System, Don State Technical University, Rostov-on-Don 344000, Russia*
[2]*Institute of Mathematics, Mechanics and Computer Science, South Federal University, Rostov-on-Don 344006, Russia*

Correspondence should be addressed to Alexey Beskopylny; besk-an@yandex.ru

Academic Editor: Michael M. Khonsari

The article is devoted to the analysis of the state of the contact surfaces of the higher kinematic pair in the general case of relative motion, that is, in the presence of rolling, sliding, and twisting, which is characteristic of Novikov's circular-screw gears. The purpose of the work is to assess the impact of friction forces, the state of contact surfaces after tool processing, and the localization of the instantaneous contact spot on the level of contact—fatigue durability of gears. Power contact in the presence of geometric slippage of the mating surfaces leads to a significant change in the initial geometry and the mechanical properties of surface layers. In the existing methods of calculations of contact strength, the effect of running-in is investigated insufficiently, which leads to an incorrect result, especially for gear with high hardness of the teeth. In this work, the conditions of contact interaction close to the real requirements are studied on the basis of experimental material, numerical solution of the contact problem, determination of the terms of the contact areas of slip, and adhesion within the instantaneous spot. The shape of the instant contact spot has asymmetry and can be approximated by an ellipse with the introduction of a correction factor. The running-in period is of a plastic nature with cold deformation and reduction of the roughness of surfaces. As a result of the run-in period, the area of actual contact (tooth height) is increased by 2 or more times. It is not desirable to spread the area of contact at the area of adhesion that initiates the formation of pitting. The presence of defective surface area on the level of contact strength does not have significant influence, because of the running-in period, but increases the risk of spalling and brittle fracture.

1. Introduction

Running-in is the process of transition of the properties of the contacting surfaces of the parts from the initial to the operational. There is a change in size and macro - and microgeometry, as well as physical and mechanical properties of the material of the interacting surfaces to optimize their parameters in a relatively short period. For Novikov's transmissions, this process is of particular importance, because due to its initial point contact passes into contact on the surface, which is due to a sharp increase in the contact strength of such transmissions. However, this is a qualitative indicator. The burnishing process is influenced by many factors—geometric, kinematic, power, lubricant properties, and contact surfaces. It is clear that instantaneous contact spots, both calculated and determined experimentally, differ significantly (Figure 1).

The quasielliptic instantaneous contact spot (IPC) obtained from the initial geometry by the numerical solution of the contact problem [1] and experimentally and numerically on models [2, 3] in the course of processing turns into a kind of curvilinear trapezoid, significantly increasing in size. Thus, for transmission, given in a peer-reviewed article, when p_0= 2013 MPa chordal arc length of the active portion of the profile of the head of the tooth L (Figure 2) due to earnings increased from (1,58...1,6) mm to more than 2,15 mm.

However, if the results of the study are evident, its mechanism is a subject of discussion. Thus, confirmed by many subsequent studies, for example [5], it is considered, in particular,

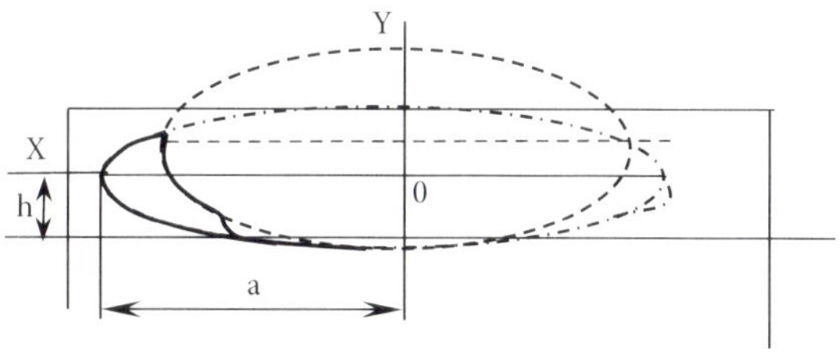

FIGURE 1: The spots of contact in Novikov's gear: (a) unprocessed single gear line (SGL), model on Plexiglas [2], and (b) computer modeling [3, 4].

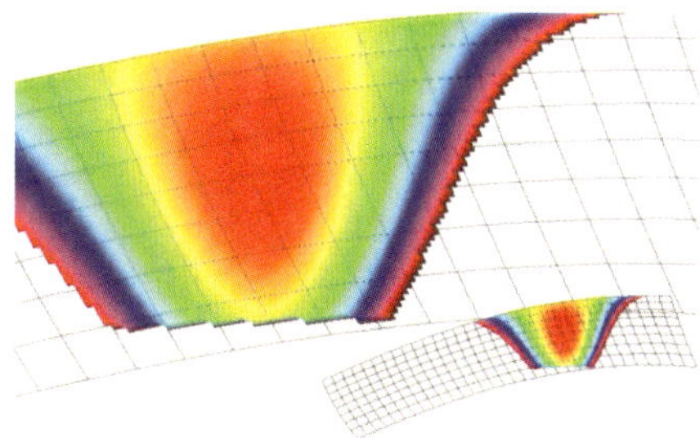

FIGURE 2: Changing the chordal length of the arc L of the active part of the Novikov gear in the process of running-in. L_0: after running-in at the moment on the output shaft T_2=1000 Nm; L_1: as a result of running-in at the time of failure—a breakdown on the tooth of the tooth fillet at T_2 = 3550 Nm with the operating time of $7.2 \cdot 10^6$ loading cycles; ρ_α: radius of curvature of the tooth; α_κ: pressure angle at the theoretical contact point; α_ρ: the minimum angle of the profile of the active part of the tooth.

a consequence of positive hydrodynamics—increase in the thickness of the lubricant film due to the increased speed of movement of the contact area of the teeth in stable rolling conditions. The opposite point of view is presented in [6], where good workability is due to "bad" hydrodynamics due to a sharp increase in the reduced radius of curvature of the contacting surfaces. In confirmation, it is declared of "low transmission Novikov resistance to scoring." The latter is doubtful because there are no cases of bullies for more than half a century of experience in using Novikov's gears (with different initial contours) and the references given in [6] also do not say anything about the low luminosity of such transmissions, especially about their "congenital defect." When operating in the boundary friction mode on the contacting surfaces, there would be traces of adhesive action or microcutting, but the worked surfaces are carbonized and mirrored. The opinion of Yakovlev [7] on the plastic character of run-in under conditions of a stable elastic hydrodynamic contact, when the total wear is several times sufficient for a qualitative run-in, is sufficient (in the cited paper up to 5), and there are fewer scallops of initial roughness.

Running-in process experiments were conducted on a pin-on-disk tester in [8]. The attractors of friction vibration were investigated [7] by the chaos theory, and the evolvement mechanism of the friction vibration chaotic attractors in the running-in process was analyzed. The experimental results indicate that the friction vibration has a stochastic nature, and the chaotic attractor in the phase space is an always open trajectory with a specific hierarchy and structure.

Running-in attractor was investigated as a stable and time-space ordered structure formed in running-in process in [9]. To establish prediction models of the running-in attractor, orthogonal experiments were performed by sliding pins against a disc during the running-in process. The models proposed in [9, 10] make it possible to predict the operating conditions, properties of tribological contact, and surface roughness, which gives a reference to the operating mode.

The dual-disk model testing concept, shown in [11], provides valuable information during running-in process on the applicability of new technologies, such as surface structuring, coatings, alternative fluids, or modern materials, in real machine elements.

A model for the simulation of wear particles formation and running-in in mixed lubricated sliding contacts was developed in [12]. The simulations based on a previously developed half-space algorithm coupled with a numerical elastohydrodynamic lubrication solver utilizing the load-sharing concept.

The effect of running-in on surface characteristics of spur gears and on their development during subsequent efficiency testing is studied in [13]. Micropitting was associated with surface asperities and their plastic deformation; higher running-in load gave more micropitting, also after identical efficiency tests. Running-in increased unequal compressive residual stresses in both profile and axial directions, while after efficiency testing they approached equal levels.

The conditions of running-in and wear are not exactly identical but interrelated [14]. Models for friction during running-in should include the effects of wear since wear affects the surface topography as well as the formation of transfer films, mechanically mixed layers, and third-body agglomerates.

2. Materials and Methods

Experimental studies [15, 16] and experience in the operation of surface-hardened Novikov gear (mainly type GOST 30224-96) showed that, in the group of medium quality (8...10 degree of accuracy), they are superior to their equivalent involute not less than 35...40%. At design pressures up to 2000 MPa, there were no cases of surface contact failure. The work surfaces were smooth, shiny, no pitting. However, many failures-end chips, fractures on the background of the development of deep contact cracks, to some extent, are the result of contact interaction and require more reliable information about the nature of the stress state in real contact.

The general case of relative motion—the combination of rolling, slippage, and rotation—is characteristic of Novikov gearing. The instant contact area is different from the ellipse. The effect of the cantilever application of the load to the tooth is obvious, but little has been studied. Therefore, the characteristics of the stress-strain state (SSS) in the contact area are determined numerically. That is justified in the research plan, but in the solution of applied engineering problems, it is advisable to enlist analytical models, which are corresponding to the practical problem only approximately, significantly simplify the solution, and enhance the universality of the results obtained. Hence, the assessment of the possibilities and conditions of using existing analytical solutions is a necessary stage in the development of engineering methods of calculation. The study of contact interaction provides the determination of the mode of lubricating action in the conditions of actual contact, determination of the characteristics of the contact area, and determination of the components of the stress-strain state (SSS) in the contact area.

Hardening thermal or chemical heat treatment (CHT) makes significant adjustments that affect the changes in the conditions of contact and the properties of the material of the contacting elements. The existing regulations of traditional involute transmissions are not applicable since the stress state in the contact area is different even for Novikov transmissions with different initial contours. This paper analyzes the results of the assessment of SSS in the contact area of gear Novikov-based solutions to problems of solid mechanics—analytical and numerical—and the results of bench testing carbonitriding transmission Novikov with the initial loop of SWG-5 (base for GOST30224-96) “Reducer” (Izhevsk).

3. Results and Discussion

3.1. Contact Mechanics Study. Tests were carried out in single-stage gearboxes CU-160 on stands with a closed flow of power, mechanical loading, the number of revolutions of the drive shaft n_1=1500 min^{-1}, and the circulating oil lubrication system MS-20. The transmissions with parameters were tested: center-to-center distance: a_w = 160 mm; module: m = 3.15 mm; number of gear teeth and wheels: z_1 = 32, z_2 = 65; angle of inclination of teeth: β = 17,28390; the bias coefficients are $x_1 = x_2 = 0$; width of the rim: b_1 = 50 mm and b_2 = 60 mm; material: steel with 0.25% C, 1% Cr, 1% Mn, and 0.2% Mo; CHT: nitrocarburizing with $h_{\text{eff}} \approx$ 0.8mm (up to He = 550HV1); finishing is not provided; accuracy on the scale of ND-4. The average roughness of the nontreated surfaces is R_a = 0.45 μm (grade of cleanliness - 7b) and the working surfaces are R_a = (0.30... 0.35) μm (cleanliness class - 9a). A total of 13 pairs were tested in the range of nominal torques on the drive shaft: T_1 = 1500 ... 2000 Nm.

The study of contact interaction provides the solution of the following tasks: determination of the mode of lubricating action in the conditions of actual contact; determination of the characteristics of the contact site—**E** and the pressure distribution—p and the components of the stress-strain state in the contact area. The solution of the first problem determines the possibility of solving subsequent ones. In the conditions of boundary lubrication, the contacting takes place according to the so-called ”contour area” that depends on the waviness of the surfaces, load, and mechanical properties of the contacting bodies and not exceeding (10-15) % of the nominal, that is, on which the bodies would touch if their surfaces had a perfectly smooth geometric shape. In hydrodynamic (liquid) lubrication, the decisive factor is internal friction in the lubricant volume. The combined effect of the rheological properties of the lubricant and the elastic properties of the bodies determines the elastic hydrodynamic lubrication regime. Characteristic mode of friction is the parameter λ, the relative thickness of the film layer between the contacting surfaces: $\lambda = h/(R_{a1} + R_{a2})$, where h is the absolute film thickness; R_{a1} and R_{a2} are an arithmetic average deviation of the profile of asperities from the baseline. Recommendations on the criterion values of λ differ, but most often the regimes for λ <2, are classified as the boundary, at λ=2...5, as elastic hydrodynamic, and at λ > 5, as hydrodynamics. Under the conditions of the elastic hydrodynamic regime, the pressure plot differs from the Hertz plot for the “dry” contact by the presence of the input zone and the perturbation at the output of the contact. The difference in the maximum pressure in the contact zone differs from that determined by the formulas, following from the theory of Hertz, and does not exceed 20 %. That makes it possible to use the dry contact dependencies in the first approximation to determine the pressure in the lubricant. The thickness of the lubricant film in the elliptic contact was determined by the known formula [17, 18]

$$h = R_x \left[(1.82 - 0.68X) \left(\frac{\mu \alpha \nu}{R_x} \right)^{0.75} \left(\frac{p_0}{E^*} \right)^{-0.25} \right] \quad (1)$$

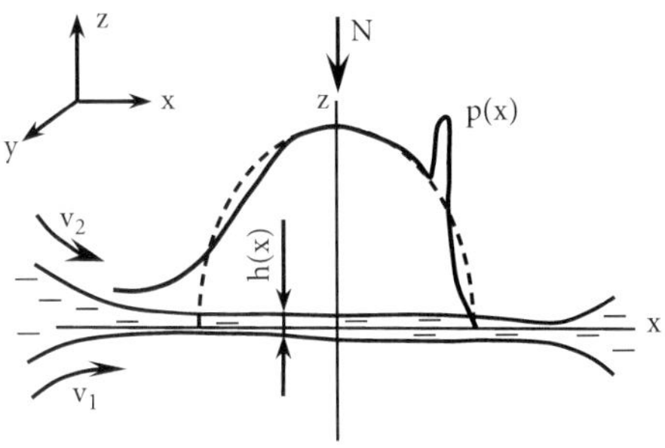

FIGURE 3: Scheme of the pressure distribution in the contact area of the cylindrical surfaces in elastohydrodynamic regime.

where $R_x = R_{x1}R_{x2}/(R_{x1} \pm R_{x2})$ is the reduced radius of curvature; R_{x1} and R_{x2} are the curvature radii of the first and second surfaces in the section passing through the direction of rolling or sliding (Figure 3, the x-axis) and a common normal to the surfaces (Figure 3, the z-axis); $X=R_{\min}/R_{\max}$ is measure of the curvature of the surfaces in the *yz* cross section; $0< X <1$. R_y is given the radius of curvature of the surfaces in the *yz* cross section; μ and a are dynamic viscosity and piezoelectric effect of the base oil at a temperature T at the entrance to the contact zone; $E^* = E_1/(1-\gamma^2)$ is reduced modulus of elasticity; γ is Poisson's ratio; $v = (v_1 + v_2)/2$ is average speed of movement of the surfaces of parts; v_1 and v_2 are speed of movement of the first and second surfaces; p_0 is the maximum pressure in contact with the Hertz.

3.2. The Parameters of the Tested Gear. Center distance is a_w= 160 mm; module is m = 3,15 mm; number of gear teeth and wheels is z_1= 32 and z_2= 65; angle of inclination of the teeth is β = 17.28390; the coefficients of the displacement is $x_1 = x_2 = 0$; the width of the bases is b_1= 50 mm b_2= 60 mm; material is steel with C=0.25%, Si=0.17-0.37%, Mn=0.9-1.2%, Cr=0.9-1.2%, Ni<0.3%, Cu<0.3%, and Mo=0.2-0.3%; chemical-thermal treatment nitrocarburizing is c $h_{\text{eff}} \approx 0.8$ mm (to He= 550HV); finish machining is not provided; the accuracy is on a scale ND-4. The average roughness of the untreated surfaces is Ra= 0.45 μm (purity class - 7b), worked surfaces are Ra= (0.30...0.35) μm (purity class - 9a). Test conditions are as follows: torque on the drive shaft: T_1 = 1500-2000 Nm; mode: 100%; drive shaft speed: n_1 = 1500 min^{-1}; lubrication with MS-20 oil.

Tested gear is as follows: R_x= 716.55 mm; R_y= 22.86 mm; p_o= 2013 MPa; E^*= 2.15e+5 MPa; γ = 0,3. Speed v is determined by [19]

$$v = \omega_1 \left\{ \left[r_1 + 0.5L_t \left(u^{-1} - 1 \right) \sin \alpha_t \right]^2 + \left[-0.5L_t \left(u^{-1} - 1 \right) \cos \alpha_t \right]^2 + \left(r_1 ctg\beta \right)^2 \right\}^{0.5} \tag{2}$$

where $r_1 = mz_1/2\cos ß$ is the dividing diameter of the gear and $u = z_2/z_1$ is the gear ratio; L_t is the pressure angle and the distance from the pole to the theoretical contact point in the end plane: v = 27.9 m/s. At a temperature of T=1000 C (almost not reached in the tests), μ =1.84e-4 Pas and a = 14.5e-7 Pa^{-1}. In this case (even without taking into account a multiplying factor X) the minimum value of the thickness of the layer h = 2.07 μm and (when $R_{a1} = R_{a2}$= 0.45 μm) λ = 2.3. Consequently, the transmissions worked in a stable elastohydrodynamic regime.

At the same time, similar calculations were carried out in accordance with the works of Hamrok-Dawson [18], but the calculated hydrodynamic parameter never fell below 2.5, which is even more favorable.

3.3. Characterization of Contact Area E and Pressure Distribution p without regard to Running-in. When the elastic characteristics of the materials of interacting bodies are equal, the jump of normal displacements determining the contact area depends only on the normal load [19], which allows separating the problem of determining *S* and *p* from the problem of determining the SSS and solving them sequentially.

Problem 1. In the orthogonal coordinate system $OX_1X_2X_3$, where X_3=0 is the common tangent plane, *E*, an unknown contact region, is lying in this plane. The problem is to find the solution of the homogeneous equilibrium equations of the elastic medium in the half-spaces $X_{3(1)} > 0$, $X_{3(2)} < 0$ satisfying the boundary conditions

$$\begin{aligned} & u^+(x_1,x_2) - u^-(x_1,x_2) \\ & \quad = \delta - r^+(x_1,x_2) - r^-(x_1,x_2) = F(x_1,x_2), \\ & p(x_1,x_2) \geq 0, \\ & (x_1,x_2) \in \mathbf{E} \\ & u^+(x_1,x_2) - u^-(x_1,x_2) > F(x_1,x_2), \\ & p(x_1,x_2) = 0, \\ & (x_1,x_2) \notin \mathbf{E} \end{aligned} \tag{3}$$

Problem 2. The solution obtained in the previous problem—the definition of the contact area and the density of the normal load [**E**, $p(x_1, x_2)$]—allows us to calculate the components of the stress state, in which under the simultaneous action of the normal $p(x_1,x_2)$ and tangent $t_{x1}(x_1, x_2)$ loads can be, based on the solution of Boussinesq-Cherutti, presented as

$$\tau_{i,j} = \varphi_2 \left(x_3, \frac{\partial V}{\partial x_1}, \frac{\partial V}{\partial x_2}, \frac{\partial V}{\partial x_3}, \frac{\partial^2 V}{\partial x_k \partial x_3}, \frac{\partial^2 V}{\partial x_i \partial x_j}, \frac{\partial^2 V}{\partial^2 x_k}, \frac{\partial^3 W}{\partial x_i \partial x_j \partial x_k}, \frac{\partial^3 U}{\partial x_i \partial x_j \partial x_k} \right) \tag{4}$$

$$\sigma_{i,j} = \varphi_1 \left(x_3, \frac{\partial V}{\partial x_3}, \frac{\partial^2 V}{\partial x_1 \partial x_2}, \frac{\partial^2 W}{\partial x_1 \partial x_2} \right), \tag{5}$$

$$i = 1,2, \quad j = 1,2, \quad k = 1,2,$$

V, W, and U are, respectively, Newtons, logarithmic Boussinesq, and biharmonic potentials of a simple layer.

Tangential loads are caused by friction forces, which are generally determined by the kinematics of the interacting

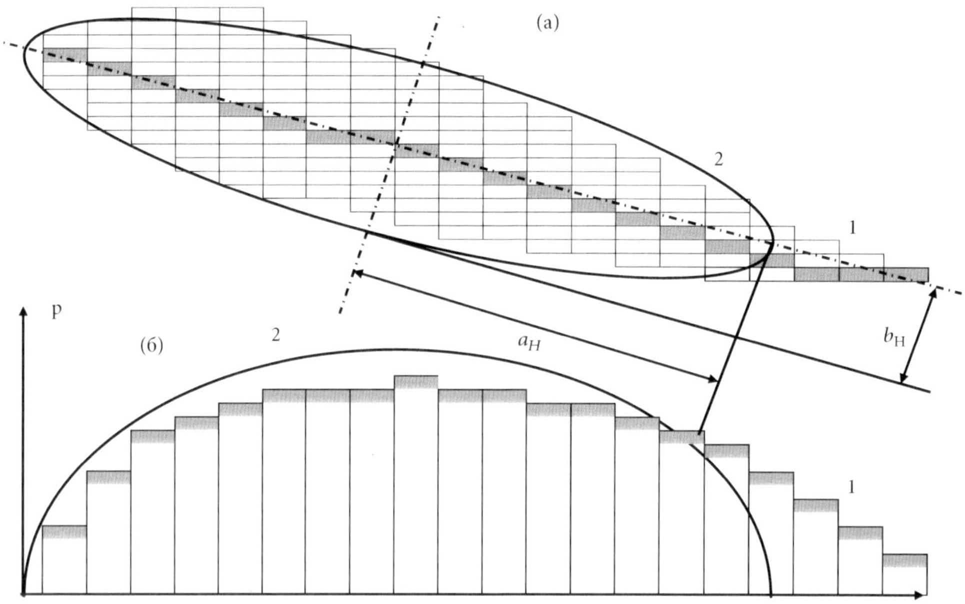

FIGURE 4: Contact area (a) and pressure distribution (b) in Novikov's cylindrical gear: 1: contact of real surfaces; 2: equivalent elliptical contact.

bodies as absolutely rigid, as well as by elastic deformations that violate the "ideal" kinematics of rolling. Thus, the instantaneous contact spot can represent either the area of total slip $\mathbf{E} = E_+$ or a combination of slip and clutch zones E_0 with zero relative slip speeds, i.e., $\mathbf{E} = E_+ \cup E_0$; this model of friction on Amonton-Coulomb is realized only in E_+.

The solution of the problem of determining the tangential loads in the contact of a higher kinematic pair (in particular, with respect to Novikov transmissions) was carried out on the basis of the variational approach, which formulated it as the problem of minimizing the Kalker [20] energy functional in Spector and Fedorenko [21]. The error in the numerical results (relative to the analytical ones) on the grid was 0.62% and 1.06% for the intensity of the octahedral stresses.

The theoretical and experimental studies carried out showed the following:

(1) The shape of the instantaneous contact spot (Figure 4(a)), obtained by a numerical solution of the contact problem, has a pronounced asymmetry and qualitatively corresponds to the results of the experiments.

(2) For the same compressive forces, the numerically obtained values of $p_{\max}$ of the maximum contact pressures are lower than when using the analytical solution for elliptical contact (Figure 4(b)). The discrepancy did not exceed 5%, and as the load increased, it decreased.

(3) But, despite the smaller values of $p_{\max}$, the equivalent stresses (in this case, the intensity of the octahedral stresses) in the central region of the contact area, determined numerically, are somewhat higher than those obtained analytically. The excess is small (up to 4%) and in the calculation methods was taken into account by the introduction of the corresponding coefficient.

(4) Under the conditions of an out-of-pole contact, the distribution of tangential frictional stresses is signposted (in the case of an elliptical contact close to an ellipsoidal one) with a certain shift of the maximum in the rolling direction (Figure 5(a)). Under these conditions (complete slippage), the frictional forces are increased and proportional to the magnitude of the friction coefficient at the contact surface, but they quickly decay as they move away from the surface.

(5) The presence within the contact area of the adhesion region with zero relative slip (which is possible with the extension of the contact to the pole zone) dramatically changes the distribution of tangential frictional stresses. On the interface between E_0 and E_+, zones of local concentration of these stresses appear (Figure 5(b)). Insignificant in absolute value, they indicate the probability of pitting.

3.4. Metallographic Analysis of Samples. Metallurgical studies checked the quality of the near-surface area of the tested gears. Fractographic analysis, metallographic analysis of the core and hardened layer, measurement of microhardness from the surface to the core, and measurement of the roughness of the contact surface were carried out. The measurements were carried out both on the working side and on the nonworking side of the tooth.

The microstructure is quite stable (Figure 6): at a depth up to 0.14 ... 0.18 mm trooco-or-latent-leaf martensite; further small-and-middle-needle martensite (points 4, partially 5), residual austenite (scores 3 ... 5). At depths of 0.4...0.6 mm are bainite martensitic structures, and deeper is bainite. Ferrite in the core is not detected, and the dark component is absent.

Despite the fact that all samples were made from the same batch steel and HTO passed simultaneously, the spread of hardness values reached 150 HV1 units within one tooth and 250 HV1 for the same points of different samples (Figure 7). Measurements were performed on both the working and the nonworking sides of the teeth at the load application site. No significant change in hardness due to run-in was detected. In

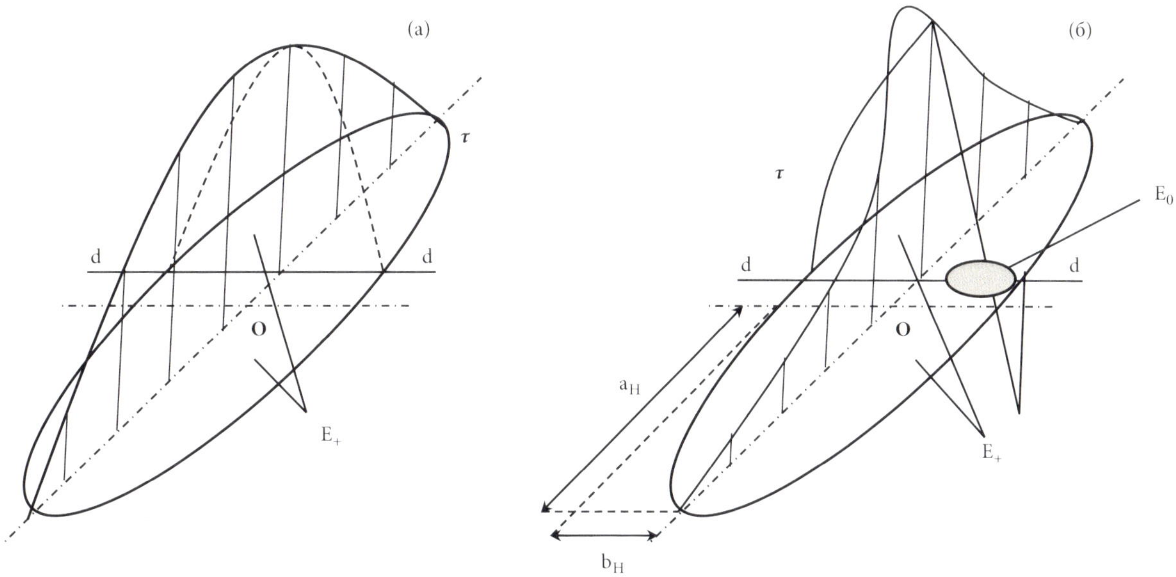

FIGURE 5: The distribution of tangential stresses over the contact surface in the Novikov transmission: (a) - $X_1 = X_2 = 0$: complete slip; (b) -$X_1 = -0.3$; $X_2 = +0.3$: slippage and grip. X_1 and X_2: the coefficients of displacement, respectively, gears, and wheels.

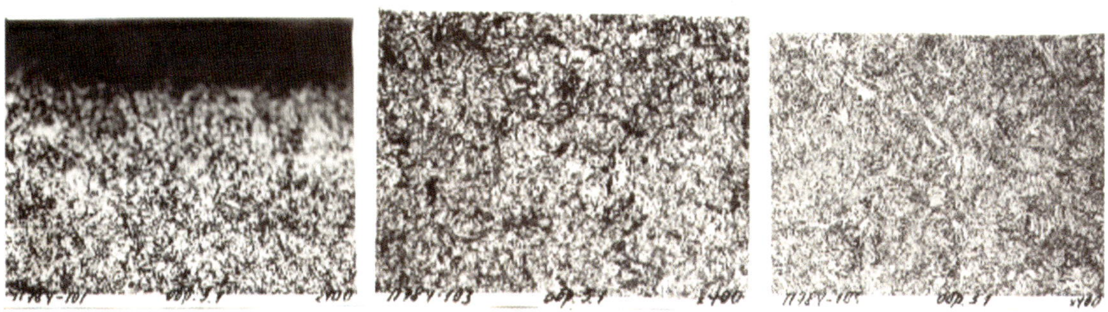

FIGURE 6: Microstructure of the reinforced layer and core (Sample 3.1). (a) near-surface zone; (b) effective area; (c) the core.

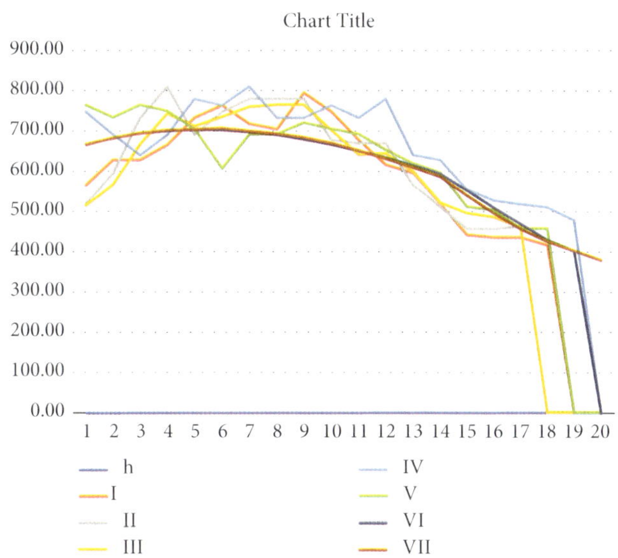

FIGURE 7: Hardness distribution along the thickness of the hardened layer. 1-3: actual distribution of hardness of samples of one batch (3 samples); 4: probabilistic curve of the actual hardness distribution (for 14 samples with a probability of 90%); 5: distribution according to the approximation.

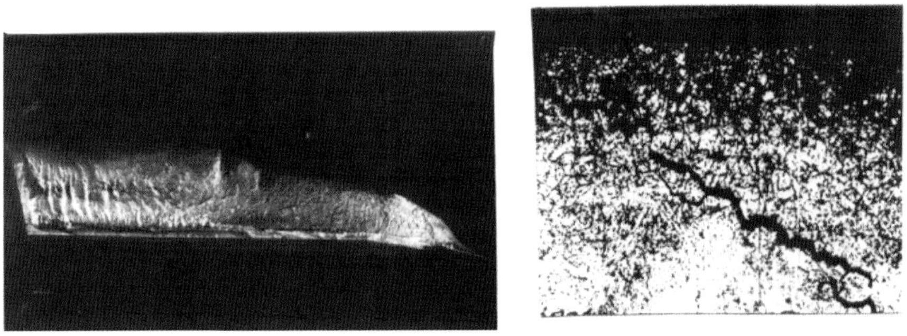

Figure 8: Surface of failure (sample 3) and sample microslip.

65% of cases hardness on the working side was slightly higher and in 20% the opposite picture was observed, i.e., there was "loosening."

The destruction of teeth in the majority of cases had the character of fatigue kinks (Figure 8), with more or less pronounced zones of slow and accelerated development of cracks developing from surface foci. Defects, witnessing about them, as a result of contact interaction are not found. An exception is sample 12, in whose section (Figure 8) a crack is found that develops from the depth at an angle to the surface of about 320. The microstructure indicates a significant overheating: in the effective zone a large needle (up to 16 micrometers) at a depth of 0.30...0.35mm (more than 100HV lower than at a depth of 0.50...0.55mm).

The qualitative side of the process of running-in is obvious, but quantitative side is practically not studied. In [22] it was proposed to introduce the coefficient of increment of teeth in height [22] for its evaluation. It is assumed that the radius of curvature of the radiused curvature Ra will be increased until the instantaneous contact patch spreads along the height of the tooth by the value Z_1K_1L. Here L is the chordal length arcs of the active portion of the tooth head and K_1 is the coefficient of reduction of the active tooth height due to the cutting of the teeth by running-in. The value of K_1 is determined by the parameters of the initial contour, the coefficient of displacement by the module, and the number of teeth of the transmission wheels formulas (11-10-11-12), table 17 of work [22]. Based on the test results, the dependence Z_1 = 0.95-3.7.10-4 (HHB-200), where HHB is the minimum hardness of the active surfaces of the teeth in the gear wheel and the wheel. For hardened and X-hardened wheels it is possible to take Z_1 = 0,82.

Considering the running-in, the dependence for determination of the normal contact stress of the gear, with the original contours type GOST 30224-96 (with the center of curvature of the head of the tooth lying outside the pitch line), takes the form

$$\sigma_H = Z_H \left[\frac{T_H}{z \cos \alpha_k} \right]^{0.687} m^{-2.063} \left(R_x\right)^{-0.312} \cdot \left(Z_1 K_1 L_p\right)^{-1.063} \quad (6)$$

where T_H is estimated torque on the wheel, Nm.

Z_H = 3,7103 is for transmissions with source contours of GOST 30224-96 type (with the center of curvature of the tooth head lying outside the dividing line), with the center of curvature on the dividing line Z_H = 3.96.103.

Rx is the reduced longitudinal radius of curvature at the theoretical contact point.

L_p is the chordal length of the arc of the active portion of the head of the tooth of the lath.

4. Conclusions

(1) Run-in significantly changes the conditions of contacting. During the running-in process the contact spot spread in height from 15...20% of the active profile to 60...80%. Almost completely neutralized defectiveness (due to CHT) is near-surface zones of the hardened layer. The cleanliness of surfaces increased by 1-2 classes.

Even the unhandled transmissions worked in a stable elastic hydrodynamic lubrication regime. The evaluation of the stress-strain state at the early stages of operation can be performed on the basis of the initial one, without taking into account the rheology of the lubricant, which greatly simplifies the calculations.

(2) Chemical heat treatment (CHT) of the experimental samples was performed at an effective depth of $h_{eff} \approx 0.25$m, which seems overestimated. The contact strength does not exert any influence, and the hardest layers of a lesser thickness will be optimal for breaking strength. Another negative point is the increase in the time of the CHT process, which leads to an increase in the austenite grain and the depth of the defective near-surface zone. The result of overheating (samples 8,12) is an increase in the surface troostomartensitic zone and the adjacent zone of medium- and coarse-grained martensite: the character of the destruction is brittle with a developed region of the dolomite. In addition, cases of a decrease in the hardness of the nitrocement layer on the working surface are noted, as compared to the nonworking layer, i.e., loosening. It is possible to recommend CHT Novikov transmissions to the effective depth $h_{eff} \approx (0.18 \ldots 0.2)$ m with mandatory control of the microstructure of the surface zone.

(3) The appearance of the area of adhesion within the contact area is undesirable. A Novikov's high-speed transmission was tested with a rectilinear (in the initial contour) near-pole section, connecting the arc sections of the head and the

tooth fillet (the root), providing the possibility of its use in the process of engagement. Despite the fact that the profile angle of this site was 340, it was struck by pitting, while the extra-pole (Novikov) areas were mirrored. When designing Novikov's adjusted gears, the bias coefficients should be assigned from the condition that the contact propagation to the pole zone be excluded [22]. It can be measured with nondestructive tests [23, 24].

(4) When designing transmissions, consideration should be given to the effect of alternative types of failure. With surface hardening, the anticipatory effect of deep contact fractures (DCF) with a fundamentally different (from the surface) model of contact fatigue life is possible. Calculations for the prevention of DCF must be carried out on the original geometry, without taking into account the burnishing, because they are the leading in the initial stages of operation (up to about 107 loading cycles).

Conflicts of Interest

The authors declare that they have no conflicts of interest.

Acknowledgments

The work was financially supported by the Russian Foundation for Base Research (Project 18-01-00715-a).

References

[1] V. I. Korotkin and N. P. Onishkov, "To the problem of contact fatigue life of surface hardened Novikov gears," *Vestnik Rostovskogo Gosudarstvennogo Universiteta Putey Soobshcheniya' (Vestnik RGUPS)*, vol. 3, p. 14, 2007.

[2] A. I. Pavlov, "Contact of convex and concave surfaces in gearing," *Herald of the National Technical University "KhPI"*, vol. 10, pp. 99–102, 2002.

[3] S. V. Lunin, "New discoveries in WN gear geometries," http://www.zakgear.com, 2001.

[4] S. Lunin, *Interactive visualization with parallel computing for gear modeling*, 2012, http://www.zakgear.com/Parallel.html.

[5] Y. Ariga and S. Nagata, "Load Capacity of a New W-N Gear with Basic Rach of Combine Prifile," *Journal of Mechanisms, Transmissions, and Automation in Design*, vol. 107, no. 4, pp. 565–572, 1985.

[6] G. A. Zhuravlev, "The fallacy of the physical basis of engagement Novikov as a reason for the limited application," *Reducers and Gears*, vol. 1, 2006, http://news.reduktorntc.ru/arc/06_1/1_06_38_45.pdf.

[7] T. Liu, G. Li, H. Wei, and D. Sun, "Experimental observation of cross correlation between tangential friction vibration and normal friction vibration in a running-in process," *Tribology International*, vol. 97, pp. 77–88, 2016.

[8] D. Sun, G. Li, H. Wei, and H. Liao, "Experimental study on the chaotic attractor evolvement of the friction vibration in a running-in process," *Tribology International*, vol. 88, pp. 290–297, 2015.

[9] Y. Zhou, X. Zuo, H. Zhu, and T. Wei, "Development of prediction models of running-in attractor," *Tribology International*, vol. 117, pp. 98–106, 2018.

[10] P. Bergmann, F. Grün, F. Summer, I. Gódor, and G. Stadler, "Expansion of the Metrological Visualization Capability by the Implementation of Acoustic Emission Analysis," *Advances in Tribology*, vol. 2017, Article ID 3718924, 17 pages, 2017.

[11] J. Moder, F. Grün, M. Stoschka, and I. Gódor, "A Novel Two-Disc Machine for High Precision Friction Assessment," *Advances in Tribology*, vol. 2017, Article ID 8901907, 16 pages, 2017.

[12] A. Akchurin, R. Bosman, and P. M. Lugt, "Generation of wear particles and running-in in mixed lubricated sliding contacts," *Tribology International*, vol. 110, pp. 201–208, 2017.

[13] D. Mallipeddi, M. Norell, M. Sosa, and L. Nyborg, "Influence of running-in on surface characteristics of efficiency tested ground gears," *Tribology International*, vol. 115, pp. 45–58, 2017.

[14] P. J. Blau, "On the nature of running-in," *Tribology International*, vol. 38, no. 11-12, pp. 1007–1012, 2005.

[15] V. I. Korotkin, N. P. Onishko, and Y. D. Kharitonov, "Stress state of Novikov gearing teeth under the conditions of real multipair engagement," *Spravochnik. Inzhenernyi zhurnal*, pp. 11–17, 2015.

[16] V. I. Korotkin and D. A. Gazzaev, "Modeling of the contact interaction of the teeth of the gear wheels of Novikov," *Journal of Machinery Manufacture and Reliability*, vol. 43, pp. 104–111, 2014.

[17] "Friction, wear and lubrication," Handbook. Moscow, Mechanical engineering, vol. 2, 1979. 358 p.<?cmd?>.

[18] B. J. Hamrock and D. Dowson, "Isothermal Elastohydrodynamic Lubrication of Point Contacts, Part III," *Journal of Lubrication Technology, Transactions of ASME*, vol. 99, pp. 264–275, 1977.

[19] A. N. Beskopylny, N. P. Onishkov, and V. I. Korotkin, "Assessment of the fatigue durability of the rolling contact," *Advances in Intelligent Systems and Computing*, vol. 692, pp. 184–191, 2018.

[20] J. J. Kalker, "Rolling with slip and spin in the presence of dry friction," *Wear*, vol. 9, no. 1, pp. 20–38, 1966.

[21] R. V. Goldstein, A. F. Zazovskii, A. A. Spektor, and R. P. Fedorenko, "The solution of spatial contact rolling problems with slip and clutch by variational methods," *Advances in mechanics*, vol. 5, no. 3, pp. 61–100, 1982.

[22] V. I. Korotkin, N. P. Onishkov, and Yu. D. Kharitonov, "Novikov gears. Achievements and development," *Machine Building*, p. 384, 2007.

[23] A. N. Beskopylny, A. A. Veremeenko, E. E. Kadomtseva, and N. I. Beskopylnaia, "Non-destructive test of steel structures by conical indentation," in *Proceedings of the MATEC Web of Conferences*, vol. 129, 2017.

[24] A. N. Beskopylny, A. A. Veremeenko, and B. M. Yazyev, "Metal structures diagnosis by truncated cone indentation," in *Proceedings of the MATEC Web of Conferences*, 2017.

Permissions

All chapters in this book were first published in AT, by Hindawi Publishing Corporation; hereby published with permission under the Creative Commons Attribution License or equivalent. Every chapter published in this book has been scrutinized by our experts. Their significance has been extensively debated. The topics covered herein carry significant findings which will fuel the growth of the discipline. They may even be implemented as practical applications or may be referred to as a beginning point for another development.

The contributors of this book come from diverse backgrounds, making this book a truly international effort. This book will bring forth new frontiers with its revolutionizing research information and detailed analysis of the nascent developments around the world.

We would like to thank all the contributing authors for lending their expertise to make the book truly unique. They have played a crucial role in the development of this book. Without their invaluable contributions this book wouldn't have been possible. They have made vital efforts to compile up to date information on the varied aspects of this subject to make this book a valuable addition to the collection of many professionals and students.

This book was conceptualized with the vision of imparting up-to-date information and advanced data in this field. To ensure the same, a matchless editorial board was set up. Every individual on the board went through rigorous rounds of assessment to prove their worth. After which they invested a large part of their time researching and compiling the most relevant data for our readers.

The editorial board has been involved in producing this book since its inception. They have spent rigorous hours researching and exploring the diverse topics which have resulted in the successful publishing of this book. They have passed on their knowledge of decades through this book. To expedite this challenging task, the publisher supported the team at every step. A small team of assistant editors was also appointed to further simplify the editing procedure and attain best results for the readers.

Apart from the editorial board, the designing team has also invested a significant amount of their time in understanding the subject and creating the most relevant covers. They scrutinized every image to scout for the most suitable representation of the subject and create an appropriate cover for the book.

The publishing team has been an ardent support to the editorial, designing and production team. Their endless efforts to recruit the best for this project, has resulted in the accomplishment of this book. They are a veteran in the field of academics and their pool of knowledge is as vast as their experience in printing. Their expertise and guidance has proved useful at every step. Their uncompromising quality standards have made this book an exceptional effort. Their encouragement from time to time has been an inspiration for everyone.

The publisher and the editorial board hope that this book will prove to be a valuable piece of knowledge for researchers, students, practitioners and scholars across the globe.

List of Contributors

Prashant Thapliyal
Department of Physics, Army Cadet College, Indian Military Academy, Dehradun 248007, India
Uttarakhand Technical University, Sudhowala, Dehradun 248007, India

G. D. Thakre
Tribology and Combustion Division, CSIR-Indian Institute of Petroleum, Mohkampur, Dehradun, India

Neelima Khare, Praveen Kumar Limaye and Dhananjay Tatyasaheb Jadhav
Refueling Technology Division, Bhabha Atomic Research Centre, Mumbai 400085, India

Kulwant Singh
Materials Science Division, Bhabha Atomic Research Centre, Mumbai 400085, India

Arundhati Bute
Laser & Plasma Technology Division, Bhabha Atomic Research Centre, Mumbai 400085, India

Navnath Kalel
Department of Mechanical Engineering, Rajarambapu Institute of Technology, Sangli 415414, India

Aoshuang Ding, Xiaodong Ren, Xuesong Li and Chunwei Gu
Department of Thermal Engineering, Tsinghua University, Beijing 100084, China

Jatinkumar V. Adeshara
Research Scholar, Department of Mathematics, H. N. G. University, Patan, 384 265, Gujarat State, India

M. B. Prajapati
Head, Department of Mathematics, HNG University, Patan 384 265, Gujarat, India

G. M. Deheri
Department of Mathematics, SP University, Vallabh Vidyanagar 388 120, Gujarat, India

R. M. Patel
Department of Mathematics, Gujarat Arts and Science College, Ahmedabad 380 006 Gujarat, India

Thomas Lohner, Andreas Ziegltrum, Johann-Paul Stemplinger and Karsten Stahl
Gear Research Centre (FZG), Technical University of Munich (TUM), Boltzmannstraße 15, 85748 Garching, Germany

Kodanda Rama Rao Chebattina and V. Srinivas
Department of Mechanical Engineering, GITAM University, Visakhapatnam, India

N. Mohan Rao
Department of Mechanical Engineering, University College of Engineering, JNTU, Kakinada, India

Andre Shihomatsu and Sergio Tonini Button
Department of Manufacturing and Materials Engineering, School of Mechanical Engineering, University of Campinas, 13083-970 Campinas, SP, Brazil

Iris Bento da Silva
Department of Mechanical Engineering, School of Engineering of Sao Carlos, University of Sao Paulo, 13566-590 Sao Carlos, SP, Brazil

Angela Maria Tortora and Deepak Halenahally Veeregowda
Application Development Laboratory, Ducom Instruments Europe B.V., 9713 GX Groningen, Netherlands

Jimit R. Patel and G. M. Deheri
Department of Mathematics, Sardar Patel University, Vallabh Vidyanagar, Anand, Gujarat 388 120, India

F. Concli
R&D Department, Bonfiglioli Mechatronic Research, Via Fortunato Zeni 8, 36068 Rovereto, Italy

A. Della Torre and G. Montenegro
Department of Mechanical Engineering, Politecnico di Milano, Via La Masa 1, 20156Milano, Italy

C. Gorla
Department of Energy, Politecnico diMilano, Via LaMasa 1, 20156 Milano, Italy

Ashutosh Kumar and S. K. Kakoty
Indian Institute of Technology Guwahati, Guwahati, India

Sergiu Spinu
Department of Mechanics and Technologies, Stefan celMare University of Suceava, 13th University Street, 720229 Suceava, Romania
Integrated Center for Research, Development and Innovation in Advanced Materials, Nanotechnologies, and Distributed Systems for Fabrication and Control (MANSiD), Stefan cel Mare University of Suceava, Suceava, Romania

F. Robbe-Valloire and T. Da Silva Botelho
QUARTZ (EA 7393), Supmeca, 3 rue Fern and Hainaut, 93407 Saint-Ouen, France

R. Progri
QUARTZ (EA 7393), Supmeca, 3 rue Fernand Hainaut, 93407 Saint-Ouen, France Retired Supmeca Professor, France

T. P. Jeevan and S. R. Jayaram
Department of Mechanical Engineering, Malnad College of Engineering, Hassan 573201, India
Visvesvaraya Technological University, Belagavi, India

M. H. Sakinah, A. K. Amirruddin and D. Ramasamy
Faculty of Mechanical Engineering, University Malaysia Pahang, Pekan, Pahang, Malaysia

M. M. Rahman, M. M. Noor and K. Kadirgama
Faculty of Mechanical Engineering, University Malaysia Pahang, Pekan, Pahang, Malaysia
Automotive Engineering Research Group (AERG), University Malaysia Pahang, Pekan, Pahang, Malaysia

Farid Al-Bender and Dominiek Reynaerts
Division of Production Engineering, Machine Design and Automation, KU Leuven, Leuven, Belgium

Rodrigo Villavicencio and Federico Colombo
Department of Mechanical and Aerospace Engineering, Politecnico di Torino, Torino, Italy

Tobias Waumans
Leuven Air Bearings, KU Leuven, Leuven, Belgium

Feng Shen, Cheng-Jin Yan, Jian-Feng Dai and Zhao-Miao Liu
College of Mechanical Engineering and Applied Electronics Technology & Institute for Advanced Mechanics in Engineering
Beijing University of Technology, Beijing 100124, China

Alexey Beskopylny and Nikolay Onishkov
Department of Transport, Construction and Road System, Don State Technical University, Rostov-on-Don 344000, Russia

Viktor Korotkin
Institute of Mathematics, Mechanics and Computer Science, South Federal University, Rostov-on-Don 344006, Russia

Index